PENGUIN BOOKS

ISAMBARD KINGDOM BRUNEL

L. T. C. Rolt was born at Chester in 1910. After his
education at Cheltenham College he embarked on an
engineering career, until he decided to turn to writing.
Since childhood he has been fascinated by the history of
engineers and engineering, and his writing reflects this
interest. His first book, *Narrow Boat*, published in 1944,
describes a journey along the English canals during
the twelve years that he lived afloat. His subsequent
biographies of famous engineers, like his writings on
railways and motor cars, show his concern to give the
story of the Industrial Revolution an imaginative and
literary shape.

He was a founder member of the Vintage Sports Car
Club and was also co-founder and first Honorary
Secretary of the Inland Waterways Association. He
founded the Talyllyn Railway Preservation Society, of
which he was Vice-President for many years, and he was
a member of the Science Museum Advisory Council and
Vice-President of the Newcomen Society for the study of
the history of engineering and technology. He was a
Fellow of the Royal Society of Literature, and in 1965
was awarded an Honorary M.A. degree by the University
of Newcastle, followed in 1973 by an Honorary M.Sc.
from Bath University. Mr Rolt died in 1974.

Among his many publications were the biographies of
Thomas Telford and *George and Robert Stephenson* (both
published in Penguins), and two volumes of autobiography, *Landscape with Machines* and *Landscape with
Canals*.

L. T. C. ROLT

Isambard Kingdom Brunel

with an Introduction by R. A. Buchanan

PENGUIN BOOKS

PENGUIN BOOKS

Published by the Penguin Group
Penguin Books Ltd, 80 Strand, London WC2R 0RL, England
Penguin Putnam Inc., 375 Hudson Street, New York, New York 10014, USA
Penguin Books Australia Ltd, 250 Camberwell Road, Camberwell, Victoria 3124, Australia
Penguin Books Canada Ltd, 10 Alcorn Avenue, Toronto, Ontario, Canada M4V 3B2
Penguin Books India (P) Ltd, 11 Community Centre, Panchsheel Park, New Delhi – 110 017, India
Penguin Books (NZ) Ltd, Cnr Rosedale and Airborne Roads, Albany, Auckland, New Zealand
Penguin Books (South Africa) (Pty) Ltd, 24 Sturdee Avenue, Rosebank 2196, South Africa

Penguin Books Ltd, Registered Offices: 80 Strand, London WC2R 0RL, England

www.penguin.com

First published by Longmans Green 1957
Published in Pelican Books 1970
Reprinted in Penguin Books 1985
Reprinted with an Introduction 1989

7

Acknowledgements

The author wishes to express his thanks to the following who either provided material to illustrate this
book or granted reproduction permission:

The Librarian of Bristol University for providing copies of pictures numbered 5 and 19.
F. C. Ferguson, Esq. for supplying number 26. The *Illustrated London News* for permission to
reproduce numbers 21 and 25. The Museum of British Transport for permission to copy and
reproduce numbers 2, 13 and 15. Lady Noble for her kind permission to copy and reproduce numbers
1, 3, 4, 5, 6, 7 (photograph Desmond Tripp Studios), and 19. Radio Times Hulton Picture Library for
permission to reproduce number 22. The Director of the Science Museum for permission to reproduce
numbers 8, 9, 10, 11, 12, 14, 16, 17, 23, 24. Messrs Joseph Westwood Ltd and A. E. Tawse for the
loan of material from which number 20 was reproduced.

Printed in England by Clays Ltd, St Ives plc
Set in Linotype Georgian

> The intellect of man is forced to choose
> Perfection of the life, or of the work,
> And if it take the second must refuse
> A heavenly mansion, raging in the dark.
>
> When all that story's finished, what's the news?
> In luck or out the toil has left its mark:
> That old perplexity an empty purse,
> Or the day's vanity, the night's remorse.

W. B. YEATS

Contents

Plates

Introduction

WHEN Tom Rolt died at the age of sixty-four in 1974, he had already established for himself the reputation of being the outstanding popular historian of engineering history and biography in the twentieth century. This reputation was based upon a clutch of historical accounts of canals, railways and motor cars, some general works of mechanical engineering and institutional history, and a number of biographies of distinguished British engineers, including I. K. Brunel, Thomas Telford, George and Robert Stephenson, and James Watt. Of these, the first and most successful was that published in 1957 as *Isambard Kingdom Brunel: A Biography*. This has gone through many impressions in both hardback and paperback form, but no new edition has been prepared and the text has never been modified, even though what may be called 'Brunel studies', which have been greatly stimulated by the success of Rolt's book, have made significant advances and raised several serious questions about some of his interpretations and judgements. It is, therefore, necessary to explain to the reader the thrust of these new developments, and to justify the decision to proceed with a further printing of the book without any substantial changes to the text as Rolt left it.

Tom Rolt brought to engineering history an unusual combination of skills. He was both an engineer and a gifted author. The British engineering profession has not been distinguished by a high proportion of members who have taken an active interest in the history of their skills, but there have been a few who have taken a vigorous part in associations such as the Newcomen Society for the History of Engineering and Technology, of which Rolt became a Vice-President. Of these few, however, an even smaller proportion has displayed the sort of verbal felicity which makes a popular author; but Rolt was one of this select group. After an engineering training in Staffordshire and Gloucestershire, and some difficult years in the 1930s attempting to make a living for himself as a motor mechanic, all of which has been vividly recorded in the first volume of his biography, *Landscape with Machines*, Rolt took the major decision to devote his life to writing. To this end he bought a canal boat and set out with his first wife, Angela, to write about the canals and the lives of the people who lived and worked upon them. The Second World War interrupted his plans but did not seriously deflect him from his objective. His first book, *Narrow Boat*, was published in 1944, and was a brilliant evocation of life on the declining canal system of Britain as seen from the cabin of *Cressy* during these years. Other works soon followed, and quickly made his name as a sensitive author with a touch of romantic melancholy and a gift for robust descriptions of people and places. The livelihood of anybody relying upon their penmanship is generally precarious, except, possibly, for journalists, but Tom Rolt managed to sustain himself and his second wife, Sonia, with whom he brought up their two sons, by deriving a steady income from his books. It was for him an eminently satisfying achievement.

Canals were the first preoccupation of Rolt as an author, and his writings helped to stimulate a vigorous movement to preserve the remaining network, which took the form of the Inland Waterways Association. It was the same when he turned to writing about railways and motor cars: interest in these subjects burgeoned and Rolt found himself involved in a range of preservation societies, whether it be for the Tal-y-llyn Railway in North Wales, or for the owners of Bugatti cars. He never shirked from the practical consequences of his romanticism, but recognized that if the delightful modes of transport of the recent past were to be preserved for posterity it was necessary for men and women who felt strongly about them to work together for this objective. In his later years he became a committed enthusiast for the pursuit of industrial archaeology, and was elected as President of the new Association for Industrial Archaeology shortly before his death.

When Tom Rolt started to write engineering biography, he did so with the same commitment and with a similar result in so far as his writing generated an enthusiastic response amongst readers who found their own interest in the subject aroused and who looked to him for help and guidance. It was as a biographer of engineers that Rolt achieved the fullest expression of his talent for putting engineering achievements across to the lay reader in a way which made them vivid and fascinating. In this respect he was a worthy successor to Samuel Smiles, who had laid the foundations of engineering biography in the nineteenth century. Smiles is probably most remembered for his best-selling work of 1859, *Self-Help*, the title of which encapsulated the mid-Victorian philosophy of individualism whereby deserving men made good through their own hard work and application. But in the process of expounding this moralistic view of life, Smiles – who was himself a Scottish medical man turned journalist and railway secretary – came to see the pioneering British engineers of the eighteenth and nineteenth centuries as archetypes of self-help, and devoted much effort to gathering biographical evidence about them. The result was the three-volume book *The Lives of the Engineers*, which he completed in 1862, and subsequently he went on to write other biographies of engineers both living and dead. Smiles thus established engineering as a subject worthy of detailed biographical treatment, and set a very high standard for it.

Smiles wrote *The Lives of the Engineers* when British engineering was in its heroic age. The triumvirate of great railway builders, Robert Stephenson, I. K. Brunel and Joseph Locke, had all died recently, and the public at large was deeply impressed by the men who had covered the land with iron rails and steaming locomotives, who had built bridges, factories, harbours and lighthouses, and who had launched iron-built steamships upon the oceans. The sense of euphoric progress induced by these novelties had been symbolized in the Great Exhibition of 1851, with its visionary prefabricated structure of cast iron and glass, the Crystal Palace, and the knighthoods conferred on the men who built it were deemed to be well deserved. Robert

Stephenson had received a state funeral in Westminster Abbey in October 1859. In eulogizing this successful new professional group, therefore, Smiles was not at a loss for eligible material. He could afford to be very selective, and concentrated his attention on a dozen or so leading engineers from the seventeenth-century precursors to George and Robert Stephenson. But it is surprising that, even within such limitations, he excluded I. K. Brunel. Perhaps it was because he regarded Brunel as a foreigner or as having received too much help from a talented father to be regarded as a model of 'self-help', although on both counts he was wrong because Brunel was thoroughly anglicized despite the French origins of his father, Marc Brunel; and while the help of his father was of inestimable benefit to I. K. Brunel, his was only one of several engineering dynasties at this period, and Smiles did not exclude Robert Stephenson from his pantheon on account of the help that he received from his father. It has also been suggested that, as a railway company employee, and a standard-gauge railway at that, Smiles regarded it as undiplomatic to give too much attention to the broad-gauge Great Western Railway and its engineer. Rather more likely by way of explanation is the fact that Brunel's family was excessively protective of the great man's reputation after his premature death, and probably made access to his papers difficult. Whatever the reason, the fact is that Smiles did not include Brunel in the select company of those whose biographies he wrote. His passing references to Brunel appear to be well disposed to the engineer, but he never tried to write his life-story.

More surprising than Smiles's selectivity is the way in which he determined the field of engineering biography for most of his successors. Time after time, students of the subject have returned for their material to the heroic age as encapsulated by Smiles. There have been plenty of prominent engineers since the 1860s, and autobiographies and biographies have been written by them and about them, but almost without exception these have been dull and uninspiring expositions. On the other hand, when H. W. Dickinson and other able authors amongst the pioneers of the Newcomen Society turned their attention to engineering biography in the 1920s and 1930s, they chose their subjects exclusively from the heroic age with enduring studies of James Watt, Richard Trevithick and others of the same period. Likewise, Sir Alexander Gibb turned to Thomas Telford as the focus of his lively biographical work, and Lady Celia Noble wrote her delightful family account of the two Brunels. Then in the 1950s Tom Rolt followed the same course: the subjects of all his works of engineering biography were dead by 1860. He is on record as having indicated the biographical potential of later generations of engineers, but for his substantive works he stuck closely to the period delineated by Smiles. It has been the same with most subsequent students in the field: they have continued to return to the heroic age for their major studies, and later periods have continued to be largely disregarded.

The perception of an heroic age in British engineering is thus enshrined in the bibliography of engineering biography, and L. T. C. Rolt was completely traditional in his adoption of this interpretation. It is, indeed, an interpretation which is easily justified in terms of the way in which British historians have understood the development of the Industrial Revolution. The well-established consensus of historical opinion is that this process of rapid and continuing industrialization began in the eighteenth century, and that it began in Britain. Such was the dynamic impact of the process that it enabled Great Britain to achieve leadership in world trade and industry and, through the mediation of the British navy and the expansion of the empire, an unprecedented degree of world political dominance also. It was the fact that the engineering novelties of iron and steam represented this exhilarating sense of national greatness which gave them an extraordinary poignancy in the middle decades of the nineteenth century. For a short period engineering achievement was appreciated not only as an end in itself but as a potent symbol of national supremacy. It could not last because other countries were quick to learn the lessons of industrialization, and some of these such as the United States of America had much greater resources than those available to Britain. In the second half of the nineteenth century Britain was overtaken by other countries in many areas of industrial and commercial activity and ceased to enjoy the unchallenged supremacy which had characterized the years around the Great Exhibition. Engineers continued to build larger and more impressive structures than ever before, and to adopt powerful new technologies in the shape of electricity and the internal combustion engine. After 1860, however, continental and American engineers came to take an increasingly prominent share of such innovations, and Britain had to concede leadership in one area after another. Small wonder, therefore, that the heroic age of British engineering seemed to recede into the past together with the age of British industrial supremacy: the first was a product of the latter, and the implicit understanding of this relationship determined the template of British engineering historiography.

Even though Rolt accepted this traditional interpretation of the role of the heroic engineers without question, he departed from Smiles's example by adopting Isambard Kingdom Brunel as the subject for his first and most outstanding engineering biography. There had been a substantial biography of Brunel produced by his elder son, Isambard Brunel, in 1870, and this had been supplemented by Lady Celia Noble's study in 1938. Both these earlier works had drawn extensively on the private and business papers preserved by the Brunel family, and although sadly incomplete these comprise an astonishingly wide and revealing view of Brunel. Happily, when Rolt embarked on his study, most of the drawing books and business letter books had been presented to the University of Bristol Library by Lady Noble and her son, Sir Humphrey Noble, and he was able to have full access to these and to the more personal papers still in the possession of the family. It is important to note that he was the last scholar to have had the advantage of such unlimited access, because although the family have subsequently passed other material to Bristol, there are some items, particularly the diary kept by Brunel between 1827 and 1829 (which Rolt described as 'only recently discovered') and the two volumes of Brunel's journal covering the years from 1830 to 1834, that are not available for study. This is much to be regretted, and it must be hoped that the family will be persuaded of the wisdom of placing all the surviving documents in the care of the University of Bristol – as only such a scholarly archive can take best care of them – and thus ensure full regard for the reputation of their illustrious ancestor. But so far it must be said that Rolt had unrivalled access to the remarkably extensive collection of papers left by I. K. Brunel, and that the biography which he produced is largely the result of his immersion in this archival material.

The result is a biography of exceptionally high historical and literary quality. The idea of writing the book appears to have been given to Rolt by his friend David Cape of the publishing house Jonathan Cape, to whom he pays a tribute for 'his infectious enthusiasm and determination' in the Acknowledgements at the end of the book. In the event, the book was published by Longmans rather than Cape, but the enthusiasm certainly seems to have been infectious because Tom Rolt achieved that insight only rarely attained even by the best biographers of getting inside the character of his subject and appearing to see the world through his eyes, and thus illuminating his deepest thoughts and motivations. The clue to this achievement can be found in the brilliant Epilogue with which Rolt concluded his study. There he suggests that Brunel was 'the last great figure . . . of the European Renaissance', a man who, despite his superficial gaiety and wit, was driven by a profound melancholy which was fuelled by a 'doubt and pessimism which might have driven weaker natures to apathetic despair or to orgies of self-indulgence', but which drove him instead 'into a fury of creative activity'. There can be little doubt that Rolt felt a resonance here, not only with Brunel as a fellow engineer, but also with his own melancholy stoicism about society and human fate. He was unable to replicate it with his other subjects: for George and Robert Stephenson he evinced sympathy, but not identification; while for Thomas Telford he confessed that he was unable to penetrate the central reserve of his character. But with Brunel he achieved a symbiosis of minds between biographer and subject which is extraordinary and ensures an outstandingly high quality of biographical accomplishment.

Rolt begins his study of Brunel with a literary anecdote – a story of Charles Macfarlane, Victorian raconteur, portraying his encounter with the Brunel family in 1829. It sets the scene beautifully for an account of the relationship between I. K. Brunel and his father, the talented *émigré* French engineer who had sought refuge in Britain and married Sophia Kingdom in 1799. They had two daughters and then their only son, Isambard Kingdom, who was born on 9 April 1806. The second chapter describes Marc Brunel's Thames Tunnel project, on which the young I. K. Brunel was almost killed in January 1828. During his long convalescence, Isambard came to know well an enterprising group of Bristol merchants and industrialists, through whom he became associated with the scheme to bridge the Clifton Gorge, with improvements in the 'Floating Harbour' at Bristol (so called because it was in effect an enclosed dock in which vessels remained afloat at all stages of the tide) and, most significantly, with the ambitious scheme for a railway between Bristol and London which was authorized as the Great Western Railway Company in 1835, with the 29-year-old I. K. Brunel as its first engineer. Rolt tells the epic story of the construction of the G.W.R., with Brunel's commitment to the broad gauge and to the abortive atmospheric system which he introduced on the South Devon part of the network, in the second section of the study (Book II), and then devotes the third part (Book III) largely to Brunel's career as a marine engineer, from his design for the *Great Western* as an extension of the G.W.R., through the breath-taking innovations of the *Great Britain*, to 'that great leviathan' the *Great Eastern*, 'the final hazard' which destroyed its creator. The parts are thus overlapping rather than strictly chronological, but the biography follows a natural development in time, with the main events of a hyperactive and colourful career as an engineer being interleaved with fascinating sections on Brunel's personal relationships with his family, his wife, his friends, and his professional assistants and rivals. The narrative is then rounded off in the superb Epilogue which places Brunel in his time and ties up many of the loose ends which have aroused the curiosity of the reader.

The whole story is told with delightful felicity and elegance, so that it appears as a unified work of art. This cohesive quality of the text makes it impossible to edit it in any significant way without disrupting the balance of the narrative, so no such disruption has been attempted in this edition. The reader is invited to enjoy a distinguished work of biographical scholarship as it flowed from the pen of its author. There are, nevertheless, some observations about the text of which the reader should be aware. In the first place, although it is proper to describe Rolt's *Brunel* as a work of scholarship, it is not an academic treatise. Rolt lists his sources at the end of the text, but he does not provide any footnotes and the reader is not helped in finding the sources of specific pieces of information. It would be pedantic, even if it were possible, to attempt to provide such references now, particularly as some of the personal papers available to Rolt are not at present accessible, so no such attempt has been made. Secondly, it has to be noted that the success of Rolt's biography helped to stimulate interest in the study of I. K. Brunel, so that a considerable amount of scholarship has derived from it, casting new light on several aspects of Brunel's career. We now have a fuller account of his work on the Bristol Docks, for instance, and of his work with timber structures and wrought iron, as well as a better understanding of his theoretical calculations and working practices. Readers who wish to pursue these themes may find the appendix to the Bibliography of interest. And thirdly, even though most subsequent work has served to endorse Rolt's judgements, there have been a few serious criticisms. Two of these are worth mentioning, one fairly slight and the other substantial.

The slighter criticism is that of the econometric historian Gary Hawke, who took Rolt to task in his book *Railways and Economic Growth* (1970) for ridiculing Professor Dionysius Lardner. Rolt is certainly scathing in his comments on 'the egregious Doctor Dionysius Lardner', who made dire predictions about the performance of trains in the Box Tunnel and about steamships attempting to cross the Atlantic Ocean. In so far as Lardner was an able statistician, whose compilations are of inestimable value to the railway historian, Hawke is right to remind us that he was no idiot. But apart from Lardner's own achievement in bringing his career as a university professor and scientific popularizer to a premature and slightly ridiculous end by a bigamous elopement, it must be admitted that he represents the sort of man that British engineers love to hate: the highly theoretical and somewhat pompous academic who is prepared to adumbrate abstract propositions as grounds for judgements which can then be faulted in practice. There is a long tradition of such theoretical interventions in engineering, and Rolt's judgement is that of an engineer and will still find a warm response amongst engineering readers. After all, Lardner was proved to be wrong by the skill of Brunel's practical judgement, although it should not be forgotten that his theoretical competence was greater than that of most of his contemporaries. A similar tension occurred between Brunel and the Oxford geologist Professor Buckland about the geological soundness of the Box Tunnel, but the debate between them was more constructive than that with Lardner and did not receive such strong treatment from Rolt.

The more serious criticism arises from Rolt's account of the struggle to build, launch, and equip the *Great Eastern*. The series of disasters associated with the great ship present a problem to a biographer of Brunel, coming as they do at the end of a career marked by outstanding successes in overcoming obstacles. Rolt developed an ingenious theory to explain them, which he claimed had emerged unexpectedly 'only after a most careful investigation of all the available facts', whereby the fault for all the miscarriages with the *Great Eastern* was attributed to the 'strange character' of John Scott Russell. Russell had been Brunel's chosen partner in the venture, having already established his reputation as an innovative ship-designer with his own shipyard on the Thames in which the great ship could be built.

According to Rolt, Russell became so jealous of Brunel that he malevolently undermined his authority and precipitated the crises in the construction and financial affairs of the enterprise. There is a dramatic cohesion about this thesis which makes it particularly attractive to anybody concerned with defending the unsullied character of Brunel's reputation as a great engineer, because it lifts the responsibility for the comparative failures of the *Great Eastern* project off his shoulders and places them firmly on those of Scott Russell. Admirers of Brunel and of Rolt have thus found it easy to accept. However, it is necessary to remember that, whatever its attractions may be, it is only a speculation based upon a particular interpretation of the available documents. It is, moreover, a speculation which is very damaging to Scott Russell, whose reputation also deserves to be considered. This point has been made very forcibly by Russell's modern biographer, George S. Emmerson, in his book *John Scott Russell: A Great Victorian Engineer and Naval Architect* (1977), and in contributions to the journal *Technology and Culture*. Russell has been less fortunate than Brunel in that very few of his personal papers have been preserved and there was no previous biography to give his side of the story. But Emmerson has produced a meticulous piece of scholarship with such material as is available, and has demonstrated that the Rolt thesis is seriously flawed. The vituperative spirit with which he attacks Rolt is regrettable, and he fails to carry conviction in his attempt to restore Russell's reputation against what he regards as the unfair attacks of contemporaries and historians: Russell remains, at the end of his account, a strange and shifty operator. But on the crucial point Emmerson argues effectively that Russell had too much to lose for himself and his family to be guilty of the destructive malevolence which Rolt imputes to him. What, then, went wrong, first with Rolt's interpretation, and secondly with the *Great Eastern* project?

It is not possible to do justice to either of these questions here, but brief answers can be made to both. As far as Rolt's version is concerned, it seems likely that it suffered from over-exposure to the rich material in the Brunel archive. In particular, he was considerably influenced by the notes made by Henry Brunel, the second son of I. K. Brunel, who was trained as an engineer and who prepared some comments for his brother's biography of their father. Henry seems to have had little doubt about the culpability of Scott Russell and, without any countervailing evidence, Rolt was ready to accept this judgement as the basis for his own exposition. The view of the Brunel family, however, was, to say the least, biased: overwhelmed by the premature death of the great man, they were ready enough to find a scapegoat for the comparative failure of his later years. Rolt's mistake was to accept this evidence without reservation, and it is now clear that such reservations must be made.

If the Rolt/Brunel version of the travails over the *Great Eastern* must be discounted as being unfair to Russell, what did cause the problems? It is not necessary to resort to a conspiracy theory to explain the clash between Brunel and Scott Russell. Of course, conspiracies do occur, and explanatory theories then have to take account of them. But in this abrasive and damaging relationship as it developed between 1854 and 1859, other causes can be discerned. There was, for instance, the striking difference of approach the two men had towards the enterprise. Brunel went into it as the chief engineer, expecting to direct the complete engineering undertaking in the same way in which he controlled the railways for which he was responsible, while Scott Russell behaved like the traditional shipbuilder who expected to be given a brief and then be left to construct the ship. In other respects such as in keeping accounts and supervising subordinates, it is clear that the two men had very different habits of work.

These differences of approach put the two men on a collision course which was then reinforced by clashes of temperament and professional jealousy – the jealousy being present on both sides. Scott Russell never saw himself in the role of a subservient assistant to Brunel: in his view he was the builder who had contracted to construct the *Great Eastern* for the Eastern Steam Navigation Company to a design discussed and agreed with Brunel, which incorporated Scott Russell's principles of waveline design. He expected, and had the right to expect, something more in the nature of a partnership in the enterprise than Brunel was willing to permit him. Right from the beginning Brunel showed possessive jealousy about the great ship, being hyper-sensitive to any comment about her amongst his fellow-engineers. Scott Russell responded by trying to define his share in the venture, and the urbanity and blandness with which he did so seem merely to have stirred Brunel to greater agitation. These personal differences exacerbated and were exacerbated by the unprecedented managerial, financial and technical prob-lems posed by the size of the undertaking. The whole project was one of enormous complexity, but an explanatory hypothesis along these lines can account satisfactorily for most of the misadventures of the *Great Eastern* affair, and also for the most remarkable of all the facts in the case – namely that, despite all the traumas and personal tragedies involved in its construction, the ship *was* completed satisfactorily and *did* enjoy a consider-able technical triumph, even though it never became a com-mercial success. Perhaps Scott Russell's ultimate offence in the eyes of the Brunel family was that he appeared to enjoy all the credit for this success after the death of the engineer. But any attempt to blame him alone for the troubled gestation of the great ship is no longer tenable.

Despite all this, it would be a mistake to tamper with Rolt's account in an effort to improve its historical reliability. Flawed as it undoubtedly is, the story of 'the final hazard' in Brunel's career is brilliantly told, and it possesses a poetic cohesion with the rest of the narrative which could be altered only at the risk of marring a great work of art and literature. The story is thus left as Rolt told it, with this warning to the reader and with a note on some of the literature expounding the necessary modifications in the appendix to the Bibliography. For Tom Rolt's biography of I. K. Brunel deserves to remain as he left it, as the outstanding work of engineering biography of the twentieth century, pro-viding a definitive insight into the greatest hero of the heroic age of British engineering, and arguably the greatest engineer of all time.

R. A. Buchanan
University of Bath
October 1988

Preface

A CENTURY is a brief span as a historian measures time. Yet between 1760 and 1860 a comparatively small group of men transformed the face of England and brought about an economic and social upheaval so vast that the life of no single person in this country remained unaffected by it. They set in motion a process of rapid technical evolution that still continues and which, spreading round the world, created international problems with which we are still vainly grappling.

Much has been written about this Industrial Revolution, as we call it, but the effect of most of this writing is curiously impersonal and therefore unreal. It is as though the Revolution was the product of some *deus ex machina* or of some corporate act of will on the part of a whole people. For if the fathers of the Revolution are mentioned at all they appear only as shadows. Brindley, Telford, the Stephensons, Brunel – their names may be celebrated, we may even know a little of what they did, but they are still only names. We do not know what sort of men they were or what impulses drove them on their momentous course. The reason is that they have been almost totally ignored by the serious biographer, whereas one could point to many a historical figure who scarcely caused a ripple on the surface of world history and yet has a bibliography to his credit which would fill a fair-sized bookshelf. Why this should be so is a question upon which I have speculated elsewhere.

Of this small group of men whose lives had such prodigious consequences Isambard Kingdom Brunel was perhaps the outstanding personality. He has his statue in marble; every boy's railway book refers to him; we may have seen his name engraved upon that great bridge at Saltash which is the gateway to Cornwall; we may know of him as the over-ambitious author of the broad gauge or of that premature giant among ships, the *Great Eastern*; perhaps we only remember him by virtue of the evocative overtones of that remarkable name, a name in which all the pride and self-confidence of an era seems to ring out like a brazen challenge. But what sort of a man was this Brunel? That is the question which this book tries to answer.

The results of such an inquiry can be disappointing; the spirit and personality of a man may prove on investigation to have been much smaller than the size of his material achievement had suggested. I was prepared for such a disappointment in this case, for although I have always admired Brunel's work my inquiry was inspired by curiosity and not by hero-worship. But the further I went the clearer did it seem to me that, large though the achievement was, the man was larger still. Brunel, in fact, was more than a great engineer; he was an artist and a visionary, a great man with a strangely magnetic personality which uniquely distinguished him even in that age of powerful individualism in which he moved. To learn something about such a man, about his private thoughts, his hopes and ambitions and about the spirit which drove him, is to know a little more about the sources from which the greatest of all revolutions derived its dynamic strength.

If I have achieved my purpose in this book of giving some substance to a distinguished but hitherto unsubstantial shade, it is due above all to the kindness of those, particularly Brunel's descendants, who so freely gave me access to private diaries, and letters not hitherto available for study and publication. Their help is suitably acknowledged in the note at the end of this book, where particulars of source material are also recorded. It was this fresh material which has enabled me to piece together for the first time the full story of the construction, launching and disastrous first trial of the *Great*

Eastern which was directly responsible for Brunel's premature death. This last episode in Brunel's life has hitherto been represented as the simple failure of a hopelessly over-ambitious project dictated by a desire for self-aggrandisement. I hope I have succeeded in dispelling this notion by showing that there was much more to it than this. The desire for self-aggrandisement was there, sure enough, and fatal it proved; but it was not on Brunel's side. Like all great men, Brunel throughout his life had jealous enemies and the last of these was the instrument of his undoing. The story of the *Great Eastern* was a tragedy of trust misplaced and betrayed. As such it has a certain epic quality which makes it a fitting – one might almost say inevitable – end to an extraordinary life.

In 1910 *The Engineer* said of him: 'In all that constitutes an engineer in the highest, fullest and best sense, Brunel had no contemporary, no predecessor. If he has no successor, let it be remembered that ... the conditions which call such men into being no longer have any existence.' If this was true in 1910, how much more true is it today!

L. T. C. R.

BOOK I

[1]

Father and Son

'A LITTLE, nimble, dark-complexioned man with a vast deal of ready, poignant wit.' So Charles Macfarlane, the author and traveller, described one of his two young companions on a journey from Paris to Calais during the bitter winter of 1829. All northern France lay snowbound, and in such circumstances a night's discomfort in the coupé of a coach followed by a channel crossing in a small packet boat was hardly calculated to promote friendship. Yet Macfarlane was obviously captivated by one whose high spirits seemed quite impervious to cold and discomfort, and who, despite his youth, was so evidently an experienced traveller. At each stopping place he called in perfect French for bundles of hay until the coupé was almost filled and the trio buried up to their necks 'like three stone Schiedem bottles packed for safe carriage'. It was some hours before Macfarlane discovered who this remarkable young man was. His name was Isambard Kingdom Brunel. He was the only son of Marc Brunel, the engineer of the great Thames Tunnel that had stirred the imagination of Europe but which now languished for lack of funds, and he was on his way home after a visit to Paris with his friend, Orlebar, a cadet at Woolwich.

The three friends celebrated their arrival at Dover by dining and wining so lavishly at the 'Old Ship' that they had only just enough money left between them to pay the coach fare to London. So their original intention of spending a comfortable night at the inn had to be abandoned. Instead there followed another freezing journey on a night coach which was only relieved by the midnight stop at Canterbury, where they were able to warm themselves at a blazing taproom fire. It speaks volumes for their vitality and hardihood that they were still laughing and joking at six o'clock the next morning when the coach finally set them down at the London Coffee House on Ludgate Hill. Here, over a farewell glass of port wine negus, Brunel invited Macfarlane to meet his parents at his home in Bridge Street, Blackfriars.

Macfarlane lost no time in accepting this invitation, and there he met his friend's two sisters, Emma and Sophia, the latter's husband, Benjamin Hawes, and Mrs Brunel, who, as Macfarlane expressed it, was obviously so 'devotedly attached to her dear old French husband'. He it was who instantly captivated the visitor. 'The dear old man had,' he wrote, 'with a great deal more warmth than belonged to that school, the manner, bearing, and address and even the dress of a French gentleman of the *ancien régime,* for he had kept to a rather antiquated but very becoming costume. I was perfectly charmed with him at this, our first meeting. . . . What I loved in old Brunel was his expansive taste and his love or ardent sympathy for things he did not understand or had not had time to learn. What I most admired of all was his thorough simplicity and unworldliness of character, his indifference to mere lucre, and his genuine absent-mindedness. Evidently he had lived as if there were no rogues in the world.' Macfarlane expressed his sympathy at the stoppage of the Thames Tunnel works, calling it 'one of the greatest and most disgraceful of European mishaps'; but Marc Brunel scarcely needed consolation of this kind for he had suffered many disappointments in his lifetime and was ever philosophical in adversity. 'Courage!' he was fond of saying. 'A man who can do something and keep a warm, sanguine heart will never starve.'

Despite his mother's English blood, young Isambard, with his olive complexion, brilliant dark eyes and sensitive hands which his natural vivacity moved to frequent expressive gesticulation, must at this time have conformed much more closely than the father to the Englishman's idea of the typical Frenchman. In the knowledge of Marc Brunel's background of engineering achievement and contrasting his calm benignity, his courtliness and poise with an ebullience which could so easily be mistaken for shallowness, it was not unnatural that Macfarlane should have awarded the palm to the older man. 'I had liked the son,' he writes, 'but at our very first meeting I could not help feeling that his father far excelled him in originality, unworldliness, genius and taste; perhaps also in those eccentricities which cottoned with mine.' When these words were written the post of engineer in charge of the tunnel works had been young Brunel's only opportunity to prove his mettle, and Macfarlane could never have guessed that the ready wit and the gaiety concealed a fire and a power which would drive him, undeterred by repeated disappointments, to achieve within a decade a degree of fame and fortune such as his father had never enjoyed.

Only in one respect is Macfarlane's revealing portrait of Marc Brunel misleading. By repeatedly referring to 'old Brunel' he suggests a benevolent patriarch whereas at the time of writing Marc was only sixty-one and had twenty more years of life before him. Indeed, his famous son was destined to outlive him by only ten years. Nothing reveals the contrast between their two characters more clearly than this: that while the father lived to a peaceful old age the son sacrificed his life to his great achievements.

Although his origins and his adventurous career were very different from theirs, in character Marc Brunel resembled those engineers, Brindley, Rennie, Telford and George Stephenson, whom Samuel Smiles delighted to honour. Like them, Marc was self-taught, a born craftsman with a flair for invention. He was an unselfconscious man, and those qualities of simplicity, unworldliness and natural dignity which Macfarlane so much admired in him are the natural attributes of the craftsman in any age. From these qualities sprang his 'indifference to mere lucre' and an implicit trust in his fellow men which was so frequently misplaced. Hence his repeated disappointments; hence, too, the fact that he so often failed to secure an adequate return for his labours.

His son was a bird of very different feather. To begin with he could not be called a self-made man, a fact which, knowing Samuel Smiles's addiction to the doctrine of self-help, may serve to explain his almost complete exclusion from that worthy's massive *Lives of the Engineers*. As no single man imparted more momentum to the greatest social revolution in all history than did Isambard Brunel, this omission is not otherwise explicable. Certainly, unlike the characters in Smiles's *Lives,* he had no initial handicaps of birth or upbringing to overcome. By the time he reached adolescence his *émigré* father had already achieved honour and distinction as an engineer, moved freely in good society and was able, despite financial vicissitudes, to give his son an excellent education and training. Yet such advantages alone cannot make the man. The name of Isambard Brunel would not mean what it does today if he had not displayed the same characteristics of dogged persistence and an unlimited capacity for hard work which distinguish the self-taught engineers, with the addition of gifts which they lacked. For he was more than a painstaking and ingenious craftsman; he was also an artist of remarkable versatility and vivid imagination. But what most distinguished him was the force which drove him, so long as life lasted, to the utmost limit of his bent and which charged his personality with that mysterious magnetic power which so often discomfited his opponents and which drew lesser men to follow him, sometimes to prosperity but not infrequently to heavy financial loss.

It may possibly be true, as Macfarlane maintained, that Marc Brunel possessed the greater power of original invention. His son's strength lay rather in that imaginative flair which could seize upon and combine ideas in new ways wherein they became gilded with that magnificence which was to be the hallmark of all that he accomplished. He was as complex in character as his father was simple. He was acutely self-conscious and, as we shall see, the private man was a character very different from that of the cold, proud, abundantly self-confident engineer whom he impersonated to such perfection on the public stage. He had none of his father's naïve belief in the natural goodness of his fellow men. As his journal reveals, even in his youth he was accustomed to weigh the characters of his associates with a shrewdness which seldom erred. Only once in his life was this judgement to fail him, with most disastrous consequences. Once a man had passed his exacting appraisal he would find in the younger Brunel the most loyal and warm-hearted of friends, but woe betide the defaulter. He would find himself shrivelled by a power of ironic invective such as the father could never summon but of which the son was a master. For where Marc could only command affection, his son inspired awe, and the power of his personality was a weapon which could terrify as surely as it could charm.

It has become fashionable nowadays not to praise famous men but to belittle them. Biographers search assiduously for any chink in the armour of their subject which promises to promote controversy and so stimulate the sale of their work. Yet apart from such commercial considerations it has always been in the nature of little men to besmear those whom they envy, and for this reason Isambard Brunel never lacked detractors. It has often been implied that he was an exhibitionist among engineers whose grandiose schemes were conceived solely from the desire to gratify his own vanity. It has also been alleged that he sometimes claimed the full credit for works which owed much to others. The more closely the records of his career are examined, however, the more baseless do these charges become; indeed on the second charge it would be correct to say that the opposite is true.

A great man achieves eminence by his capacity to live more fully and intensely than his fellows and in so doing his faults as well as his virtues become the more obvious, with the consequence that he will often present an easy target to his enemies. It is not in freedom from faults but in the ability to transcend and master them that greatness lies. The key to that mastery is self-knowledge, and in his early years young Brunel's entries in his private journal reveal his awareness of his own weakness. He writes: 'My self-conceit and love of glory, or rather approbation, vie with each other which shall govern me. The latter is so strong that even of a dark night, riding home, when I pass some unknown person, who perhaps does not even look at me, I catch myself trying to look big on my little pony.' And again: 'I often do the most silly, useless things to appear to advantage before those whom I care nothing about.' Knowing this weakness, he learned to guard against it in later life. Pride remained, but whatever imagination suggested and pride drove him to undertake was never embarked upon or divulged to others until its every detail had been subjected to the cool, critical scrutiny of the accomplished technician which he so soon became. So reason tempered imagination and pages of close calculation decided whether the first inspired sketch should be pursued or discarded. If a project failed, it was usually because he overestimated the ability of his fellows and of the technical equipment of his day to realize his design. A man of the highest courage, if he was wrong, he was the first to admit it, nor would he ever commit others to any hazard to which he would not commit himself. On the last and greatest hazard of all he staked both life and fortune.

Despite the profound difference of character between them, the debt which the son owed to his father was none the less great. He undoubtedly inherited some of Marc's characteristics: his artistic sense and his gift of draughtsmanship; his refusal to admit defeat when confronted by some technical problem and, most important of all, his scrupulous attention to detail. Over and above all this he undoubtedly benefited greatly at the outset of his career from the lustre which his father had already given to the name of Brunel and from the help, the example and the experience of the

distinguished friends whom his father had made in society
and in the engineering profession. We can appreciate this
better in the light of Marc's life-story and the training which
he was able to give his son.

Marc Isambard Brunel was born in the little village of
Hacqueville in the rich Vexin plain of Normandy on 25 April
1769. Here his family had held the 'Ferme Brunel' as
tenants in unbroken male succession for three hundred and
thirty years. They had a status equivalent to that of reasonably prosperous yeomen farmers in England, and, as the
roomy stables and coach-houses of their farm still reveal,
they held the hereditary privilege of controlling the posting
arrangements of the district. As was so often the case in
England, while eldest sons succeeded fathers, the younger, if
they showed sufficient intelligence, were trained for the
priesthood – and this was the vocation assigned to Marc.
With this intent he was sent first to the neighbouring college
of Gisors and later to the seminary of St Nicaise at Rouen,
where he unwillingly wore the black soutane. For young
Marc was not at all drawn to the church. While he showed
remarkable aptitude in drawing and mathematics he hated
Latin and Greek, and during his holidays he haunted the
village carpenter's shop. It was quite obvious that he had a
strong creative bent which in England, even at this date,
might have been warmly encouraged. But the influence of
the industrial revolution had not yet touched rural Normandy and his father, Jean Charles, could see no future for
such talent other than some occupation so menial as to be
beneath the family dignity. Consequently he tried by every
means in his power, including cruel punishments, to force
the boy into the traditional vocation. It was the principal of
St Nicaise who encouraged Marc's gifts and who finally persuaded Jean that his son was not destined for the priesthood.
Leaving St Nicaise he went to lodge with a friend and kins-
man of the Brunels, François Carpentier, the American
Consul at Rouen and a retired sea captain, who undertook to

train him for service in the Navy with the help of M. Dulagne, an authority on hydrography.

Apart from its architecture, which he delighted to draw,
there was little to be found in Rouen to feed Marc's enthusiasm. But one day when he was exploring the waterfront
his imagination took wing at the sight of two enormous cast-
iron cylinders which had just been landed on the quay. His
excited questions brought the answer that they were part of
a new 'fire' engine for pumping water and that they had
come from England. 'Ah!' he exclaimed. 'Quand je serais
grand, j'irai voir ce pays-là.' Little did he know then the
vicissitudes and adventures that lay between this wish and its
fulfilment.

Having obtained a commission from the Minister of
Marine with the help of M. Dulagne, Marc Brunel joined
the corvette *Le Maréchal de Castries* and sailed to the West
Indies as a 'Volontaire d'honneur' in 1786. Just as apprentices
were wont to produce a 'masterpiece' on completing their
training so, as proof of his ability, Marc had made for himself a small ebony quadrant which he used throughout his
six years' service at sea.

The storm clouds of the Revolution lay heavy over France
when Marc, his ship paid off, found himself in Paris in January 1792. He was, and indeed remained for the rest of his life,
a staunch royalist, and as a result of a bold but imprudent
remark in a café he was forced to flee the city, finding temporary refuge with his old patron, Captain François Carpentier, in Rouen, where Royalist sympathies were still
strong and where he became a member of the loyalist guard.
It was here that he met his future wife, Sophia, for the first
time. The daughter of William Kingdom, a Plymouth naval
contractor, she was an orphan, the youngest of sixteen children, and she had been sent to learn French with the Car-
pentiers by her eldest brother and guardian. Sophia was an
extremely personable and accomplished young woman and
her graces immediately captivated the susceptible Marc. At

first his advances were foiled by Sophia's zealous chaperon, Mme Carpentier, but the young royalist, living in imminent danger and involved, as a member of the guard, in frequent clashes with the republican mob, cut a romantic figure not easy to resist. Love thrives in such an atmosphere of risk and tension, and one night, after a particularly serious clash in the streets of Rouen, Marc and Sophia declared their love for each other.

The royalists of Normandy had earlier suggested that the King and his family should take refuge in the Château Gaillard, that great fortress built by Richard Coeur de Lion which towers above the Seine. Had Louis accepted this advice instead of making his ill-judged and fatal flight to Varennes, French history might have been different, and Marc Brunel might never have come to England. As it was the full fury of the Jacobin Terror followed the imprisonment and death of the King. Shortly after Marc's engagement the insurrection of Normandy and Brittany against the Jacobins provoked ferocious reprisals, and it became obvious that for Marc to remain any longer in France meant almost certain death. Through the good offices of François Carpentier a passport was obtained for him on a false pretext from the American Consulate. A fall from his horse on the road from Rouen would have foiled his escape had he not been picked up by a passing traveller and taken to Le Havre, where he boarded the American ship *Liberty* bound for New York on 7 July 1793. Even then his safety was not assured, for it was not until the ship was challenged by a French frigate searching for runaways that Marc discovered to his horror that he had lost his precious passport in the accident. In the two hours available to him he managed to forge a document so convincing that it passed the scrutiny of the French boarding party without question. One can only imagine his feelings as he presented it for examination with the ink scarcely dry!

His landing in New York on 6 September marked the beginning of a six years' stay during which, characteristically, he acquired considerable reputation as architect and engineer but secured little or no return for his services. His first venture was to set out with two fellow *émigrés* through what was then practically virgin territory to survey the country in the neighbourhood of Lake Ontario. It was on the return from the expedition that Marc fell in with the American, Thurman, who engaged him to survey the course of a projected canal to link the Hudson with Lake Champlain. This meeting with Thurman marked another turning point in his life, for it encouraged him to abandon his dream of rejoining the French Navy when order was restored in his country and to become instead a professional engineer.

His next exploit was to submit the winning design in a competition for a new Congress Building at Washington. This design was never realized on the score of expense, but in modified form it subsequently took shape as the Palace Theatre, New York, which was destroyed by fire in 1821. Finally, after he had assumed American citizenship, he was given the post of Chief Engineer of New York, in which capacity he designed a new cannon foundry and advised on the defences of Long Island and Staten Island.

In 1798 Marc Brunel was invited to dine with Major-General Hamilton, the British aide-de-camp and secretary at Washington, in order that he might meet an exiled fellow countryman, M. Delabigarre, who had just arrived from England. Talk of Britain's victories at Cape St Vincent and Camperdown and of her expanding naval power re-awakened the desire to visit England which Marc had first felt when he had seen as a boy the beam-engine cylinders on the quay at Rouen. Moreover, during this talk of the sea and ships, M. Delabigarre happened to mention the method of manufacturing ships' blocks which was then a monopoly of Messrs Taylors of Southampton. When it is realized that in those great days of sail a seventy-four-gun ship of the line required no less than 1,400 blocks, the importance of this item of equipment and the effect of its high cost and limited supply on naval expansion will be appreciated. It was this chance conversation which prompted Marc to evolve improved block-making machinery and to journey to England with the object of realizing his designs. His decision was certainly influenced by another more personal consideration. He had maintained a correspondence with Sophia Kingdom, who, having been imprisoned for a time in a French convent, had managed to return to England. Sophia had recently acknowledged in a most encouraging manner his gift of two miniatures, a portrait of her which he had painted from memory and a self-portrait, both of which are still in the possession of his descendants.

Staking his future on his block-machinery designs, Marc sailed from New York on 20 January 1799 and landed at Falmouth in March. At this time a Frenchman was regarded by officialdom in England with as much suspicion as a visitor from behind the Iron Curtain might be today, but Marc was favoured in two ways: by the reputation he had gained during his six years in America, and by an introduction from his friend General Hamilton to Earl Spencer of Althorp, then First Lord of the Admiralty under Pitt. In Lord and Lady Spencer the Brunels were to find most loyal and devoted friends whose influence was to prove invaluable to both father and son. Notwithstanding these advantages, however, Marc could never pass the gates of Portsmouth Dockyard without a special permit, even when he was superintending the erection of his own machinery there.

It seems fairly certain that Marc's drawings of his block-making machinery made a contribution to British engineering technique much greater than the machines they represented. For it is safe to assume that he had mastered the art of presenting three-dimensional objects in a two-dimensional plane which we now call mechanical drawing. It had been evolved by Gaspard Monge of Mezieres in 1765 but had remained a military secret until 1794 and was therefore unknown in England. But to produce beautiful drawings of what were, at that time, intricate machines calling for very precise workmanship was one thing; to get them accepted and to find someone capable of making them in a strange country was quite another. Once again, however, fortune favoured him.

Marc proceeded straight from Falmouth to London, where his Sophia was living with her eldest brother, who, like his father, was a naval contractor. Here, after six years' exile, the lovers were reunited. He had not been in London long before he fell in with a fellow *émigré* by the name of de Bacquancourt to whom he mentioned the problem of finding someone who could construct his machines. From this chance encounter began his association with the man who was without doubt the greatest mechanic of the age. For de Bacquancourt advised Brunel to call at a little shop in Wells Street, off Oxford Street, where he found, working with a single journeyman assistant, Henry Maudslay.

Whereas the names of James Watt, Trevithick, Telford, the Rennies, the Stephensons and the Brunels have become household words, insufficient credit has been given to the great mechanics, to the men who supplied the practical 'know-how', who designed and built the machines and evolved the workshop techniques without which the schemes of the engineers could never have taken workable shape. Although Maudslay learnt his craft in the workshop of Joseph Bramah, himself no mean engineer and inventor, it is Maudslay and not Bramah who deserves to be known as the father of the modern machine shop. Maudslay's slide rest lathe, his screw-cutting lathe and planing machine introduced standards of accuracy in machining hitherto unknown. And because they were self-propagating they initiated a process of improvement and ever-increasing precision which has continued down to the present day. But Maudslay not only begat a new race of machines, he was also the father of a school of engineers. Joseph Clement, James

Nasmyth, the inventor of the steam hammer, Joseph Whitworth, Richard Roberts, the founder of the famous firm of
Sharp, Roberts & Company, these were only a few of the
distinguished men who learnt their craft in the school of
Maudslay. Soon after his first meeting with Marc Brunel,
Maudslay moved to a larger workshop in Margaret Street,
Cavendish Square. By 1810 this in turn had become too small
and so he set up his plant in a disused riding school which he
had bought in the marshes of Lambeth. Thus was founded
the works of Maudslay, Sons & Field, perhaps the most
famous engineering firm of the nineteenth century and one
with which the Brunels, father and son, were to be closely
associated.

Maudslay was a consummate craftsman, and Smiles has
left a delightful picture of him in his private workshop in his
old age, working at his vice to the sound of one of the many
musical boxes which he loved, and repeatedly snuffing himself or pausing to illustrate his ideas with rapid sketches
chalked on the work-bench. After his death, one of his old
workmen said of him: 'It was a pleasure to see him handle a
tool of any kind, but he was *quite splendid* with an eighteen-
inch file.' No engineering craftsman could wish for a higher
tribute.

This, then, was the man to whom Marc brought his drawings, and after several meetings, as mutual confidence grew,
Maudslay agreed to produce models of the block machines.
The problem which remained was how to finance their full-
scale manufacture. First, Marc's future brother-in-law,
Kingdom, wrote to the existing manufacturer, Samuel Taylor, of Southampton, on his behalf, but the latter replied:
'My father has spent many hundreds a year to get the best
mode, and most accurate, of making the blocks, and he certainly succeeded; and so much so, that I have no hope of
anything ever better being discovered, and I am convinced
there cannot.' After this unqualified rebuff Marc decided to
make use of his introduction to the Spencers, and through

them his designs were brought to the notice of Sir Samuel Bentham, Inspector-General of Naval Works. Bentham was himself an accomplished engineer who, in anticipation of the expiry of the Taylors' contract, was already planning an improved block-making plant for Portsmouth Dockyard. He at once appreciated the value of Marc's designs, and it was eventually arranged that Henry Maudslay should make the new machines while Marc superintended their installation at Portsmouth.

On 1 November 1799 Marc married his Sophia at the church of St Andrew's, Holborn, and the couple settled in a small house at Portsea. Marc never failed to remember his wedding anniversary, for he made an ideal match which was his great solace in all the misfortunes and disappointments which later beset him. At the age of seventy-six he would write in touching acknowledgement: 'To you my *dearest* Sophia I am indebted for all my successes.'

Maudslay laboured for six years on the block machinery, and there were innumerable delays, difficulties and needless troubles before the plant was finally finished and working satisfactorily. Meanwhile the children were coming to Marc, first Sophia, destined to marry Benjamin Hawes and to become known as 'Brunel in petticoats', and then the weaker and milder Emma who became the wife of the Rev. George Harrison. The Brunels were still at Portsea in 1806 when Marc wrote in his journal: 'On the 9th of April, and at five minutes before one o'clock in the morning my dear Sophia was brought to bed of a boy.' This was Isambard, his only son.

Marc's claim to the sole invention of the block machinery has been constantly disputed. After Sir Samuel Bentham's death his widow claimed for him the major share of the credit, while others have asserted that Henry Maudslay did much more than give shape to a set of finished designs. Yet again, others assert that insufficient credit has been given to the pioneer work of the Taylors, whose machines at Southampton were by no means so crude as Marc Brunel's biographer, Richard Beamish, implies. Nevertheless the facts remain that the block machinery established Marc's reputation in England and that, after considerable difficulty, it is true, he eventually obtained from the Admiralty the sum of £17,000 for his work, a figure which disappointed him and was considered inadequate by Lord Spencer. It was claimed that with his machines six men could do what sixty had done before. Certainly they represented what was perhaps the first example of fully mechanized production in the world and they soon became a showpiece for visiting notabilities. 'Machinery so perfect appears to act with the happy certainty of instinct and the foresight of reason combined,' wrote one admiring visitor, Maria Edgeworth; while Tsar Alexander I was so impressed that he tried to persuade Marc to return with him to St Petersburg and gave him, as a token of friendship and esteem, the magnificent ruby encircled with diamonds which is still in the possession of the family. The Tsar's invitation was to stand him in good stead later.

Despite Marc's dissatisfaction, this first English enterprise brought a greater financial return than his later ventures. When his work at Portsmouth was completed he decided to sink all his capital in a sawmill at Battersea, and the family moved from Portsea to No. 4 Lindsey Row, Chelsea. He had also obtained another Admiralty contract for installing improved timber-handling and sawing machinery at Chatham Dockyard and was full of enthusiasm for a new scheme for manufacturing army boots by machinery. This last had occurred to him after seeing, at Portsmouth, the wretched state of the troops returning from Corunna. One feature of his work at Chatham was later to be turned to far more celebrated purpose by his young son. This was a wide-gauge railway for conveying logs from the hoists to the sawmill and it consisted of wrought-iron rails continuously supported on longitudinal timber baulks with cross-ties.

For a time all went well. The Battersea sawmills were completed and yielded a handsome profit. The bootmaking machinery was approved and the Government advised him to lay down a plant with an output sufficient to supply the army's entire needs. Marc accepted this as a guarantee, built the factory and its machines, manned it, at his own suggestion, with disabled ex-soldiers, and was soon turning out boots at the rate of 400 pairs a day. Although the Government had been ready enough to persuade him into the venture, they proved much less eager to purchase his output and in the sudden peace which followed the battle of Waterloo he found himself landed with a vast stock of boots which the army authorities would not take. To this heavy financial loss was added the consequences of another disaster. On the night of 30 August 1814 his Battersea sawmills were practically destroyed by fire. Most unfortunately the outbreak coincided with another serious fire on Bankside which had engaged nearly all the available engines. Marc was at Chatham when he heard the news and at first he was not in the least downcast but at once began preparing plans for rebuilding the mills. It became apparent later, however, that while he had been preoccupied with his other concerns his partner at Battersea had allowed the finances of the mills to become hopelessly involved. Capital had been squandered as income and, although Marc did succeed in re-equipping the mill, in the words of his biographer, 'no amount of ingenuity could call back the balance in his banker's books'. As a result of these two misfortunes Marc's personal finances went from bad to worse until, on 14 May 1821, he was arrested for debt and confined in King's Bench prison whither his faithful Sophia accompanied him.

The existing law was such that Marc Brunel was by no means the only eminent man to become familiar with the inside of a debtor's prison, but for years afterwards this episode in his career was taboo in the family. If it had to be mentioned, it was referred to darkly as 'the Misfortune'. Not unjustifiably, Marc felt that the Government had let him down badly over the bootmaking enterprise, and it was on this score that he appealed for help to his influential friends. But the negotiations were protracted, and as the weeks went by the anxiety and strain began to tell heavily upon him. 'My affectionate wife and myself are sinking under it,' he wrote despairingly to Lord Spencer. 'We have neither rest by day nor night. Were my enemies at work to effect the ruin of mind and body, they could not do so more effectually.' It was only when he threatened to take advantage of Tsar Alexander's invitation and remove to Russia that action was forthcoming. Through the good offices of the Duke of Wellington he was granted £5,000, which enabled him to obtain his discharge, upon the condition that he would remain in England.

This short sketch of Marc Brunel's life down to his release from prison in July 1821 has included only those exploits which shaped his career and so affected the fortunes and movements of his family. The range of his activities after his coming to England was in fact much wider and displays his remarkable powers of invention and his versatility both as a civil and as a mechanical engineer. He invented a stocking-knitting machine, improved printing machinery and carried out experiments in the use of paddle wheels and screw for marine propulsion. In the civil engineering field he designed bridges and docks and was also responsible for the design of Liverpool's first floating landing stage. He carried out government work at Woolwich as well as at Chatham and advised on many of the more important engineering projects of the day. Very few of his works seem to have brought him an adequate financial return, and the impression gained is of a man plunging into any new enterprise with such whole-hearted enthusiasm for its technical possibilities that he could not stop to think about money, to protect his ideas or to weigh the probity of his associates. Nevertheless in spite of his financial difficulties he managed to give his only son an excellent education.

Young Isambard began to display his talent for drawing when he was only four years old, and by the time he was six he had mastered his Euclid. Such a precocious display of inherited talent obviously delighted Marc so that he determined to foster it to the limit of his means. He first sent him to Dr Morell's boarding school at Hove, where the boy amused himself in his spare time by making a survey of the town and sketching its buildings, just as his father had done in his own youth at Rouen. Marc always insisted that this drawing habit was as important to an engineer as a knowledge of the alphabet, and it was undoubtedly in this way that both father and son developed such extraordinarily acute powers of observation. So uncanny was their gift for instantly detecting an error in design or workmanship that sometimes it seemed to the uninitiated that they must possess second sight. 'You would not venture, I think, on that bridge unless you would wish to have a dive,' was Marc's comment after one glance at a section drawing of M. Navier's first suspension bridge over the Seine at Paris, and a few days later he heard the news that the bridge had collapsed. On another occasion when he was passing a newly erected storeshed at Deptford with Richard Beamish he suddenly quickened his pace, exclaiming in some agitation: 'Come along, come along; don't you see, don't you see?' and in answer to his friend's puzzled query he pointed to the building: 'There! Don't you see? It will fall!' The next morning the building was in ruins. In exactly the same fashion young Brunel astonished the other boys by successfully predicting the fall of part of a new building which was going up opposite his school at Hove.

Holidays at Lindsey Row before 'the Misfortune' must have been pleasant indeed. No. 10 was a reminder of the past splendours of the Chelsea village waterfront, for it was a part of old Lindsey House which had been built by Sir Theodore Mayerne, physician to James I and rebuilt by Lord Lindsey, Lord Great Chamberlain to Charles II. The portion occupied by the Brunels included the original great staircase, while spacious rooms with wide fireplaces of marble and porphyry made an ideal setting for Christmas or Twelfth Night parties when the young people danced and their elders played whist. The two Miss Brunels had just left school by this time and are described as 'models of everything young ladies should be', but their young brother was far less inhibited. Whenever a party or some game or charade was in the air he was the ring-leader; he loved to walk along the top of the garden wall so that he could joke and gossip with the Miss Mannings who lived next door. In summer there was swimming in the river from the steps below the house, and exciting excursions to town, then still remote from the rural quiet of Chelsea, made almost invariably by boat.

When his son reached the age of fourteen, Marc Brunel sent him first to the College of Caen in Normandy and later to the Lycée Henri-Quatre in Paris, which was at that time famous for its mathematical teachers. Finally, his Continental education was completed by a period of apprenticeship under Louis Breguet, maker of chronometers, watches and scientific instruments. This last move reflects the soundness of Marc's judgement. He could not have found a finer or more critical nurse for his son's mechanical talents. What Henry Maudslay was to machine-tool making, Abraham Louis Breguet was to watchmaking, and he is regarded by most horologists as the supreme craftsman in this field. Moreover, like Maudslay, he was the father of a school of craftsmen who in after life were proud to inscribe themselves as 'Élève de Breguet'.

The boy may already have acquired some interest in clocks and watches from Sophia, for her mother had been sister-in-law to Thomas Mudge, the inventor of the lever escapement and one of the greatest horologists of his day. When Mrs Kingdom, with her large family, was widowed at Plymouth, Mudge was working there on his marine chronometer, so he very probably helped her. One of Sophia Kingdom's treasures which her son inherited and passed on to his descendants was one of the only two bracket clocks with lever escapement which Mudge ever made. He would value this legacy all the more highly through having served so great a craftsman as Breguet.

He was lucky to have known such a master, for Breguet was nearly at the end of his life. Born at Neuchâtel in 1747 he died in 1823, leaving his business to be carried on by his son, Louis Antoine. The exacting standards of workmanship upon which Brunel insisted throughout his life and which are revealed in the superb quality of his timepieces and of his engineering instruments, many of which he made or repaired himself, were undoubtedly formed at this time. That the old watchmaker thought highly of his pupil is evident in the letter which he wrote to Marc about him on 1 November 1821. 'Je sens,' he writes, 'qu'il est important de cultiver chez lui les heureuses dispositions inventives qu'il doit à la nature, ou à l'éducation, mais qu'il serait bien dommage de voir perdre.' Living in straitened circumstances and with the memory of her recent sojourn in the King's Bench prison all too vividly in mind, Sophia Brunel was not unnaturally opposed to the idea of her only son following his father's hazardous profession, but such a tribute and such advice from so eminent a master as Breguet silenced her objections and her husband, needless to say, was delighted.

During his three years in the Paris of the Bourbon Restoration, Brunel was subjected to another formative influence besides the craftsmanship of Louis Breguet. This was the fashionable admiration for the art and the architecture of the *Grand Siècle* which sharpened that sense of grandeur, of great occasion, which would so distinguish all his work.

In 1822, when he was sixteen, he returned to England and began to work with his father in his little office at No. 29 Poultry which was staffed by a single clerk. That he also spent a lot of his time in that great school of engineers Maudslay, Sons & Field of Lambeth is clear from a letter he was to write to Henry Maudslay's son many years later in which he referred to 'your firm, with which all my early recollections of engineering are so closely connected and in whose manufactory I probably acquired all my early knowledge of mechanics'. Breguet and Maudslay – two greater mentors could not have been found in all Europe.

So gifted was young Brunel and such was the intelligence and enthusiasm which he brought to bear upon his father's projects that in spite of his youth he rapidly became more of a trusted and able partner than an assistant. He was thus able to play a major role when, in 1824, Marc Brunel embarked upon his greatest adventure – the boring of a tunnel under the Thames.

[2]

Ordeal by Water

WHEN Marc Brunel planned his tunnel beneath the Thames no sub-aqueous tunnel existed if we except certain mine workings such as those of the celebrated Botallack mine near St Just in the farthest west of Cornwall. In May 1798 Ralph Dodd, the eccentric engineer of the Grand Surrey Canal, advocated a 900-yd tunnel between Gravesend and Tilbury but no work was done. In 1802 the idea was revived by a Cornish engineer, Robert Vazie, who proposed a tunnel from Rotherhithe to Limehouse. This time considerable capital was raised and an Act of Parliament passed incorporating the Thames Archway Company, whose proprietors held their first meeting at the Globe Tavern in Fleet Street on 25 July 1805.

Vazie's plan was to drive a small pilot tunnel or driftway under the river which would serve as a drain during the construction of the tunnel proper. Under his direction a shaft 11 ft in diameter was sunk at a distance of 315 ft from the river in the neighbourhood of Lavender Lane, Rotherhithe. Such was the difficulty experienced owing to the repeated influx of land water that by the time a depth of 42 ft had been reached the capital had been exhausted. When more money had been raised Vazie sank a further 34 ft at the reduced diameter of 8 ft, only to encounter a quicksand. This filled the shaft with water and the proprietors with alarm and despondency. The advice of John Rennie and William Chapman was sought, but as these two eminent engineers failed to agree, or to give the unhappy Company any constructive guidance, no less a man than Richard Trevithick was called in on the recommendation of his friend Davies Giddy and Vazie. The former, under the name of Davies Gilbert, was later to play a part in Brunel's career. At first Trevithick worked as resident engineer under Vazie's direction, but the latter was subsequently dismissed, to his great and understandable disgruntlement, and Trevithick assumed sole charge. Working with a picked team of Cornish miners, then the most experienced men in the world at this class of work, Trevithick cleared the shaft and began to excavate the driftway. Treacherous quicksands were repeatedly encountered through which water poured into the workings, yet in six months the heading had been driven over 1,000 ft out of the total length of 1,200 ft. This was a heroic feat, for the drift was only 5 ft high by 3 ft wide, so that the appalling working conditions in this confined space under the constant threat of death may be vividly imagined.

The heading had actually reached low tide mark on the Limehouse shore when, on 26 January 1808, under the stress of an abnormally high tide, a terrific break-through occurred which quickly overpowered the pumps. The miners were submerged up to their necks as they retreated and Trevithick himself, who was the last to leave, narrowly escaped with his life. He was still undaunted, however. Clay flung down in the bed of the Thames above the breach reduced the flow, while to effect a permanent repair he proposed constructing a coffer dam of sheet-piling. The directors of the Archway Company were not so stout-hearted; work came to a standstill while they called in more consultants in the hope, which was happily frustrated, of making Trevithick the scapegoat for their misfortunes. Finally William Jessop pronounced 'that it was impracticable to make under the Thames a tunnel of useful size by an underground excavation'. Trevithick countered this by proposing to excavate the bed of the river from above by means of a series of coffer dams and to lay within this excavation a sectional cast-iron tunnel like a huge pipeline. Such an idea was typical of this great, far-sighted engineer, but it was not adopted and the Thames tunnel works were abandoned. It was left to the

Americans in this century to prove the soundness of Trevithick's proposal when the Detroit river and San Francisco Harbour tunnels were both built successfully by his method.

Marc Brunel must have followed Trevithick's heroic efforts with keen interest, but apparently what first led him to ponder the possibilities of underwater tunnelling was the problem of effecting a crossing of the Neva at St Petersburg, where spring ice floes threatened to carry away the piers of any bridge. Although nothing came of the Neva plan it resulted in the most celebrated of all his inventions – the tunnelling shield which he patented in 1818. The principle of this shield occurred to him when he observed the tunnelling action of the destructive 'ship-worm' *Teredo navalis* in ships' timbers while he was working at Chatham Dockyard.

In 1823 I. W. Tate, one of the promoters of the old Archway Company, learnt of Marc's invention and urged him to discuss his ideas with some influential friends. So impressed were they that a meeting was held at the City of London Tavern on 18 February 1824 where a considerable sum was raised by subscription. A bill to incorporate a Thames Tunnel Company was promoted and received the royal assent in the following July. Marc was appointed engineer at a salary of £1,000 a year. He was also awarded the sum of £5,000 for the use of his patent with the promise of a further £5,000 when the tunnel was completed. In order to be nearer the work, Marc shut his little office in the Poultry and moved with his family from Lindsey Row to No. 30 Bridge Street, Blackfriars. Poor Sophia, she cannot have relished this removal to a neighbourhood which was then considered so unfashionable. But such upheavals are often the lot of an engineer's wife, and at least she fared better than her predecessor Jane Trevithick, who complained bitterly when she was uprooted from her native Camborne and planted in Rotherhithe in what her son described as 'a gloomy situation near the mouth of the driftway'. Sophia made the best of a bad job and transformed No. 30 into an oasis of comfort and elegance which astonished her visitors.

The site selected for this new attempt was about three-quarters of a mile west of Trevithick's abandoned driftway from a point near St Mary's, Rotherhithe, to the Wapping shore near the junction of Wapping Lane with Wapping High Street. It had been decided to drive the tunnel at a level which would bring the crown of the arch only 14 ft below the bed of the Thames at its deepest point, for by doing so the geologists had assured Marc, after making numerous trial borings, that he would find a stratum of strong blue clay and avoid the quicksands which had beset Trevithick at a deeper level.

As in the case of the previous attempt, the work began with the sinking of a shaft, this time at Cow Court on the south bank, but the scale of this shaft and the method of sinking it were vastly different. It was constructed above ground, a great cylinder of brickwork 50 ft in diameter and 42 ft high so securely braced by iron tie rods secured to cast-iron rings called 'curbs' at top and bottom that it formed a completely rigid unit. The plan then was to excavate the ground from within and beneath it so that it would sink into position by its own weight.

The opening ceremony of the undertaking was performed by the chairman of the Company, William Smith, M.P., on 2 March 1825, Marc Brunel laying the first brick and young Isambard the second. This was followed by the customary junketings. While the bell-ringers of St Mary's worked with a will a company of two hundred sat down to what was described as 'a sumptuous collation'. A model of the tunnel in sugar decorated the table, while a bottle of wine was ceremonially laid aside for the similar banquet which would crown the success of the venture. Mellowed by a combination of good wine and optimistic speeches, few of those present could have guessed how many years were to pass before that bottle would be broached.

The shaft was built to its full height in three weeks, each bricklayer laying a thousand bricks a day. This done, the work of excavating the ground beneath began, the shaft slowly sinking at an average rate of 6 in. a day. Not since the days of the Cherry Garden had Rotherhithe seen the rank and fashion which now flocked to watch the progress of this unique operation which Marc Brunel had so carefully planned. The company included the Dukes of Cambridge, Gloucester and Northumberland, Prince Leopold, Marc's old friend Lord Spencer, the Duke of Wellington and Lord Somerset. The last two gentlemen ventured to descend the shaft, where they were much struck by the grinding noise, reverberating between the walls, as the iron curb which formed its base slowly sank under the weight of nearly 1,000 tons of masonry.

On 6 June the shaft reached its full depth and the work of underpinning and removing the iron curb was begun. At the foot of the shaft, below tunnel level, a large reservoir was constructed to receive the water draining from the workings, and above it the steam-driven plunger pumps were installed. Meanwhile Henry Maudslay had been building Marc's famous tunnelling shield in his works at Lambeth and the time had now come for its installation. In the light of subsequent events, some description of this machine is essential.

It consisted of twelve massive cast-iron frames 21 ft 4 in. high by 3 ft wide, their feet connected by ball-joints to broad iron shoes which rested on the floor of the excavation. Six of these frames occupied each of the twin arches of the tunnel and as they were divided into three storeys they formed altogether thirty-six working cells, each of which was to be manned by one excavator. Above the heads of the men working in the uppermost cells pivoted plates, called 'staves', upheld the roof of the excavation and so corresponded to the shoes at the bottom. Similar staves protected the sides of the excavation, while rollers were interposed between each frame and its fellow to assist their relative movement. To the

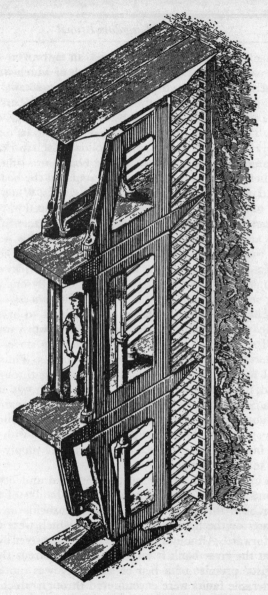

One Section of the Brunel Tunnelling Shield
(from Richard Beamish's *Life of Sir Marc Isambard Brunel*, 1862)

working face each frame presented an unbroken tier of heavy oak planks or 'poling boards' each of which was held against the face by two jacks called 'poling screws' which butted against recesses in the main vertical members of the frames and which, as designed by Marc, had a travel of only $4\frac{1}{2}$ in. The method of working, which seems to us today incredibly slow and laborious, was as follows. Each workman in the frames removed only one poling board at a time, excavated behind it to a depth of $4\frac{1}{2}$ in., replaced the board and secured it with the now extended poling screws. When he did so the butt ends of the poling screws were placed in recesses provided for them in the uprights of the adjacent frames. When all the ground had been excavated in this way, the frame was thus free to be moved forward $4\frac{1}{2}$ in. by horizontal jacks bearing against the newly completed brickwork behind at top and bottom of the arches. One by one all the poling screws were then retracted, and the process of excavation began again. Bricklayers worked back to back with the excavators in the frames, and the brickwork was thus carried right up to the tails of the protecting staves which together formed the shield round the frames. Thus Marc Brunel had solved the problem of tunnelling through soft ground, for, except for a space of $4\frac{1}{2}$ in. which was exposed when a frame had just been moved, the whole excavation was continuously supported. Behind the shield there followed a timber staging, mounted on wheels and equipped with a hoist which facilitated the removal of spoil and the supply of materials to the bricklayers.

The great shield was completely installed and began its laborious advance on 28 November 1825, baulks of timber having been thrown across the shaft to provide an initial purchase for the horizontal jack screws which were to propel it forward. Although the excavation was as yet over 100 ft from the river bank it was very soon apparent that the geologists' promise of a bed of solid clay was quite false. Considerable faults were encountered through which water poured into the workings. However, it was a case of slow but sure until the directors, dissatisfied with the rate of progress, insisted upon putting the men on piecework and doubling the travel of the poling screws. This meant that each frame would in future excavate twice as much at each 'bite'. Marc protested in vain that piecework meant careless work and that the lengthening of the screws was extremely risky because twice the previous area would be left unsupported when a frame was moved forward.

Meanwhile the volume of water which was encountered revealed the falsity of another economy upon which the directors had insisted. Following the precedent established by Trevithick, Marc had proposed cutting a driftway below the main tunnel to carry off the water, but this was refused. If Marc had had his way in this, subsequent disasters would have been considerably mitigated and the working conditions would have been more tolerable. Owing to the downward inclination of the tunnel (a fact not made clear in contemporary section drawings which show it perfectly horizontal) as many as forty men had to be employed with bucket-pumps clearing water from the shield whence it flowed through an iron pipe to the reservoir below the shaft. Sand in the water caused constant trouble with these pumps, and the men in the lower cells of the frames were often working up to their knees in water. Moreover, because the Thames at this period was little better than an open sewer, the stench at the working face was appalling. Working under such conditions it is not surprising that throughout the whole period of construction sickness claimed a much heavier toll than accident. This was a danger which Marc had not bargained for and against which his shield could offer no protection. In one horrible form of this 'tunnel sickness' men were quite suddenly struck blind and some, even if they survived, which many did not, never recovered their sight.

When the work began William Armstrong, assisted by

young Isambard, held the post of resident engineer under
Marc, but the shield had not begun its advance before Marc
was taken seriously ill, while in the following April Armstrong broke down and resigned his post. So it came about
that while Marc Brunel was author and stage director, it was
his son who, in the role of engineer in charge, took over at
short notice the leading part in an engineering drama which
excited the wonder of Europe. The fact that he was barely
twenty years of age occasioned little remark; yet today, when
an educational fetish prolongs childish irresponsibility far
into adolescence, it seems almost incredible that such an immense burden of responsibility should have been laid on
such young shoulders and that it should have been carried
with such distinction. Throughout all the hazards which beset the works, Marc's constant anxiety was tempered by his
pride in his son and his admiration for his remarkable
powers.

And what an amazing drama it was, this stubborn
struggle between man and earth which went on relentlessly,
month after month, year after year in the darkness under
the Thames. Always dramatic, and sometimes tragic, upon
one occasion it became sheer fantasy. Isambard Brunel
threw into the work all that unsparing energy which was to
distinguish his whole life. For as much as thirty-six hours at
a time he would not leave the tunnel, pausing only for a
brief cat-nap on the wooden stage behind the shield. For fear
that he would work himself to death he was given three assistants, Beamish, Gravatt and Riley. The tunnel soon
claimed Riley; he died of fever in January 1827 after only two
months' work. Beamish, too, was struck down. He suddenly
became blind in one eye and never fully recovered his sight.
Isambard also suffered a short spell of illness in the autumn
of 1826 but was soon back at his post as indomitable as
ever.

As the great shield crept slowly under the river, so the
wonder grew, and when 300 ft of the western arch had been

completed the directors of the tunnel company began to recoup their fortunes by admitting sightseers at a shilling a head. A barrier was erected to prevent them approaching the shield and here they gathered in awe-struck crowds to peer through the gloom at the men who toiled in the gas-lit tiers of the frames. To Marc these visitors were an added anxiety and he protested in vain against their admission. For with that foresight which was almost prevision, both father and son had measured the appalling risk they were running and guessed that disaster was imminent. Instead of the continuous bed of clay which they had been so confidently promised, they had encountered faults where only gravel separated them from the waters of the river. 'During the preceding night,' wrote Marc Brunel in his diary on 13 May 1827, 'the whole of the ground over our heads must have been in movement, and that, too, at high water. The shield must, therefore, have supported upwards of six hundred tons! It has walked many weeks with that weight, twice a day, over its head! Notwithstanding every prudence on our part a disaster may still occur. May it not be when the arch is full of visitors!'

The debris which fell through into the workings from the bed of the river – pieces of bone and china, an old shoe buckle and finally the sheaf of a block and a shovel – left no doubt as to the state of the ground above. Everything depended on the shield which alone prevented immediate collapse and the movement of a frame became a most hazardous operation. First there was the risk of a collapse at the working face when the thrust of the poling screws was transferred to the uprights of the adjacent frames, and then when the frame was shifted forward there was exposed that perilous gap between the new brickwork and the tail of the staves. Marc's diary records his growing apprehension as conditions became more treacherous: 15 May 'The water increased very much at 9 o'clock. This is very *inquiétant*! My apprehensions are not groundless.' 18 May 'Visited by Lady

Raffles and a numerous party. I attended Lady Raffles to the frames, most uneasy all the while as if I had a presentiment . . .'

High water in the river was always the time of danger and on the evening of the 18th as the tide was making, Beamish, who had relieved Isambard Brunel at the frames, put on his waterproof, sou'wester and mud boots in the expectation of a wet night. The tide was making fast and more and more water poured through the shield. Nos. 9 and 11 stages were ready to be worked forward and the screws were being moved on the top boards of No. 11 when suddenly Beamish heard Goodwin, one of the best of the picked corps of miners who manned the frames, cry out for help. Beamish at once called for assistance from the men in the next frame, but before they had time to reach Goodwin a tremendous torrent of water roared like a mill sluice out of No. 11, bowling over a bricklayer named William Corps who would have been swept off the stage behind the frames had not Beamish managed to cling to him. Beamish wanted to make a final attempt to enter No. 11 frame but it was impossible. Rogers, one of the men from No. 9, caught his arm, shouting above the thunder of the water: 'Come away, sir, come away; 'tis no use, water's rising fast.' At the visitors' barrier he encountered Brunel and both men turned for a last look at the shield. 'The effect', wrote Beamish afterwards, 'was splendid beyond description; the water as it rose became more and more vivid from the reflected lights of the gas.' Then a great wave advanced down the tunnel, carrying with it a chaos of empty cement barrels, boxes and timber spars. It struck a small office building which had been erected midway between the shield and the shaft; there was a deafening crash, and a blast of air as the building collapsed and then all the lights went out. Someone shouted 'The staircase will blow up,' and Brunel's voice could be heard ordering the men to ascend the shaft as quickly as possible. Scarcely had the last man cleared it than the lowest flight of the spiral stairway in the shaft was swept away. No sooner had the soaked and exhausted group gained the surface than a faint cry for help was heard from below. The voice was recognized as that of Tillett, the old engineman who had gone below to repack his pumps. Without an instant's hesitation Brunel called for a rope, seized it and slid down one of the iron ties of the shaft into the tumultuous darkness. Miraculously, he managed to find the old man and to secure the rope round his waist as he struggled in the water. When both had been hauled safely up the shaft a roll-call was held. Not a man was missing.

Now that the disaster, so long dreaded, had actually happened without loss of life Marc Brunel was able to write in his diary next day: 'Relieved as I have found myself, though by a terrible catastrophe, of the worst state of anxiety that I have been in for several weeks past, I had a most comfortable night.' The day following being a Sunday, Marc noted: 'The Rotherhithe Curate in his sermon today, adverting to the accident, said that it was "but a just judgement upon the presumptuous aspirations of mortal men ... etc!" The poor man!'

The Brunels were not men to sit down with folded hands and bewail their misfortune. Only twenty-four hours after the disaster Isambard descended to the bed of the river in a diving bell borrowed from the West India Dock Company to take stock of the damage. There he found a great depression which, as he suspected and Thames watermen afterwards confirmed, had been caused by gravel dredging. At the bottom of this hole he was able to step from the footboard of the diving bell and stand with one foot on the tail of the shield staves and the other on the brickwork of the arch. Bags of clay laced with hazel rods to form a bond were flung into this breach and a raft loaded with 150 tons of clay was sunk. For a while the pumps began to master the water in the tunnel, but the following day there was a further inundation and it was found that the raft had tilted over on the ebb tide. Marc, who had been persuaded into the raft expedient against his better judgement, now decided to raise it and rely upon the clay bags alone. The raft was successfully grappled with chains, afer which iron rods were laid across the gap between the shield and the brickwork to form a bed for the clay bags.

In the course of these operations Isambard Brunel had gone down in the diving bell with an assistant named Pinckney when Beamish, who was standing on the bell barge, was horrified to see the footboard of the bell suddenly float to the surface. For a moment he feared the worst until a tug on the communicating cord urged him to haul in. Pinckney had foolishly let go his hold on the bell in stepping out on to the treacherous ground near the breach and it had given way beneath him. Brunel had promptly extended his leg below the mouth of the bell and Pinckney had just managed to cling to his foot. In the ensuing struggle the footboard broke away, but Brunel succeeded in hauling his companion back into the bell and brought him safely to the surface. He then continued his descents quite unperturbed by this hair-raising incident.

By 11 June 19,500 cu. ft of clay had been flung into the hole and the pumps had cleared the shaft and the first 150 ft of the tunnel by the 25th. Two days later Brunel decided to inspect the rest of the tunnel by boat and picked as his companions those miners who had been the last to leave the frames on the day of the catastrophe. Clad only in bathing trunks and armed with bull's-eye lanterns they glided slowly into the darkness watched by an anxious group of miners who had assembled at the bottom of the shaft. In the lowest part of the tunnel the water was so deep that they were able to propel the boat by pushing against the tunnel roof until it eventually grounded on the huge mound of silt which had been washed through the breach. Over this they crawled until they succeeded in reaching a point directly above the stage where they could see that the shield, though entirely filled with silt, was still intact. Here Brunel called for three cheers, and the answering cheers from the foot of the shaft echoed back to them through the darkness.

In his private journal Brunel afterwards recorded his impressions of his adventures in the diving bell and of this first hazardous voyage through the drowned tunnel:

What a dream it now appears to me! Going down in the diving bell, finding and examining the hole! Standing on the corner of No. 12! The novelty of the thing, the excitement of the occasional risk attending our submarine (aquatic) excursions, the crowds of boats to witness our works all amused – the anxious watching of the shaft – seeing it full of water, rising and falling with the tide with the most provoking regularity – at last, by dint of clay bags, clay and gravel, a perceptible difference. We then began pumping, at last reaching the crown of the arch – what sensations! ...

I must make some little indian ink sketches of our boat excursions to the frames: the low, dark, gloomy, cold arch; the heap of earth almost up to the crown, hiding the frames and rendering it quite uncertain what state they were in and what might happen; the hollow rushing of water; the total darkness of all around rendered distinct by the glimmering light of a candle or two, carried by ourselves; crawling along the bank of earth, a dark recess at the end – quite dark – water rushing from it in such quantities as to render it uncertain whether the ground was secure; at last reaching the frames – choked up to the middle rail of the top box – frames evidently leaning back and sideways considerably – staves in curious directions, bags and chisel rods protruding in all directions; reaching No. 12, the bags apparently without support and swelling into the frame threaten every minute to close inside brickwork. All bags – a cavern, *huge, misshapen* with water – a cataract coming from it – candles going out ...

While all these perilous operations were going on, Marc's worries were not lessened by the fact that eminent visitors still insisted upon inspecting the works regardless of the danger. One of these was Charles Bonaparte, and among his entourage was Sir Roderick Murchison, the geologist, who left an amusing account of the party's experiences, which it is interesting to contrast with Brunel's. 'The first operation we underwent,' he writes,

(one which I never repeated) was to go down in a diving-bell upon the cavity by which the Thames had broken in. Buckland and Featherstonehaugh, having been the first to volunteer, came up with such red faces and such staring eyes, that I felt no great inclination to follow their example, particularly as Charles Bonaparte was most anxious to avoid the dilemma, excusing himself by saying that his family were very short-necked and subject to apoplexy, etc.; but it would not do to show the white feather; I got in, and induced him to follow me. The effect was, as I expected, most oppressive, and then on the bottom what did we see but dirty gravel and mud, from which I brought up a fragment of one of Hunt's blacking bottles. We soon pulled the string, and were delighted to breathe the fresh air.

The first folly was, however, quite overpowered by the next. We went down the shaft on the south bank, and got, with young Brunel, into a punt, which he was to steer into the tunnel till we reached the repairing shield. About eleven feet of water were still in the tunnel, leaving just space enough above our heads for Brunel to stand up and claw the ceiling and sides to impel us. As we were proceeding he called out, 'Now, gentlemen, if by accident there should be a rush of water, I shall turn the punt over and prevent you being jammed against the roof, and we shall then be carried out and up the shaft!' On this C. Bonaparte remarked, 'But I cannot swim!' and, just as he had said the words, Brunel, swinging carelessly from right to left, fell overboard, and out went the candles with which he was lighting up the place. Taking this for the *sauve qui peut*, fat C.B., then the very image of Napoleon at St Helena, was about to roll after him, when I held him fast, and, by the glimmering light from the entrance, we found young Brunel, who swam like a fish, coming up on the other side of the punt, and soon got him on board. We of course called out for an immediate retreat, for really there could not be a more foolhardy and ridiculous risk of our lives, inasmuch as it was just the moment of trial as to whether the Thames would make a further inroad or not.

A similar exploit ten days later ended in tragedy. Two directors of the Tunnel Company, Martin and Harris, insisted on inspecting the works in the same way. They were accompanied by Brunel's assistant, Gravatt, and two miners, Dowling and Richardson. They had reached the deep water when Martin stupidly stood up, struck his head against the roof of the tunnel and fell backwards, capsizing the punt. The only swimmers in the party were Gravatt and Dowling, who managed to return and fetch another boat. They found the two directors clinging to the plinth of one of the side arches between the twin tunnels, but there was no sign of Richardson and his body was subsequently recovered by dragging. He was the first to die by accident in the tunnel. In fairness to Marc and Isambard, they only agreed to such exploits under duress.

The labour involved in pumping the water out of the lower part of the tunnel and clearing the frames of silt was immense, and as it was carried out under the constant threat of a further inundation, nerves became frayed as the men listened or peered through the gloom in constant dread of the sudden roar and glitter of falling water. One night in July a man named Rogers was on duty at the shaft while Fitzgerald, the foreman bricklayer, was acting as night watchman in the frames. A sudden shout of 'Wedges, clay, oakum! – the whole of the faces be coming in – coming altogether!' sent the horrified Rogers rushing down the tunnel to find Fitzgerald fast asleep on a bed of straw on the stage but in the throes of a nightmare.

The picked miners who manned the great shield were not easily rattled, but panic was apt to spread like wildfire amongst the Irish labourers who were employed at the pumps and in clearing the tunnel. When danger threatened, their first instinct was to put out the lights in the old belief that under cover of darkness the water could not find and overwhelm them. Brunel recorded one such incident in his journal:

At two o'clock in the morning of the 17th October, Kemble, the overground watchman, came stupefied with fright to tell me that the water was in again. I could not believe him – he asserted that it was up the shaft when he came. This being something like positive, I ran without my coat as fast as I could, giving a double knock on Gravatt's door on my way. I saw the men on the top, and heard them calling earnestly to those who they fancied had not had time to escape; nay, Miles had already in his zeal thrown a long rope, swinging it about, calling on the unfortunate sufferers to lay hold of it, encouraging those who could not find it to swim to one of the landings. I instantly flew down the stairs. The shaft was completely dark. I expected at every step to splash into the water. Before I was really aware of the distance I had run, I found myself in the frames in the east arch. Nothing whatever was the matter, but a small run in No. 1 top, where I found Huggins (the foreman of the shift) and the *corps d'élite*, who were not even aware that anyone had left the frames.

Returning to the shaft, Brunel heard from Miles a more picturesque version of the affair. 'I seed them there Hirishers a come a tumblin' through one o' them small harches like mad bulls,' he said, 'as if the devil picked 'um – screach of Murther! Murther! Run for your lives! Out the lights! ... My ears got a singin', Sir – all the world like when you and me were down in that 'ere diving-bell – till I thought as the water was close upon me. Run legs or perish body! says I, when I see Pascoe ahead o' them there miners coming along as if the devil was looking for 'im. Not the first, my lad, says I, and away with me – and never stopped till I got landed fair above ground.'

In August Brunel's father again fell seriously ill, and in September Beamish had an attack of pleurisy and was off duty six weeks. The irruption had interfered with the ventilation in the tunnel and the air became so foul that a black deposit formed about the nostrils of the men, who frequently collapsed with violent attacks of giddiness and vomiting. Yet, spurred on by Brunel's unconquerable determination, the work went forward until, by November 1827, the whole tunnel and the great shield had been completely cleared and restored to order.

It was to celebrate this hard-won triumph over disaster that Brunel staged the most fantastic spectacle in the history of the tunnel. He resolved to entertain his friends to a banquet under the river. The side arches were hung with crimson draperies and a long table brilliantly lit by gas candelabra when, on the night of Saturday 10 November, a company of fifty sat down to dine to the strains of the uniformed band of the Coldstream Guards. When the time came for the toasts there was a dramatic moment when Brunel's friend Bandinel of the Foreign Office rose to propose the health of Admiral Sir Edward Codrington, one of the tunnel promoters. Waving aloft a copy of that night's *Gazette Extraordinary* he announced Codrington's victory over the Turkish fleet at Navarino, a victory that ensured that liberation of Greece from Turkey for which, three years before, Byron had thrown away his life at Missolonghi. 'In that battle,' Bandinel declared, 'the Turkish power has received a severer check than it has ever suffered since Mahomed drew the sword. It may be said that the wine-abjuring Prophet conquered by water – upon that element his successors have now been signally defeated. My motto, therefore, on this occasion, when we meet to celebrate the expulsion of the river from this spot is – 'and he raised his glass, 'Down with water and Mahomed – wine and Codrington forever!' There was tremendous applause.

In organizing this remarkable entertainment, Brunel had not forgotten his *corps d'élite*, and in the adjoining arch one hundred and twenty miners sat down to the feast. The proceedings came to an end after their chairman had presented Brunel with a pickaxe and shovel and called for 'three times three and a bumper'. It had been a great occasion but the triumph was to be shortlived.

Throughout November and December there was constant anxiety as the great shield continued its advance through treacherous ground which showed no sign of improvement. In January the fear of another break-through became acute, and on the 8th of the month both father and son were on tenterhooks when Don Miguel, the Pretender to the throne of Portugal, and a large company insisted upon examining the shield. How well founded their fears were was proved all too clearly on the morning of 12 January, when disaster struck again with even more tragic force.

Brunel was down in the frames and Beamish was in the little office at the top of the shaft making out tickets authorizing the issue of warm beer and gin to the miners who were coming off shift. Suddenly the watchman burst into the office shouting, 'The water is in! The tunnel is full!' Beamish at once rushed to the shaft to find that this time it was no false alarm. The workmen's stairs were crowded with panic-stricken men. Seizing a crowbar Beamish battered down the door of the visitors' stairway and began to descend. He had not taken many steps before a great wave of water surged up the shaft towards him and out of it, to his horror, rose the inert form of Brunel. Beamish gathered him into his arms, carried him up the stairs and laid him on the ground, where for some while he could only murmur the names 'Ball', 'Collins', over and over again. Meanwhile complete disorder and terror reigned at the top of the shaft. The roar of the water was deafening as half-drowned men staggered up the stairs and others were carried up insensible.

Brunel, with the two miners Ball and Collins, had been removing some of the side shores in No. 1 frame when the ground suddenly bellied towards them and a tremendous column of water followed which instantly swept them out of the frames and extinguished all the lights. Brunel was soon in water up to his waist. Then his leg became trapped under a baulk of falling timber from the stage and it was only by a great struggle that he managed to free it. Calling repeatedly for Ball and Collins to follow him he made for the shaft, only to find the stairs completely blocked by struggling men. He was trying to reach the visitors' stairs through the west arch when the immense wave, which actually lipped the top of the shaft, caught him and bore him up to safety. When the roll was called, Ball and Collins were found to be missing and so were four other miners who had been trying to climb the shaft by a ladder when they were swept away by the recoil of the wave.

Although it was obvious that Brunel was seriously ill and that his leg was giving him acute pain he remained, as usual, quite undaunted and refused to leave the works. He at once ordered the diving-bell and the bell barge to be got ready, and as he was unable to walk he lay on a mattress on the barge, from which he directed the diving operations, and would not be taken home until he had learned the extent of the cavity in the river bed.

On this occasion 4,500 tons of clay were dumped into the river before the work of clearing the tunnel could begin. In this work Brunel played no part. After a brief spell of re-cuperation at Brighton he looked forward to returning to work, but he indulged himself too freely there and suffered a relapse. It was discovered that in addition to his damaged knee he had internal injuries of such gravity that his holiday was followed by a long spell on his back at his home in Blackfriars. While he was laid up he recalled his terrifying experience in the tunnel and set it down, for no eye but his, in the private journal which he always kept under lock and which was not to be opened until after his death. This account gives us what is probably the most revealing glimpse we have of his strange nature, a nature in which the sense of drama was so strong that it could become totally absorbing and exclude all sense of personal danger or fear. While the body struggled, jerked like a puppet by the instinct for self-preservation, it is as though the spirit of the artist in him became wholly detached, coolly observant and rapt by the grandeur of the tragic play.

'Here I am in bed at Bridge House,' he writes.

I have now been laid up quite useless for 14 weeks and upwards, ever since the 14th January. I shan't forget that day in a hurry, very near finished my journey then; when the danger is over, it is rather amusing than otherwise – while it existed I can't say the feeling was at all uncomfortable. If I was to say the contrary, I should be nearer the truth in this instance. While exertions could still be made and hope remained of stopping the ground it was an excitement which has always been a luxury to me. When we were obliged to run, I felt nothing in particular; I was only thinking of the best way of getting us on and the probable state of the arches. When knocked down, I certainly gave myself up, but I took it very much as a matter of course, which I had expected the moment we quitted the frames, for I never expected we should get out. The instant I disengaged myself and got breath again – all dark – I bolted into the other arch – this saved me by laying hold of the railrope – the engine *must* have stopped a minute. I stood still nearly a minute. I was anxious for poor Ball and Collins, who I felt too sure had never risen from the fall we had all had and were, as I thought, *crushed* under the great stage. I kept calling them by name to encourage them and make them also (if still able) come through the opening. While standing there the effect was – *grand* – the roar of the rushing water in a confined passage, and by its velocity rushing past the opening was grand, *very grand*. I cannot compare it to anything, cannon can be nothing to it. At last it came bursting through the opening. I was then obliged to be off – but up to that moment, as far as my sensations were concerned, and distinct from the idea of the loss of six poor fellows whose death I could not then foresee, kept there.

The sight and the whole affair was well worth the risk and I would willingly pay my share, £50 about, of the expenses of such a 'spectacle'. Reaching the shaft, I was much too bothered with my knee and some other thumps to remember much.

If I had been kept under another minute when knocked down I should not have suffered more, and I trust I was tolerably fit to die. If, therefore, the occurrence itself was rather a gratification than otherwise and the consequences in no way unpleasant I need not attempt to avoid such. My being in bed at present, tho' no doubt arising from the effects of my straining, was *immediately* caused by me returning too soon to a full diet at Brighton:

had I been properly warned of this, I might now have been hard at work at the tunnel. But all is for the best.

At the time he wrote this, he still hoped to return to his post as engineer in charge, but it was not to be. The accident marked a turning point in his career. Seven years would pass before work on the tunnel was resumed and by then he would be much too fully occupied with projects as yet undreamed of for him to play his old role again. When he had fully recovered, after a spell of convalescence in the West Country, he took a holiday abroad and it was on his return that Charles Macfarlane met him.

[3]

The Years of Frustration

DURING Brunel's long illness, his father himself superintended the immense task of once more restoring the tunnel works to order. This he eventually succeeded in doing,
but meanwhile funds were running low and disputes and
divisions of opinion had broken out between the directors of
the Tunnel Company. Just as their predecessors of the old
Archway Company had tried to make a scapegoat of Richard
Trevithick, so now some of them, including the Chairman,
attempted to lay the blame for their misfortunes on Marc
and lent a willing ear to the numerous budding engineers
who bombarded the Company with impracticable schemes
for completing the tunnel. Marc was deeply hurt, talked of
going abroad and ultimately, when the post had become
purely nominal, resigned as engineer to the Company.

Even while his energies had been fully absorbed in the
tunnel works, Brunel had still found time to sit by the
fireside in his Rotherhithe lodging in the small hours, wondering what the future held for him, building what he called
his '*châteaux d'Espagne*' and recording his musings in his
secret journal. 'What will become of me?' he asks, and then
his dreams begin. He will build a fleet of ships and storm
Algiers; build a new London Bridge with an arch of 300-ft
span; build new tunnels at Gravesend and Liverpool and 'at
last be rich, have a house built, of which I have even made
the *drawings*, etc., be the first engineer and an example for
future ones'. After all, he reflects, Pitt became Prime Minister at twenty-two and, 'I may be said to have almost built this
tunnel, having been active resident engineer. What Castles!
... What a field – yet I may miss it.'

Now that work on the tunnel had been suspended his musings took a less sanguine turn, as his enforced inactivity allowed him too much time to brood over the future. In the early hours of the morning of 7 May 1828 he lay in his bed at Bridge House 'smoking some excellent canaster' and soliloquized as follows:

Here are these directors damning the Tunnel as fast as they well can. If they go on at this rate, we must certainly stop. . . . Where the devil money is to come from in that case, I know not. . . . Where then will be all my fine castles? – bubbles! Well, if it was only for myself I should not mind it. I fear if the Tunnel stops I shall find all those flattering promises of my friends will prove friendly wishes.

The young Rennies, whatever their real merit, will have built London Bridge, the finest bridge in Europe, and have such a connexion with government as to defy competition. Palmer has built new London Docks and thus without labour has established a connexion which ensures his fortune, while *I* – shall have been engaged on the Tunnel which failed, which was abandoned – a pretty recommendation.

I have nothing after all so very transcendent as to enable me to rise by my own merit without some such help as the Tunnel. It's a gloomy perspective and yet bad as it is I cannot with all my efforts work myself *up* to be *down* hearted. Well, it's very fortunate I am so easily pleased. After all let the worst happen – unemployed, untalked of, *penniless* (that's damned awkward) I think I may depend upon a home at Benjamin's. My poor father would hardly survive the tunnel. My mother would follow him. I should be left alone – here my invention fails, what would follow I cannot guess. A war now, I would go and get my throat cut and yet that would be foolish enough – well 'vogue la galère', very annoying but so *it* is; I suppose a sort of middle path will be the most likely one – a mediocre success – an engineer sometimes employed, sometimes not – £200 or £300 a year and that uncertain: well, I shall then have plenty to wish for and that always constituted my happiness. May I always be of the same mind and then the less I have the happier I shall be.

I'll turn misanthrope, get a huge Meerschaum, as big as myself and smoke away melancholy – and yet that can't be done without money and that can't be got without working for it. Dear me, what a world this is where starvation itself is an expensive luxury. But damn all croaking, the Tunnel must go on, it shall go on. By the by, why should not I get some situation, surely I have friends enough for that. Q. – Get a snug little berth and then a snug little wife with a little somewhat to assist in house-keeping? What an interesting situation!

No luxuries, none of your enjoyments of which I am tolerably fond? – Oh horrible – and all this owing to the damn'd directors who can't swallow when the food is put into their mouth. Here is the Duke of Wellington speaking as favourably as possible, offer-ing unasked to take the lead in a public meeting and the devil knows what, and they let it all slip by as if the pig's tail was soaped.

By the summer the 'dam'd directors' had settled their differences and roused themselves sufficiently to organize a public meeting at the Freemasons' Tavern on 5 July which was attended by the Dukes of Cambridge, Wellington and Somerset and many other public men. But England was in the depths of the slump which, after a brief boom, had fol-lowed the peace of Waterloo, and, despite the most eloquent appeals, the sum of money subscribed fell far short of the amount required. In August the directors ordered the shield to be bricked up. In what became known as 'the visitors' arch' a large mirror was placed against this wall and the tunnel became a peep-show for sightseers.

'The Tunnel is now *blocked up* at the end and all work about to cease,' wrote Brunel. 'A year ago I should have thought this intolerable and not to be borne; now it is come, like all other events, only at a distance do they appear to be dreaded.'

In this misfortune the public displayed their habitual fickleness and what had so lately been an object of awe and wonderment now became a target for every cheap-jack humorist. Some wag christened the tunnel 'the Great Bore', while Thomas Hood, in an 'Ode to M. Brunel', suggested that he should turn it into a wine cellar:

I'll tell thee with thy tunnel what to do;
Light up thy boxes, build a bin or two,
The wine does better than such water trades,
Stick up a sign, the sign of the Bore's Head:
I've drawn it ready for thee in black lead,
And make thy cellar subterranean – thy Shades!

Notwithstanding the reverses and the taunts, father and son both continued to hope that work on the tunnel would eventually be resumed, pinning their faith to the efforts which were being made by the Duke of Wellington and others to secure financial aid from the Government through the Exchequer Loan Commissioners. These negotiations dragged on sporadically until 6 December 1831, when we find Brunel writing in his diary what he evidently feared was the tunnel's epitaph, for the heading 'Tunnel' is enclosed in a broad frame of black ink:

Tunnel is now, I think, *dead*. The Commissioners have refused on the ground of want of security. This is the first time I have felt able to cry at least for these ten years. Some further attempts may be made – but – it will never be finished now in my father's lifetime I fear. However, nil desperandum has always been my motto – we may succeed yet. *Perseverantia.*

After the first disaster in the tunnel he had written, 'Altho' to others I appear in such cases rather unconcerned and not affected (pride), the internal anguish I felt is not to be described.' So, at the very outset of his career, pride armoured him against self-betrayal. No wonder that Charles Macfarlane, during that bitter winter's journey from France in 1829, should never have suspected that beneath his travelling companion's air of light-hearted unconcern and his unquenchable gaiety, ambitious dreams and self-doubt, extravagant hopes and bitter despondency struggled for mastery. For when Brunel landed at Dover on that occasion he must have realized that, in an age when reputations were usually made or lost in early manhood, the next few years must decide whether he would become a mediocrity or 'the first engineer and an example for future ones'. *Perseverantia!* He threw himself into the task of establishing his reputation as an engineer with his usual unsparing energy, but many disappointments lay in store for him before the tide of fortune suddenly turned and swept him to fame.

A blow only less severe than the closure of the tunnel works was the abandonment of what Brunel refers to cryptically as the 'gaz' experiments. The idea behind these experiments had first occurred to his father, but he contributed so much towards its development that he came to look upon it as his own brain child. Put simply, the scheme was to generate gas from carbonate of ammonia and sulphuric acid and pass it into two surface condensers which were alternately heated and cooled and which communicated through expansion vessels and valves with a power cylinder. When the gas in one condenser was held in its condensed state by passing cold water through the condenser tubes and the other was heated by the circulation of hot water, the difference in pressure between the two vessels was thirty-five atmospheres. This was the power which the Brunels endeavoured to harness and which they believed might supersede the power of steam. The 'gaz engine' became one of Brunel's favourite *'châteaux d'Espagne'*. He saw himself taking charge of a large works for its manufacture, while the ships which he dreamed of launching against Algiers were to use this new motive power.

So promising did the idea seem that the experiments were heavily subsidized, even the Admiralty making a grant. The technical problems which had to be solved in order to translate theory into practice were immense and that they *were* solved is extraordinary when we think of the very limited metallurgical knowledge existing at that time. The gas condensed at a pressure of no less than three hundred atmospheres, while pipes and pipe joints had to be made to withstand pressures of 1,500 lbs per square inch. This in an age when, in steam engineering, a pressure of 50 lbs per sq. in. was often considered dangerously high. Cast iron was obviously useless and a type of gunmetal was eventually evolved for the pressure vessels. After the closure of the tunnel the apparatus was erected in the abandoned works at Rotherhithe and here Brunel, with an assistant named Withers, patiently pursued his experiments. 'Here I am at Rotherhithe,' he wrote on 4 April 1829, 'renewing experiments on gaz – been getting the apparatus up for the last *six months*!! Is it possible? A 1/40 of the remainder of my life – what a life, the life of a dreamer – am always building castles in the air, what time I waste!'

Brunel's notes and sketches suggest an apparatus just about as safe as a ticking time-bomb and the most remarkable thing about the whole venture is that he and Withers failed to blow themselves up. At long last, however, Brunel was forced to admit defeat. On 30 January 1833 he wrote:

Gaz – After a number of experiments I fear we must come to the conclusion that (with carbonic acid at least) no sufficient advantage on the score of economy of fuel can be obtained.
All the time and expense, both *enormous*, devoted to this thing for nearly 10 years are therefore *wasted*. . . . It must therefore die and with it all my fine hopes – crash – gone – well, well, it can't be helped.

While the tunnel works were closed and these experiments were being pursued to their fruitless conclusion, Brunel, with the help of his father and his friends, was energetically seeking engineering commissions and becoming increasingly independent of, and divorced from, his father's activities. He continued to refer to Bridge House, Blackfriars as his home but his visits there became less and less frequent. He travelled widely about the country in the course of his work, when he was in London he took most of his meals either at the Athenaeum or at the Barge House, Christ Church,

Lambeth, where lived his sister Sophia and his brother-in-law and dearest friend Benjamin Hawes. Until his marriage, the Barge House became his second home.

While his father was making surveys for the improvement of the Oxford Canal and the Medway Navigation, one of Brunel's first commissions was to carry out certain drainage works at Tollesbury on the Essex coast and we find him catching the Yarmouth mail coach to Kelvedon and proceeding thence by hired chaise to Tollesbury. The work included the provision of a pumping engine and what he describes as a 'siphon' at the sea wall. At that time, thanks to the working economies effected by Richard Trevithick and other Cornish engineers, the pumping engines at the Cornish mines were still supreme and it was doubtless with the object of gaining experience and information that he secured introductions from John Taylor to his son Richard at the Consolidated Mines and to Captain Nick Vivian at Wheal Towan. Of the former, Taylor writes in September 1830: 'Our engines there are the largest in the County and though no one comes up to that at Wl Towan we think considering their size, what they have to do and the number at work, our duty is very good.'

Earlier in the year an accurate working model of a 24-in. beam engine ('brass for brass, iron for iron') had been built and a contract for the engine and siphon placed with Maudslay, Sons & Field. Finally, Brunel went in search of a barge capable of conveying the machinery to Maldon. 'Maldon barge would not take it,' he notes. 'No Essex barge to be found.' Such were the transport difficulties which beset engineers in the days before railways.

Another and more important commission which came his way in November 1831 was for the construction of a new dock at Monkwearmouth (Sunderland). This involved the construction of a tidal basin and an extensive locked harbour. In this he was destined to be disappointed, for his first ambitious designs were rejected by Parliament and it was not until 1834 that an Act was passed which authorized another and much more modest scheme.

What strikes the reader very forcibly about Brunel's journal during this period of his life is the facility with which he managed to travel about England. We usually assume that the people of England in pre-railway days remained firmly rooted and that for them a journey of fifty miles or so was a momentous event. This may have been so for the majority, but Brunel, perpetually on the move, shows us how much ground an active young man could cover in the heyday of the turnpikes provided he could put up with little sleep, and could afford the high fares. Thus he first heard of the Monkwearmouth Dock Scheme on 14 November, booked his place in the northbound coach forthwith, and set off in the late evening of the next day. He breakfasted at Grantham, dined at York at 4.40 p.m. and arrived at Newcastle at 2.30 a.m. on the 17th where he retired to bed at the Queen's Head, rising soon after 8 a.m. to meet the Dock Surveyor. The journey had cost him £7 9s. 6d. That day he sat up till midnight 'smoking, etc.', with the Dock Committee whom he describes as 'shrewd clever fellows – but a rum set'.

When he was on his travels he never missed an opportunity of visiting any work of architectural or engineering interest which might lie on or near his route. Thus he spent some time examining Durham Cathedral when he drove over there in a chaise with two members of the committee to deposit the Dock plans and, for a reason which will appear later, he visited the Scotswood suspension bridge over the Tyne and made a number of detailed sketches. When he finally left Newcastle on 2 December, he embarked upon a regular sightseeing tour, albeit a very purposeful one. His first objective was Stockton where he walked out to inspect the first suspension bridge in England which carried the Stockton and Darlington railway over the Tees. With this he was not impressed. His diary features an alarming sketch showing the floor of the bridge deflected 12 in. by the weight of two coal waggons and the comment: 'Wretched thing . . . the floor creaks most woefully in returning.' His judgement was correct, for shortly after his visit the floor of the bridge had to be propped up on piles and in 1844 the Stephensons replaced it with a cast-iron structure.

Before posting to Darlington to dine he visited Hartlepool and wrote: 'A curiously insulated old fishing town – a remarkably fine race of men. Went to the top of the church tower for a view.' Arrived at Darlington there was no pausing for sleep. After dinner he caught the night mail coach for York, reaching there at 5.45 a.m. and taking the connecting mail for Hull, where he inspected the docks, taking dimensions of the entrance locks and pausing en route to see Beverley Minster which he thought 'very fine'. Another overnight journey must have followed, for by early morning of the next day, Sunday, 4 December, he had arrived in Manchester to spend the day with his friends the Hulmes.

It was on the following morning that Brunel made what must have been his first railway journey, for his journals record no earlier experience and the Liverpool & Manchester Railway had been open to the public for only a little over a year. 'Went by first railway coach to Liverpool', he writes. '1 hr 25 mn on the road and 2 hrs 15 mn between Hotels' (i.e. stations). Interleaved in the diary at this page is a small sheet of paper bearing a series of wavering circles and lines and the inscription: 'Drawn on the L. & M. railway 5.12.31.' On this Brunel has added the following significant note: 'I record this specimen of the shaking on Manchester railway. The time is not far off when we shall be able to take our coffee and write while going noiselessly and smoothly at 45 miles per hour – let me try.' Evidently he already cherished a new *'château d'Espagne'* – the dream of building a railway which would be smoother, faster and better than anything so far conceived.

No sooner had the opening of the Liverpool & Manchester deprived the Leeds & Liverpool and Bridgwater Canals of their freight traffic monopoly than the latters' tolls came tumbling down and evidently the railway had not yet mastered this opposition. 'Went to the Railway establishment,' he writes. 'Nothing doing apparently in the way of conveyance of goods.'

After breakfasting, visiting Liverpool Docks and the new Custom House which he criticizes severely ('what an extravagant waste of strength in the massive corners and spires – very inferior stone'), he went on to Chester by the eleven o'clock coach. Here he saw the new Grosvenor Bridge over the Dee on which his father had advised and which was now nearing completion. This obviously filled him with admiration for he calls it: 'A most beautiful, bold and grand work – decidedly the finest and largest in the world.' He then returned home via Wrexham, Shrewsbury, Wolverhampton and Birmingham. Leaving Chester at 6 p.m. he was in Birmingham at 6.30 a.m. the following morning, where he made a business call, and was back in London that night. His devious journey of 528 miles had cost him exactly £18.

Even more disappointing than the Monkwearmouth Dock scheme was the fate of the proposals for a new dry dock and other works in the Navy Yard at Woolwich. He spent much time surveying and making trial borings at Woolwich and in meetings with the Navy Board but all to no purpose. The plans were ultimately abandoned.

Yet another unfortunate enterprise was the design and construction of an observatory at Kensington for Mr (later Sir) James South. This included a revolving dome which was completely bisected by a telescope aperture having mechanically operated shutters. Both dome and shutters were constructed by Maudslay, Sons & Field. He devoted a great deal of time and pains to this observatory. One night, while the dome was under construction and covered with tarpaulin, a gale sprang up and Brunel, who was at Bridge House, feared that his work would be carried away. After walking several times on the Blackfriars Bridge to judge the force of the wind, he decided at midnight to go to Kensington where he arrived in the small hours and swarmed over the palings to reach the observatory. A strolling constable, obviously attracted by such suspicious behaviour on so wild a night, soon found himself helping to make all secure, which done Brunel wrapped himself up in his wet cloak and spent the rest of the night on a bed of shavings on the observatory floor.

The completion of the observatory was celebrated on 20 May 1831 when a company of forty-five sat down to dine on the lawn and Brunel, in his own words, 'lionized it till six'. His small triumph was very short-lived. The work had exceeded the original estimate, and Sir James South tried to justify his refusal to pay by finding fault with a building whose design he had originally approved. A wretched dispute developed which dragged on until the following spring when the *Athenaeum* published an anonymous article, obviously inspired if not written by Sir James, in which the observatory dome and its shutters were described as 'an absurd project' which 'had no other object than the display of a *tour de force*, and was an effort to produce effect on the part of the architect'. Brunel, deeply hurt and incensed by this attack, contemplated a libel action, but his friends, although they took his part in the affair, advised him against any form of retaliation.

Among these friends was Edward Blore, the eminent architect who was responsible for Scott's Abbotsford and for Prince Worontzow's fabulous Tudor castle in the Crimea. Blore acted as mediator between Brunel and Sir James, although the former's journal does not reveal that his efforts met with any success. At this time Blore, who was a friend of both the Brunels, was about to begin his work on Buckingham Palace, or Buckingham House as it was then called, and he invited them to inspect the building. John Nash had previously begun rebuilding and enlarging the old town house of the Earls of Buckingham, but on the accession of the Prince Regent, Blore succeeded him. Nash's work has been described as his least satisfactory achievement and judging by Brunel's outspoken comments on his visit, this would certainly appear to be correct. 'An extraordinary, iniquitous, jobbing, tasteless, unskilful, profligate waste of money,' he writes. 'Walls without foundation, ornament without meaning – job, job, job. . . .' Two years later, after another visit to the Palace he wrote of Blore's work: 'Do not like the alterations to the garden front at all – it is quite spoiled – all the rest very good.'

These must have been discouraging times for Brunel. He had abandoned his secret journal for a diary concerned with day-to-day happenings which affords us only an occasional hint of the frustrations he must have felt. While Telford was lauded for his great suspension bridge across the Menai; while the Rennies' splendid new London Bridge was opened by the King with great acclaim; while the Stephensons rode unchallenged on the crest of the new railway wave with bills for the Grand Junction and London & Birmingham Railways already before Parliament, what had he to show? An uncompleted tunnel, a useless 'gaz engine', two abortive dock schemes of little moment and the miserable dispute over the observatory. He had, in fact, as we shall see, already laid the foundations for his success, but this was not evident to him as yet. For the present he had to accept the crumbs let fall by more successful engineers while he endeavoured, with the help of influential friends, to obtain a footing in the new railway world. Had these early attempts succeeded, not only Brunel's life story, but the whole history of Britain's railways might have been very different. For in these momentous years the map of that system which was to bring about the greatest social and economic revolution which the world had ever known was being determined by chance introductions and by small informal gatherings in tavern or coffee house.

So early as January 1830 Brunel applied for the post of engineer of England's first coast-to-coast railway – the Newcastle & Carlisle, which had secured its Act in May 1829. His application was supported by recommendations from Lord Lonsdale and that staunch family friend Lord Spencer, but James Lock, one of the promoters, damned him with faint praise in a letter to his friend Howard in which he wrote: 'We cannot have Stephenson without the leave of the Directors, so we had better look elsewhere. I perceive that Lord Carlisle has a strong inclination for Brunel who I suppose is a clever young man. . . .' The post was given to Francis Giles on the understanding that if he proved unsatisfactory it would be offered to Brunel. This must have been a bitter blow to his pride, for Giles was a canal engineer of mediocre ability, who had already damned himself when the Liverpool & Manchester Bill was in Committee by expounding at length on the impracticability of George Stephenson's proposed crossing of Chat Moss.

In November of 1830 his father's old friend Charles Babbage, the inventor of the calculating machine, gave Brunel an introduction to Mr Whateley of Birmingham, solicitor to the proposed Bristol & Birmingham Railway. He took coach to Birmingham and put up at the 'Hen & Chickens', where he met Whateley, who took him out to see the improvements on the Birmingham Canal which Thomas Telford had recently completed. Commenting on the cutting spanned by the graceful cast-iron Galton bridge, by which Telford had straightened Brindley's original tortuous line near Smethwick, Brunel writes: 'The excavation and deep cutting is prodigious . . . the whole merit of the work appears to be its grandeur.' That evening he dined with the railway promoters. 'Something may at last turn up of this,' he says, but once again he was doomed to disappointment. Although he carried out some preliminary survey work, railway interest in the Midlands was wholly absorbed by the Grand Junction and the London & Birmingham Railways, with the result that the Bristol railway scheme perished for lack of support. When it was revived as the Birmingham & Gloucester in 1836 it was not Brunel's but William Moorsom's route which was adopted.

Brunel's character was of that finely tempered, resilient quality which flexes easily under misfortune but never breaks. He could, and undoubtedly did during these years of difficulty and repeated disappointment when fate seemed implacably against him, plumb depths of despondency unknown to less sensitive, self-conscious and artistic natures. Yet he never lost faith in himself. Once one project on which he had pinned his hopes had failed he would rapidly recover from the blow, dismiss it from his mind and concentrate upon the next with undiminished energy. This unshakable faith in himself, though he sometimes suspected it to be the sin of pride, schooled him, during this time of adversity to hide his feelings behind a bold front of self-confidence and enthusiasm which impressed everyone he met and which, aside from his remarkable abilities, contributed more than anything else to his ultimate success. Moreover, pride and ambition never drove him to make the fatal mistake of refusing any commission as too humble for his notice or because it appeared to be a blind alley offering no opportunities for advancement. Thanks to those acute powers of observation which he acquired under his father's tutelage, everything he undertook contributed something of value to that store of experience which was the secret of his versatility.

Thus in the spring of 1833 when he was fully occupied with concerns of much greater promise he did not scorn the invitation of the proprietors of the Fossdyke Navigation to make a survey and report on the state of their waterway as a result of criticisms levelled against it by the boat owners. With railways already in the ascendant, such an invitation was no open sesame for an ambitious engineer, yet Brunel accepted immediately, took coach for Lincoln and set to work with his usual energy and thoroughness, cramming more work into two days than most men would have accomplished in a week. He travelled by packet boat up the

Fossdyke to the Trent Lock at Torksey and down the Witham taking soundings from which he prepared cross-sections of the channel. He also took dimensions of all the boats lying in Brayford Mere. Not content with this he hired a chaise next day and drove to Torksey, returning on board the loaded keel *Industry* drawn by two horses. Little difficulty was experienced on the Fossdyke, but on entering the Witham the keel stuck fast at High Bridge and Brunel notes that they 'were obliged to drag her all through with the windlass'. This episode is particularly notable in one respect. One of the proprietors of the Navigation who came to meet Brunel on his arrival at the 'Black Boy'[1] was Colonel Sibthorp, member for Lincoln, arch opponent of free trade, railways, or indeed anything else which might threaten to disturb a country gentleman's status quo. It must certainly have been the only occasion on which the great engineer appeared sitting on the same side of the table as the author of the famous phrase: 'I would rather meet a highwayman, or see a burglar on my premises than an engineer.'

After a stay of four days, Brunel returned to London, but not until he had spent some time examining the 'exceedingly beautiful' cathedral of Lincoln. A fortnight later he was back in Lincoln to deliver his report and this time he made a special Sunday pilgrimage by gig and postchaise to Boston to see and climb the famous 'stump'. When he calls it 'the most complete, perfect and beautiful thing I have seen', it is as if, in surveying the high majesty of this incomparable tower, Brunel had paused for a moment on the brink of his own mighty achievements to salute his mediaeval masters. For on this April afternoon in 1833 his hour was not far off.

So far, England's engineering genius had been nursed either in her mining and manufacturing districts or in London. The North had been the birthplace of canals and rail-

1. The 'Black Boy' stands on Castle Hill. It lost its licence in 1922 and is now the headquarters of the Castle Hill Club.

ways, but many a gifted provincial engineer had made his reputation only after migrating to London like Dick Whittington with no more assets than the skill of his hands and the pack on his back. It was therefore one of fortune's strange twists that Brunel, a Londoner since early boyhood and enjoying many advantages, should fail to find any adequate outlet for his skill either in London or in the North. His fortune lay in the West Country. It was the merchant venturers of Bristol who first gave him the chance to prove his mettle and it was a gift which he never forgot. These citizens of Bristol had realized the changes which the new railways were bound to make to the trade routes of England and became alarmed by upstart Liverpool's threat to their trade. Already Liverpool had supplanted Bristol as England's second city and port. So they began to plan countermeasures and in Isambard Brunel they found the man to carry them out.

[4]

Opportunity at Bristol

I⊤ is imposible to read history or biography without being
struck by the momentous consequences of trivial events. We
are constantly reminded that life is a great and unpredictable adventure to which we awake each morning; that
the most humdrum or apparently wasted day may afterwards be seen in recollection to mark a significant turning-
point in our lives. So it is with Brunel. Had he not been
guilty of over-indulgence immediately after his accident in
the tunnel his future might have taken a different course
and the great works with which his name will always be associated might never have been built.

After he had recovered from his long illness at Bridge
House his parents rightly agreed that, where their high-
spirited son was concerned, Regency Brighton was certainly
not the best place for convalescence. The sea air was all
very well, but there were too many counter-attractions. So
he was sent to Clifton, where the wealthy but eminently respectable merchants of Bristol had taken up their abode on
the limestone heights above the city and where he was unlikely to find any dangerous distractions. But for Brunel, unlike the sophisticated and jaded bucks who followed in the
Regent's extravagant wake, the pursuit of pleasure was
never a drug; his delight in his powers of observation ensured that he was never bored with himself, and in Clifton he
found a place which was very much to his taste.

Throughout the length and breadth of England it would
be hard to find a landscape so well calculated to appeal to
Brunel's romantic temperament, to his love of drama and his
sense of grandeur as the fantastic gorge which the Avon has
carved through the limestone escarpment which barred its

way to the Bristol Channel. The precipitous crags of white limestone, capped, as appropriately as a folly tower in a nobleman's park, by the little observatory; the black maw of the Giant's cave adding a suggestion of Gothick gloom; the admirable foil of Leigh woods, thickly clothing the Somerset slopes opposite, and lastly, far below, the silver skein of the tidal river that bore to Bristol Port the shipping of the seven seas; here was every ingredient that his fancy could desire. So Brunel spent his convalescence sketching and clambering about the gorge while the tall ships came and went with the tides, sails furled, and hauled by the hobblers, as they had done since Bristol began. It is easy to imagine his excitement when he heard of the proposal to bridge the Avon gorge and to invite engineers to submit designs. Here indeed was a project after his own heart. Let him design a bridge worthy of such a setting, a triumphal arch that would leap from lip to lip of Bristol's seaway in one sheer and splendid span!

In 1753 a wine merchant of All Saints, Bristol, by the name of William Vick bequeathed the sum of £1,000 to the Society of Merchant Venturers for the purpose of building a bridge at Clifton. It was to accumulate at compound interest until it reached a total of £10,000 which the optimistic Mr Vick fondly supposed would be sufficient. Nothing further happened until 1765, when a certain William Bridges produced for the consideration of the city fathers what must surely be the most remarkable design for a bridge ever conceived. It is difficult to know whether to hail this Mr Bridges as the prince of eighteenth-century eccentrics or as the far-sighted anticipator of 'High Paddington', or the Empire State Building. His design consisted of a single immense arch of stone 220 ft high and 180 ft span carrying a roadway 50 ft wide and 700 ft long. This involved equally immense haunches or abutments but these did not daunt the designer in the least. He proposed that they should house twenty dwelling houses, a lighthouse, toll-house, chapel, granaries, a corn exchange, coal and stone wharves, a general market, a water mill, cotton and wool manufactories, a marine school, a library, museum and subscription rooms, vertical windmills, a watch house, out-offices, stabling, clock turret and belfry. Moreover he ends this catalogue with 'etc., etc.,' although it is hard to think of much else he could have included unless it was a prison, a hospital or barracks for a regiment. Nothing more is heard of this early essay in vertical town planning if only because poor Mr Vick's legacy would not at that time have gone very far towards its realization.

By 1829 the Vick legacy totalled £8,000 and a committee was formed to decide what should be done. Land was bought in Leigh Woods and designs were invited, the idea being to raise additional money by subscription. Brunel decided that the site called for a suspension bridge and he lavished upon his competition designs infinite pains and exquisite draughtsmanship so that they became not merely engineering drawings but works of art. Nor did he fail to back them with the necessary technical experience. When still in his teens he had assisted his father on the drawings of the Île de Bourbon suspension bridges for the French Government. Now he sought his father's advice and also consulted Maudslay's partner, Joseph Field. He visited the Menai and devoted two days to a minute examination of Telford's bridge, while, as we have seen, after his first designs had been submitted he also visited the Scotswood and Stockton suspension bridges. In his diary at this period, too, he has pasted the cutting of a newspaper account of the collapse of the Broughton suspension bridge near Manchester. The rhythmic tread of a company of troops marching over the Broughton bridge had set up such a violent harmonic motion that a pin in one of the suspension chains broke and the bridge collapsed at one end.[1] In this way Brunel learnt what to emulate, what to improve and what to avoid.

1. Hence the army rule, often ridiculed as a superstition, that troops should break step in such circumstances.

On the closing date of the competition he submitted four designs for bridges situated at different points along the Avon gorge within the limits stipulated by the Committee. Their spans varied between 870 and 916 ft, all greater than any suspension bridge so far built. The first two included lofty suspension towers, while in the third both bridge approaches were in tunnel and the suspension chains were anchored to the rocks above. His fourth and favourite design, which he refers to as the Giant's Hole from its proximity to the cave, had the longest span. It called for a tunnel at the Clifton end and an arched defile at the other. Above them, like two castle keeps frowning at each other across the ravine, two squat but massive castellated redoubts would carry the suspension chains. Brunel was right in preferring this last design on practical and economic as well as aesthetic grounds because it avoided the need either for massive abutments or for tall suspension towers by making full use of the natural advantages of the precipitous and rocky site. For the towers of his alternative designs he had used the Gothic style, explaining in his accompanying notes that he had modelled them on features of Lancaster Castle and the gateway of Christ Church, Oxford.

Because the Bridge Committee did not feel competent to do so, they invited Thomas Telford to judge the merits of the designs submitted. He rejected the lot. His criticism of Brunel's designs, whether inspired by professional jealousy or the timidity born of advancing years (he was over seventy) we cannot know, was that his spans were too great. The maximum safe span for a suspension bridge, he ruled, was 600 ft, which happened to be the span of his bridge over the Menai. Any bridge of greater dimensions than this could not, declared Telford, offer sufficient lateral resistance to wind pressure. Brunel had in fact allowed for the effect of wind pressure, notably by the use of extremely short suspension rods at the centre of the span, thus bringing the chains down almost to the level of the platform, by trans-

verse bracing and by the addition of inverted chains as introduced by his father in his Île de Bourbon design. These
inverted chains may be clearly seen in the drawing of the
Giant's Hole bridge.

Telford having thus made mincemeat of all the aspiring
bridge builders there was obviously nothing for the Bridge
Committee to do but invite the master himself to submit a
design. Telford accepted and proceeded to perpetrate the one
truly monstrous aberration of his long career. His design was
almost as peculiar as William Bridge's eighteenth-century
effort. It showed a three-span suspension bridge, centre span
360 ft and side spans 180 ft each, supported upon two enormous piers rising sheer from the floor of the gorge and decorated in florid Gothic so that they somewhat resembled
Beckford's fantastic tower at Fonthill. But in the incongruity
of the ironwork connecting their belfry-like pinnacles they
easily surpassed Fonthill in their folly and were, unlike William Bridge's thickly populated abutments, empty and useless shells.

It is difficult to credit that such a design could have come
from the same sure hand that had reared the tall aqueduct at
Vron Cysyllte, spanned so gracefully the Conway and the
Menai, and planned to bridge London's river with a single
mighty span of iron. Only senile decay can explain and excuse it. Nevertheless, it was at first received with acclaim.
When the Bridge Committee applied for parliamentary
powers, Telford's plan was deposited, while thousands of
copies of it were printed and sold to the public.

Bearing in mind his other disappointments at this time,
Brunel's chagrin when his beautiful drawings were rejected
in favour of this monstrosity may be easily imagined. He
wrote a typically spirited letter to the Bridge Committee in
which he said: 'As the distance between the opposite rocks
was considerably less than what had always been considered
as within the limits to which Suspension Bridges might be
carried, the idea of going to the bottom of such a valley for

the purpose of raising at a great expense two intermediate supporters hardly occurred to me.' He went on to
refer to 'the reflection which such timidity would cast upon
the state of the Arts in the present day'.

Eventually both the public and the Bridge Committee had
second thoughts about the virtues of Telford's design, and in
October 1830 it was decided to hold a second competition for
which twelve designs were submitted. These included Telford's, which the great man disdained to alter, a very remarkable Gothick erection by a gentleman named Capper which
appeared to be in an advanced stage of Romantic Ruin, a
new version of the Giant's Hole design by Brunel. In this he
had compromised, against his own judgement, with popular
opinion by featuring a massive abutment on the Leigh
Woods side of the river, thereby reducing the length of suspended roadway to 630 ft. There can be no doubt that his
original design was perfectly feasible and the Bridge Committee were to pay dearly for their timidity, for the construction of this abutment alone would cost nearly
£14,000.

The twelve designs were whittled down to four which were
submitted to the judges, Davies Gilbert (late Giddy, Trevithick's friend and past President of the Royal Society) and John
Seaward.[2] It was now Telford's turn to taste the bitter cup,
for he was not among these four finalists who were: W.
Hawks, J. M. Rendel, Captain S. Brown and I. K. Brunel. In
the report which they issued from Blaise Castle on 16 March
1831 Gilbert and Seaward placed the contestants in the following order of merit: Hawks, Brunel, Brown, Rendel. They
stated that their stipulation had been that the load on the
suspension chains must not exceed 5½ tons to the square inch
and that with the exception of Brunel's specification all the

2. John Seaward (1786–1858) was of the firm of Seaward & Capel of
Canal Ironworks, Millwall. He was a marine-engine builder of distinction, but it is difficult to understand why he was selected to judge
a bridge competition.

chains were much too heavily stressed. They objected to his design, however, on three other grounds: the use of single pins for the chain links, the method of attaching the suspension rods and of anchoring the chains to the rocks. Now the mention of single pins alludes to the practice hitherto adopted by Telford and others of using short links between each of the main links of the suspension chains, a method which in no way added to their strength or their safety factor. By dispensing with these short links and using main links each 16 ft long, nearly double the length of those used on the Menai bridge, Brunel had designed a chain of equal strength which was at once lighter and cheaper to construct. He felt equally confident about the other points which had been criticized and was therefore not prepared to accept an unjust decision without protest. He was already at Clifton when the result was announced, and he at once arranged through the Committee's solicitor, Osborne, to meet the judges at Blaise Castle. From the entries in his diary at this date he does not appear to have held a very high opinion of those who had sat in judgement on his work: 'D. Gilbert came down with *Seaward* to assist him!!!!!! Seaward!!! ... It appears that my details are found *very bad*, quite inadmissable.' It is evident that this was Seaward's opinion and that Gilbert had allowed himself to be swayed by it, for when Brunel had explained his design to Gilbert: '... he returned to his original opinion viz approval of all the details. Oh *quel homme!*', he adds, 'P. R. S.!!!' That afternoon he was called in to meet the Bridge Committee and writes in triumph: 'D.G. having recanted *all* he had said yesterday I was formally appointed and congratulated very warmly by everybody.' He had scored his first success.

A letter to Benjamin Hawes ten days later reveals Brunel at the height of his form and the Bridge Committee evidently completely under his spell:

'Of all the wonderful feats I have performed since I have been in this part of the world,' he writes,

I think yesterday I performed the most wonderful. I produced unanimity amongst fifteen men who were all quarrelling about the most ticklish subject – taste.

The Egyptian thing I brought down was quite extravagantly admired by all and unanimously adopted: and I am directed to make such drawings, lithographs, etc., as I, in my supreme judgement, may deem fit; indeed, they were not only very liberal with their money, but inclined to save themselves much trouble by placing very complete reliance on me. They seem warm on the subject, and if the confounded election doesn't come, I anticipate a pleasant job, for the expense seems no object provided it is made *grand*.

I mentioned that the 240 foot height would produce a better effect than 220, and 240 was forthwith ordered.

The old prospectus was then attacked, and of course, condemned as *rather out of character*. It was agreed, however, that changing Telford into Brunel, that the *general outline was good* – and – who do you think is going to have the agreeable task of doing this? – why – I!

The 'Egyptian thing' is a reference to his latest design for the towers of the new bridge. This was an inspired piece of work. In the span of a suspension bridge all is lightness and aerial grace; its strength resides in the suspension towers and anchorages which uphold it. To make this strength manifest and thus to point the contrast between the opposing towers and the slender web of links and rods they bear, Brunel's native adaptation of an ancient monumental style was a stroke of artistic genius. Every line of these squat towers which straddle the roadway with firm-planted feet is eloquent of their purpose, a purpose admirably emphasized by the simple monoliths of stone atop the chain anchorages and by the sphinxes which, crouching above the chain rollers, eye each other over the depths of the gorge. It is only when we study this design that the absurdity of Telford's Gothic piers can be fully appreciated. Having decided that his bridge needed the strength of supporting piers, Telford architecturally contradicted himself by applying to these piers, like a false veneer, a style which, because it was evolved to convey the maximum sense of lightness possible in stone, robbed them of that strength.

Although William Beckford had employed Wyatt to realize his own Gothic fantasies at Fonthill, he was none the less a man of taste who could appreciate the absurdity of Telford's towers or Capper's Gothick ruin. The previous autumn Acramon, a mutual friend, had sent Beckford Brunel's Egyptian sketches and Beckford had commented: 'A truly grand and noble design – how superior the style of the gateways to florid barbarisms!' Thus encouraged, Acramon took Brunel to call upon Beckford at his house in Lansdown Crescent, Bath. Evidently both the man and his fabulous collection of works of art deeply impressed Brunel for he wrote the following account of his visit in his diary:

> Went to Beckford – well received – an agreeable, gentlemanly, well-informed man, talking a great deal, evidently very warm and always in motion. His house a pattern of elegance – splendour rendered agreeable and unostentatious by purity of taste and well studied luxury in the highest degree – paintings, gems and articles of vertu crowded in costly cabinets and on beautiful tables. He entered warmly into the bridge affair admiring much the giant's hole plan and praising strongly the architecture I had adopted – approving of Egyptian but condemning in strong terms all the others.[3]

Brunel planned to case his Egyptian towers with plaques of cast-iron bearing designs illustrating every phase in the construction of the bridge including the mining, casting and

3. To this glimpse of the riches of Beckford's house in Bath, Mr Sacheverell Sitwell adds a curious postscript. Many years later, when its fortunes had declined and it had become a boarding house it was visited by an elderly clergyman, a friend of Verlaine and a considerable connoisseur of painting. He took a fancy to a picture in the living room and bought it for £2. It was a painting by Giovanni Bellini worth, probably, £30,000. What is so strange is that such a painting should have survived, its value unrealized, in a house known to have contained a major art collection.

forging of the iron-work. He made sketches for these and
proposed that his friend John Callcott Horsley, R.A., should
prepare the final designs. In his autobiography, Horsley
writes: 'He made very clever sketches for some of these proposed figure subjects, just to show what he intended by them.
I remember a group of men carrying one of the links of the
chainwork, which was excellent in character. He proposed
that I should design the figure subjects, and he asked me to
go down with him to Merthyr Tydvil and make sketches
of the iron processes. We accomplished our journey, and all
the requisite drawings for the intended designs were made.'
Alas, Brunel's Egyptian towers were destined never to be
built and search has failed to reveal either his sketches or
Horsley's drawings.

On 18 June 1831 the Bridge Committee decided to commence work by clearing the site on the Clifton side and on
the 21st a very odd little ceremony took place. While workmen began digging on Clifton Down a public breakfast was
held at the Bath Hotel, after which the gentlemen and their
ladies walked to the site, where they assembled in a circle
round a pile of stones which had been dug out. Brunel then
entered the circle, picked up a stone from the heap and
handed it to Lady Elton, who, holding it in her hand, made a
short speech. This was followed first by a deafening discharge of cannon which had been mounted on the rocks just
below and then by the distant strains of the National An-
them played by a band of the Dragoon Guards who were
down in the gorge. Colours were then run up on a flagstaff
erected for the occasion while Sir Abraham Elton delivered
the usual flowery oration. 'The time will come,' he said, turning to Brunel, 'when, as that gentleman walks along the
streets or as he passes from city to city, the cry would be
raised, "There goes the man who reared that stupendous
work, the ornament of Bristol and the wonder of the age."'
This provoked loud cheers, and the toast 'Success to the
undertaking and to the conductor' was drunk in sparkling

champagne, the 'humbler classes being regaled with a barrel of beer'.

This junketing was the more ludicrous because it was premature, for if the truth had been told the Bridge Committee was not in a position to put the work in hand with any confidence. As was so often to be the case with Brunel's schemes, finance, which was always a secondary consideration to him, was to be the stumbling block to the realization of his Clifton Bridge. Although he had written so optimistically that expense seemed no object to the Bridge Committee, in fact it should have been apparent both to him and to the Committee that the money in hand was quite insufficient. In October 1830 the total subscribed, including the original Vick legacy, was £30,500 and by the following March, renewed appeals had only added to this capital the paltry sum of £500 whereas it was estimated – inadequately as it proved – that the bridge would cost at least £45,000. The main reason for this poor financial support was the troubled state of the country. Brunel's reference, in his letter to Hawes, to the 'confounded election' is a misleading understatement. There was no ordinary pre-election uncertainty in the country. England was nearer bloody revolution in 1831 than at any other time in her history. All the miseries and bitter resentments that had been gathering as a result of trade depression and agricultural enclosure had become focused upon one issue – Parliamentary reform, and the rejection of the Reform Bill by the Lords threw the country into a ferment of unrest. Just as the Bridge Committee had resolved to launch a fresh appeal for funds in October 1831 all thoughts of the bridge were forgotten in the catastrophe of the Bristol Riots.

Of the extraordinary and terrible events of 29, 30 and 31 October and their tragic sequel Brunel was himself an active witness as he was staying in Clifton at the time. The City authorities had feared that the opening of Quarter Sessions would be marked by disturbances and had appealed to the

Secretary of State for military aid. Accordingly one troop of the 3rd Dragoon Guards and two troops of the 14th Light Dragoons had been sent to Bristol under the command of Lt.-Colonel Brereton and were quartered there when the Recorder arrived on the morning of 29 October. Rioting at once broke out, a large mob launching an attack against the Mansion House in Queen Square. The Riot Act was read and the troops were called but Colonel Brereton would give them no order to disperse the mob and they therefore stood by while the rioters broke into the building and inflicted much damage. That night the rioters turned their attention to the Council House but here Captain Gage, who was in command of a troop of the 14th Dragoons, upon his own initiative ordered them to charge and the mob was dispersed before any damage had been done. Picquets were then posted at both buildings and all became quiet until 8 a.m. next morning (Sunday) when Colonel Brereton unaccountably withdrew these guards, thereby inviting the fresh attack which was not long in coming. The Riot Act was again read, and an urgent call was made for troops while the Mayor and some of his councillors, who were in the Mansion House, made their escape over the roof-tops. The arrival of Colonel Brereton with the 3rd Dragoons was greeted by cheers from the mob. Once again he gave no order to his men to defend the building but was seen chatting to the rioters and shaking hands with some of them. The appearance of Captain Gage with his troop evoked a very different response – a storm of hissing and booing accompanied by showers of stones and scrap iron. Brereton thereupon ordered all troops to withdraw and proceeded to make an Ethelred-like compact with the rioters that if they would disperse he would withdraw the 14th Dragoons from the City. It was on his orders that the 14th Dragoons marched out to Keynsham at noon. The report was quickly circulated that the soldiery had been put to flight, and from that moment until 5 a.m. next morning there was mob rule in Bristol.

It was at this juncture that Brunel came down into the city from Clifton. 'Heard . . . that the 14th were gone,' he wrote in his diary. 'Could hardly believe it. Went to the Mansion House. Found it nearly deserted. It had been broken into again and sacked.' Arming himself with the back of a broken chair, he plunged into the building and with the assistance of two friends, Alderman Hillhouse and a Mr Roch, set about salvaging pictures and plate, carrying them out over the roof and into the Custom House.

Meanwhile the trail of devastation had begun. At two o'clock the prison was broken open and fired. An hour later the new gaol was attacked, 150 prisoners liberated and the Governor's house fired while Colonel Brereton and his men looked on. The Toll Houses, the Lawford's Gate Bridewell and the Mansion House were next sacked and burned and then, at seven o'clock, the rioters advanced upon the Bishop's Palace. Once again Brereton and his troop were present but made no attempt to intervene and at 9.30 p.m. he marched them away to their quarters, whereupon the mob promptly set fire to the Palace.

Between the hours of ten and eleven, Captain Codrington, commanding the Doddington troop of Yeomanry, reported to Colonel Brereton at his lodgings. Having informed Codrington that his services were not required and sent him and his men out of the city by the way they came, the Colonel retired to his bed while the skies flared an angry red over Bristol. A desperate appeal from the Mayor at midnight was ignored. The Colonel was not to be disturbed. At 4.30 a.m. he was roused from his slumbers by Alderman Camplin, Captain Warrington of the 3rd Dragoons and two others, by which time the whole of the north and west sides of Queen Square were either ablaze or reduced to smouldering ruins. The Colonel's visitors insisted that he must call out his men and also recall the 14th Dragoons from Keynsham. This was done. When, at five o'clock in the morning, the troops marched on to the square they made a concerted charge upon the rioters, but whether they did so on Brereton's orders or whether they were so shocked by the scene of devastation which greeted them that they acted spontaneously is not clear. Suffice it to say that the mob was at once dispersed and peace at last restored.

What was Brunel doing that night? Was he one of Alderman Camplin's two companions? It is very likely, for where there was adventure or danger he was invariably to be found in the thick of it. He reserved two pages in his diary under the heading 'riots', but unfortunately he never entered up the account of his experiences after the episode at the Mansion House and the pages remain a tantalizing blank. But that he was sworn in as a special constable and played an active part throughout that dramatic night is revealed by his evidence at the subsequent trial of the unfortunate Mayor, Charles Pinney, in the Court of King's Bench on a charge of neglect of duty. Pinney was acquitted but, as the following extract from Brunel's evidence shows, the trial disclosed a very curious state of affairs.

'I believe,' asked Counsel, 'you were actively engaged in rendering what assistance you could during all the time of the riots?' – 'I was.'

'You were at every place, I think, where the mob was?' – 'No, not at the prisons; I was at the Mansion House and the Palace.'

'Had you an opportunity of observing who the persons were that were engaged with this mob?' – 'Yes.'

'Did you observe, also, the conduct of the multitude that were not actually engaged?' – 'I did.'

'Did the multitude in general assist, by their shouts and presence, the mob?' – 'At first they were indifferent, allowing the mob, and always making way for them; and towards dark, certainly, a great number of them shouted.'

'What were the sort of shouts?' – 'When the military came, they shouted "The King and Reform", but before that I cannot describe the sort of shouts.'

'Those shouts of "King and Reform" came from the multitude in general?' – 'Yes.'

'Were you at Bristol upon the Monday, when the special constables began to be formed?' – 'I was.'

'Did you observe amongst any body of special constables, any of the persons you had seen active in the riots upon the preceding days?' – 'Several.'

'Was there any particular body in which you recognized a great number of those persons?' – 'No, I did not recognize any particular body, but some three or four who had been exceedingly troublesome upon the Saturday night in front of the Mansion House; I recognized one in particular.'

'At what time, on the Monday, did you see those persons who had been actively engaged in the riots acting as special constables?' – 'They joined a party with whom I was at half past seven.'

'The three or four persons whom you saw acting as constables upon the Sunday – what did they do upon the Saturday?' – 'One, in particular, rescued a prisoner twice from my hands.'

'When you saw him acting as a special constable upon the Monday, did you inquire his name?' – 'No, I did not; I spoke to him, and reminded him that I had seen him the night before.'

'Did you mention him to any magistrates?' – 'I mentioned the circumstance but not his name.'

With rioters turning constables it must have been difficult indeed for the authorities to distinguish the sheep from the goats, and the extent of public sympathy for the rioters is revealed by this evidence. It also reveals another thing: the harsh laws of the period defeated their own object, for they induced even those who, like Brunel, were on the side of law and order to withhold the names of their antagonists.

Subsequently, Brunel was again summoned as a witness, this time to attend the court martial of Colonel Brereton. On this occasion, however, he was never called, for when the court assembled for the second day's hearing it was announced that the unhappy defendant had shot himself. We can only guess at the reasons for poor Brereton's strange conduct. He was a local man and the prospect of ordering his troops to fire or to draw their swords upon his own countrymen whose grievances and miseries he doubtless appreciated only too well obviously revolted him. Yet nothing can excuse mob violence and his forbearance was ill-judged. Had he acted more firmly in the first instance, not only would such wholesale destruction have been avoided, but, at the price of a few broken heads, he would have spared far more from the horrors of transportation and the rope. For many paid this terrible price for the havoc they wrought while the Colonel slumbered and Bristol blazed.

It is not surprising that in the light of these events and the urgent need to make good the devastation of the city, all thoughts of proceeding with the Clifton Bridge scheme were forgotten. The year 1832 passed without any further move being made and Brunel occupied himself with other concerns, some of which were mentioned in the previous chapter. The summer of that year found him in a state of great despondency, at odds with himself and much discouraged by his persistent lack of success. 'Ben,' he wrote in August to Hawes, 'I have a painful conviction that I am fast becoming a selfish, cold-hearted brute. Why don't you see it and warn me and cure me? I'm unhappy – exceedingly so – the excitement of this election came just in time to conceal it.'

Almost the only reference to political issues in Brunel's writings appears in February of this year when he called upon a family named Frampton and described them as 'terrible anti-reform, anti-Catholic, anti-free-trade people'. From this it is safe to infer that he held liberal and radical views, but his future interest in parliamentary doings was confined to the passage of Bill for the works on which he was engaged and, unlike some of his contemporaries in the engineering profession, he refused invitations to stand for election. That he plunged into the election of the first reformed parliament in support of Benjamin Hawes, who successfully contested the Lambeth constituency in the Radical cause, was probably prompted by the need for some distraction which would enable him to escape from himself rather than by any ardent political sympathies. Throughout the campaign he made his home at the Barge House and supported Hawes on the platform.

With the excitement of the election over, Brunel took stock of his gloomy situation – 'So many irons and none of them hot,' he wrote. The tunnel was closed, the gas engine experiments were abandoned; Woolwich Dockyard: 'Nothing done ... no communication from the Admiralty.' Monkwearmouth Dock: 'There appears a probability of their coming to Parliament again for the North Dock, but still uncertain and my balance still unpaid.' Clifton Bridge: 'Nothing doing and no appearance of any likelihood of doing anything.' But there is now another entry: 'Bristol Docks – I am still waiting in the *expectation* of something being done but have heard nothing decisive since my report.'

One reason why Bristol lost her position as the premier provincial port was the inconvenience and inadequacy of her docks. When Liverpool already boasted enclosed wet docks the shipping of Bristol still lay in highly inconvenient mud berths in the tidal pulls of the Avon and the Froome. Had the Bristol merchant, Joshua Franklyn, succeeded in his project of building new docks at Sea Mills on the site of the ancient Roman ferry station for Caerleon at the mouth of the little River Trim, history might have been very different. As it was, the improvements made to the existing Port of Bristol in 1804 were inadequate, unsatisfactory and short-sighted in that they failed to anticipate the rapid increase in the size of merchant shipping. They consisted of converting the old sinuous course of the tidal Avon through the city into a long wet dock which became known as the Floating Harbour or simply 'the Float', the river being diverted into a new tidal cut. At the upper end of this cut the Neetham Dam was constructed by means of which the land water of the Avon could be turned into the Float. At its lower end the Float communicated with the river through the half-tidal Cumberland Basin and two sets of locks. Just above the Prince's

Street Drawbridge, which divided the Float into two parts, there was a second locked outlet to the new cut through the Bathurst Basin, while just below this bridge the river Froome provided the Float with an additional water supply.

Unfortunately, shoals soon formed in the Float, the flow through it being insufficient to scour away the silt carried into it by the two rivers. A stop gate had been installed at the mouth of the Prince's Street Bridge with the idea that first one half of the Float and then the other could thus be isolated for scouring and cleaning, but it was found difficult to do this effectively. Users of the port objected strongly to the delays thus caused, and consequently scouring was given up, mud accumulated and heavy-laden merchantmen frequently grounded in mid-channel whence they must needs be warped most laboriously up to the quays. Early in 1832 an engineer named Myhers had been consulted about the improvement of the docks but his proposals were rejected by the Dock Company on the grounds of impracticability and expense. In August, Brunel's friend Roch, who was a member of the Dock Committee, mentioned the problem to him. As a result, Brunel met the Committee and agreed to make an inspection and report. He was not impressed with what he saw. The Neetham Dam was inefficient and in bad repair, passing, even in that dry August weather, a lot of Avon water to waste into the new cut. As for the Froome it was evidently an open sewer, for he described it as 'beastly enough of itself to breed a pestilence'.

In his report, Brunel proposed a number of improvements. The Neetham Dam should be rebuilt and raised so that it would not only divert the whole of the landwater of the Avon into the Float if need be, but would also, at high spring tides, hold back tidal water which might then be passed down the Float. He recommended the construction of a large sluice at the stop gate by Prince's Street Bridge by means of which one half of the Float could be scoured from the other. For clearing the shoals he suggested, not an orthodox dredger, but a special 'drag-boat'. This would be equipped with a steam-powered winch with triple barrels, two to enable the boat to warp itself to and fro across the Float and the third to operate a scraper which would drag the mud to a point near the Cumberland Basin entrance.[4] Here there would be built a wooden culvert through which the mud could be scoured away into the river at low tide by the action of the sluices.

Like the Clifton Bridge, this plan for improving the docks was eclipsed by the Bristol riots and Brunel's report remained in a pigeon-hole in the Dock Office. At the end of January 1833, when Brunel took such gloomy stock of his prospects, his fortunes must certainly have appeared to him to be at their lowest ebb. In fact the tide was just about to turn in his favour. With the return of the first reformed Parliament a feeling of renewed confidence began to spread through the country and in a matter of days he was down at Bristol in conference with the Dock Committee. The Committee resolved to act upon his recommendations and the work was put in hand under his active supervision.

This work in the docks brought Brunel into contact with Bristol's newly appointed Quay Warden, a naval officer retired on half pay by the name of Captain Christopher Claxton. Although Claxton was the older man by sixteen years, a close friendship quickly developed which was to last until death, while in every enterprise in which his knowledge and experience could be of service to Brunel we shall find the Captain well to the fore. The dock works were completed in just over a year, but by that time Brunel's mind was occupied with a scheme of far greater moment.

Unknown to him, in the autumn of 1832 when his future had seemed so dark and unpromising, four Bristol merchants had foregathered in a dingy little office in Temple Backs. Their names were George Jones, John Harford, William Tothill and Thomas Richard Guppy and they had met to discuss the possibility of building a railway from Bristol to

4. The Cumberland Basin is still cleaned by this method.

London. Their leader, Guppy, was a very remarkable man.
Like Claxton he was soon to become a firm friend of Brunel's, although the precise occasion of their first meeting is
not recorded. Nine years older than Brunel, Guppy was himself an engineer by training, having served a five-years' apprenticeship with Maudslay, Sons & Field. He had then
travelled in America, studied drawing in Germany, and spent
a year at the Académie des Beaux Arts in Paris before settling down in Bristol to run the Friars Sugar Refinery in partnership with his brother Samuel. Guppy and Brunel thus
shared a very wide field of interests. Guppy was a man of
considerable wealth who is said to have invested no less than
£14,300 in the railway project. Curiously enough, however,
despite this heavy commitment and the fact that he was virtually the founder of the railway, he was not himself an
original member of the Committee which was set up as a
result of this first meeting, although he joined later. Nor was
he, as is sometimes supposed, responsible for introducing
Brunel to the venture. Once again it was Nicholas Roch who
performed that office.

It is on 21 February 1833 that the marginal heading 'B.R.',
standing for Bristol Railway, makes the first of many appearances in Brunel's diary. He had ridden down from Clifton that morning to superintend dragging operations in the
Cumberland Basin and, finding the work going well, he set
off in search of Roch and found him in solicitor Osborne's
office. Roch told him in Osborne's presence of the committee
which had been formed to consider the possibility of building a railway from Bristol to London and which had held its
first meeting exactly a month before. It consisted of representatives of the Bristol Corporation, the Society of Merchant Venturers, The Bristol Dock Company, the Bristol
Chamber of Commerce and the Bristol & Gloucestershire
Railway. These bodies had agreed to contribute towards the
cost of making a preliminary survey and he, Roch, had been
appointed to the sub-committee which had been formed to

see that this work was carried out. Who was to make the
survey? Already the names of Brunton, Price and Townsend
had been advanced and it had been suggested that each
should make a survey and that the Committee would then be
guided by whoever should submit the lowest estimate for
constructing the line. Roch then proposed that he should put
the name of Brunel forward also.

Was this Brunel's long-awaited opportunity? 'How will
this end?' he wrote in his diary that night. 'We are undertaking a survey at a sum by which I shall be considerably a
loser, but succeeding in being appointed engineer – *nous ver-
rons.*' Already, be it noted, he did not doubt that he could
secure the appointment. What, then, of his rivals? Admittedly they were not very formidable. William Brunton[5] and
Henry Habberley Price had previously come forward with a
scheme for a Bristol to London Railway in 1832 and had proposed a route via Bath, Bradford, Trowbridge, the Vale of
Pewsey, Hungerford, Newbury, Reading and Southall to 'a
vacant spot within three or four hundred yards of Edgeware
Road, Oxford Street and the Paddington and City Road
Turnpike'. But they were not favoured by the majority of
the Bristol Committee, and Brunel, when he subsequently
met Price, was not impressed by his abilities. W. H. Townsend was a local land surveyor and valuer who had surveyed
the route of the Bristol & Gloucestershire Railway and was
now in charge of its construction.

This Bristol & Gloucestershire Railway was by no means
such a grandiose affair as its title would suggest. It had obtained its Act in 1828 and on its completion in 1835 it consisted of ten miles of 4 ft 8 in. gauge line laid with cast-iron
rails on stone blocks from Cuckold's Pill on the Floating
Harbour at Bristol to collieries in the neighbourhood of
Coalpit Heath and Mangotsfield. As a result of his work on

5. William Brunton (b. 1770) began his career as a mechanic in
Boulton & Watt's Soho Foundry. He went to the Butterley Iron Works
in 1808 where he patented his curious 'Steam Horse'.

this tramway, Townsend had made something of a local reputation but he evidently had not the stature to survey and engineer a trunk line. After visiting Osborne the day after the first momentous meeting, Brunel wrote: 'Osborne was of opinion that unless I took the whole management and only left T – the nominal surveyor and a little bit of this end of the line we should never get through it.' At a subsequent meeting, Townsend himself appeared to admit his incapacity and to be prepared to work as his subordinate. 'How the devil I am to get on with him tied to my neck I know not', wrote Brunel after this interview.

It was typical of Brunel that with the first great opportunity of his lifetime dangled before him he scorned to betray his own judgement by compromising with the Railway Committee's wishes. Instead he gambled to win or lose all on one bold stroke. Although his friend Roch himself favoured the plan of competitive surveys and estimates, Brunel told him bluntly that he would not enter for such a competition because he was convinced that it was the wrong approach to the problem and would only lead to trouble later. He would agree to survey only one road from Bristol to London and that would be the best, not the cheapest. Having, at Roch's request, set these views down on paper for the consideration of the Committeee, he left for London by coach on 4 March to attend the annual meeting of the Thames Tunnel Company. He returned to Bristol by the night mail on the 6th, travelling outside, and as the coach rumbled down the Bath road through the cold darkness of that March night he must have wondered what fate held in store for him at his destination for the Committee was meeting that morning. On reaching Bristol he scribbled a note to Roch advising him of his arrival and then waited for what must have seemed an eternity in Osborne's office. At two o'clock a messenger appeared with a note from Roch. He had been appointed engineer of the Bristol Railway with Townsend as his assistant surveyor, and his presence was requested at the Council

House. There at three-thirty his appointment was formally confirmed by the Chairman, John Cave. When he went to dine with Roch next day he was told that the motion for his appointment had been carried by only one vote, the Chairman himself being against him. 'This', comments Brunel, 'was going too close – must be more active next time.'

Brunel and Townsend together agreed to make a preliminary survey for the sum of £500 and wasted no time in getting down to work. It had been suggested that the new railway should follow the course of the Bristol & Gloucestershire line out of Bristol and it is evident that with this end in view Townsend had already proposed a route. On the morning of Saturday 9 March the pair rode out of Bristol in the direction of Mangotsfield. The great adventure had begun.

Of this first day of his railway career Brunel writes: 'Started with Townsend (who as usual was late). Went up the B. & G. line and then across the country by Wick Court and over the hill into Lanbridge Valley and into the London Road by Bath. A most circuitous line and dreadfully hilly country, but I fear the only line which will take in the present railway. Dined at Bath and rode home by the lower road. This latter line, I think, offers greater facilities.' The latter line it was indeed to be, and already he had suggested Temple Meads as the best site for the Bristol terminus.

At 6.30 a.m. on Monday morning Brunel was away on his horse to Bath again, whence he proceeded on foot to prospect the difficult line through the narrow valley of the Avon, already occupied by the road and the Kennet & Avon Canal, through Claverton and Limpley Stoke. 'The side of the hill is a rotten description oolite laying on clay,' he notes. 'Many slips have occurred owing no doubt to the washing of the clay by the river and considerably assisted by the bad management of the canal. Blakewell the Canal engineer, a bigotted, obstinate *practical man* says the road will make the hill slip, but could not tell us why.'

Brunel subsequently surveyed in the neighbourhood of Box and Corsham before leaving Townsend in charge and returning to London where he appointed a surveyor named Hughes to assist him and first reconnoitred the proposed southern approach to London, crossing the Thames near Kingston, which was then favoured. These were such hectic days that he scarcely ever found time to sleep for more than an odd hour or so at a time. The Committee had ordered the completion of the preliminary survey by May at the latest. His days were spent in travelling about in coaches or on horseback, while by night he worked on his reports, estimates and calculations. And as if this was not enough there were his other 'irons' to be attended to: the Fossdyke survey, the Bristol Dock works and so on.

The following entries in his diary at this time reveal a typical glimpse of the arduous life of a budding railway engineer in the year 1833:

Wedn. April 17th. – Started in the evening by Cooper's coach to the Bear at Reading.

Thurs. April 18. – Got a hack and rode out to Wo. Hill [Woodley-Hill?] according to the letter from Hughes. After some search found him – on the wrong track. Directed him as he was so far to push on to the Thames across Early Court and Upper Early and the next morning to begin again at Chapple Green and go on to Shinfield Green where I would meet him on Saturday morning.

I then rode on to Bagshot Heath and returned by a line going at the back of Easthampstead Park. My horse came down at Mitchell.

Fri. April 19th. – Up all night in the intention of going by Cooper's Coach but it did not arrive till near five and was then full. Went up by 5 o'clock Reading coach; arrived home, dressed, went to BH [Barge House].

Started for Old Company's at Piccadilly, all full, obliged to wait for the mail.

Sat. April 20th. – Arrived at Reading late. Went to bed. After breakfast went in search of Hughes. After some trouble found him at 'Black Boy', Shinfield, gave him maps. With him to Theal Road and into Pangbourne. Returned to Reading, went to Theale. Met a Mr Keeps who shewed me the new church. Returned to Reading; Hughes came in the evening. Gave him £5.0.0.

Sun. April 21st. – Went to church at the great church – Dr Millman. After church lunched. Started on horseback for Wantage – baited at Blewbury and arrived late at Wantage.

Monday April 22nd. – Started at 6 a.m. Examined the ground in the neighbourhood of Wantage – breakfasted at Streatley. Determined on the outer line winding round the undulating ground. Returned to Reading, dined, and went to Theal to meet Hughes. After waiting some time gave it up and returned.

Tuesday April 23rd. – After breakfast went in search of Hughes; after some trouble found him at the canal between Shinfield and Calcott Mills, a beautiful place this in hot weather. Gave him the line to Wantage.

Next day Brunel returned to London and went from thence to Bristol, where he spent a busy week attending meetings and also rowed up the Avon with Osborne as far as Keynsham Lock, pointing out his proposed line. Then he was off on his travels once more, back to London for more meetings on 29 April and then:

Monday May 6th. – Started by Emerald Coach to Newbury. Arrived there, mounted my horse and rode to Uffington, thence to Shrivenham. Slept there and in the morning proceeded to Swindon. Met Hughes there – found letters from Osborne requiring my immediate return to town. George [?] came; our lines nearly meet, but he has been winding round in a most curious manner. Directed him to point out his Bench Mark to Hughes at Wootton Bassett and then return over his ground to Chippenham following a line I traced for him. Rode to Hungerford; thence to Newbury. Just as I got in sight of the Castle my horse came down – cut his knees and forehead dreadfully – just scratched my knee. I never saw a horse tumble over in such an '*abandonné*' style, he dirtied himself even over the withers and croup he rolled over so far. Bled him and left him at the Castle. Returned to town in Bristol Mail.

So it went on; night journeys in crowded coaches, and hard riding across country in search of elusive and not very reliable assistants. But the preliminary survey was completed in May and presented to the Committee in Bristol. As the diary entries show, Brunel had surveyed two routes west of Reading, the Kennet Valley and Vale of Pewsey line which Brunton and Price had advocated and the more northerly route through the White Horse Vale by Swindon and Wootton Bassett. He unhesitatingly recommended the latter, but east of Reading he showed alternative routes the choice of which would depend on the site selected for the London terminus.

On 30 July matters were so far advanced that the Committee staged a public meeting in Bristol to announce details and solicit support. Brunel was asked to speak. He wrote: 'Got through it very tolerably which I consider great thing. I hate public meetings – it's playing with a tiger, and all you can hope is that you may not get scratched or worse.'

Earlier, he had warned the Bristol Committee not to air their proposals too prematurely in London and his advice had been accepted. But now it was judged that the time was ripe and on 15 August Brunel and Messrs Guppy, Bright, Cave, Tothill, Jones, Harford and Roch all set out for London by coach. Contact had previously been established with George Henry Gibbs, head of the firm of Antony Gibbs & Sons and a cousin of George Gibbs, who was a Merchant Venturer representative on the Bristol Committee and senior partner of Gibbs, Bright & Company of Bristol and Liverpool. Henry Gibbs had already formed a provisional London committee and representatives of the two parties met for the first time in Gibbs's offices at 47 Lime Street at 1 p.m. on 22 August.[6] The London representatives were Benjamin Shaw (in the Chair), Henry Gibbs, Edward Wheeler Miles, Ralph Fenwick and John Bettington. George Gibbs, Nicholas Roch,

6. MacDermot in his *History of the Great Western Railway* gives 19 August as the date of this historic meeting. My date, and the other particulars, come from Brunel's diary.

John Harford, Peter Maze and William Tothill (Secretary) represented Bristol. After a trip to Brixton in the morning on Sir Charles Dance's steam coach, Brunel joined them. His one desire now was to be allowed to get his teeth into the job and he had obviously lost patience with the endless talk of committees. He was to recover that lost virtue, perforce, however, by the time the new railway was established. 'Rather an old woman's set,' he wrote in his diary after the meeting, 'regular jobbing committee. Must hope for somebody to give them a little life and sense.'

When the meeting reassembled on 27 August that hope was fulfilled in the person of the Secretary to the London Committee who here met Brunel for the first time. His name was Charles Alexander Saunders and he was soon to become one of Brunel's most staunch friends and allies. Although not all the London directors had yet been appointed, at this meeting the prospectus was finally approved and a great railway was thus brought to birth. That greatness might as yet be hidden in the future but the infant was prophetically baptized at this meeting for against the day's entry in his journal Brunel has written for the first time the initials 'G.W.R.' The proud title was to have many imitators in after years. 'Great Western Railway'; it is not fanciful to suggest that it was Brunel's invention because the name is so entirely typical of him. While others shook their heads and pronounced that a line from London to Bristol was too long and over-ambitious, his was a mind that impatiently scorned such limitations. Already in imagination he had travelled far beyond Bristol to the farthest west of Cornwall and soon, very soon, he would span the Atlantic to New York. So the western railway it must be. As to its greatness, upon that he was already determined. To Charles Saunders he took an immediate liking and after the meeting they dined together at the Caledonian Hotel. 'An agreeable man,' he wrote. 'Proposed entrance to London at Penitentiary (Wormwood Scrubs). I think well of it.'

Now that he was the appointed engineer of a great railway company, albeit as yet unauthorized, it was obvious that Brunel's peripatetic London existence, using his father's office and sleeping at Bridge Street or the Barge House, must come to an end. He must have a London office of his own, preferably near the Houses of Parliament where much of his work would have to be done. After some search he settled upon No. 53 Parliament Street, moving in with his faithful clerk Bennett, who was to remain with him until his death. He was now faced with the immense task of superintending the detailed surveys and soon, if all went well, the actual construction of the line. Experience had shown him the disadvantages and delays incurred by attempting to carry out such work by coaches and hired hacks. He must have his own travelling carriage. So Brunel designed and had built his famous black britzska which, irreverently nicknamed the 'Flying Hearse', was to become so familiar a sight to surveyors and construction gangs along the roads and lanes of the West Country during the next few years. It was designed to accommodate his plans and engineering instruments and all creature comforts including an ample stock of the cigars which he now smoked incessantly, while the seat was arranged to extend into a couch so that he could snatch an hour or two's sleep if the occasion offered.

With his own office and a travelling carriage ordered it might be said that Brunel's affairs were looking up. So in a sense they were, yet these attributes were not so much marks of success as essential equipment for a great gamble. If the Bill for the Great Western Railway was passed by Parliament his reputation would be made and his future assured – he had no doubt of that. If it was rejected he might find himself no better or even worse off than before and his dreams of a great railway would be added to the Thames Tunnel and the Clifton Bridge on his lengthening list of broken hopes. This time, however, he was going to win, but it would be a great struggle.

The Turn of the Tide

IT had been decided that the new railway company should be represented by a board of twenty-four directors, half Londoners and half Bristolians. The last directorship having been filled on 31 August by William Unwin Sims of London, it was resolved to publish the prospectus but not, for the moment, a plan because the vexed question of the route into London was still undecided. Most of the London men were strangers to Brunel, so he took careful stock of them, recording his impressions as follows:

I think I gain ground with Mr Miles, he seems an amiable man but pig-headed. Fenwick I think is a friend. Gibbs will go with the Bristol Committee. Bettington is a jobber, but probably caring little about anything but his salary and shares. Grenfell must be humoured. Gower very doubtful – stupid enough and proportionally suspicious. Hopkins I hardly know. Simonds a hot warm-tempered Tory, just such another as K. [sic] Claxton i.e. warm friend but changeable and very capable of being a devil of an opponent.

This does not reflect a very high opinion of the London men but it is obvious that the favourable impression which Brunel had formed of their Secretary, Charles Saunders, at their first meeting was abundantly confirmed and that Saunders quickly gained his confidence, friendship and unbounded admiration. On this score, events were to prove the soundness of his judgement, for with the exception of Brunel himself no man was to do more towards setting the Great Western Railway upon its feet.

At a meeting held on Saturday 7 September it was decided to proceed with the detailed survey of the line, and because this would entail far more work than the preliminary survey,

Brunel was authorized to employ assistant surveyors additional to Townsend and Hughes. On the following Monday morning Brunel's diary begins 'Up at 5' and for the next few weeks there follows the same hectic story as before; the same interminable journeys by coach and hack (for his travelling carriage was not yet ready) in search of elusive assistants; the same nights spent in posting inns poring over plans and estimates until the candles paled and it was time to saddle up once more and range over the dew-drenched downs or through the levels of Thames-side. Those early mornings may be pictured so clearly: the white mists of September smoking from the river to drift over the water meadows; the stubbles on Moulsford Down golden in the first sunlight and the lonely horseman, his slight figure cloaked against the morning chill, riding with eyes alert for the staffs that marked the line of his new road.

Not only had Brunel to superintend the survey, he had also to conciliate the local landowners, a task which was often far from agreeable but which, surprisingly for one of so impulsive and forthright a nature, he handled with great patience, tact and success. In this his ability to size up so rapidly the characters of his fellow men obviously helped him considerably. Such pressure of work taxed even Brunel's extraordinary powers of endurance to the utmost and he confessed to one of his most trusted assistants, Hammond, that 'Between ourselves it is harder work than I like. I am rarely much under twenty hours a day at it.'

The following two diary entries may be taken as typical of his activities at this period:

September 14th. – Up at 5 a.m. Joined Place & Williams ranged on to the Island east of Caversham. Breakfasted and mounted. Called on Mr Hawks, Surveyor; appointed to be with him at 8 p.m. Rode to meet Hughes; found him in barley stubble west of cottage. Directed him how to proceed and to meet me this evening at the Bear [Reading]. Rode them to Purley Hall. Met Mr Wilder just going in; spoke to him; found him very

civil; gave him a prospectus. Rode on to Basildon Farm; left Mr
Hopkins' note and my card on Mr Stone. Rode on to Streatley;
tried in every way to find a line round instead of crossing the
river at Goring; found it impossible. On looking at the country
from the high hill south of Streatley however, it was evident
that much cutting might be saved by passing SW of Streatley
Farm and winding a little more east of Halfpenny Lane.

Returned to Reading, went in search of Mr Stone; found he
was gone. Called on Mr Symonds. Hughes came at 7½. Agreed
with him that he was to have £2.2 a day and pay his own expenses instead of £35 and charges. Pointed out to him the line he
was to follow. Took him with me to Mr Hawks to look at his
large plan, Mr H — to furnish him with a copy by tomorrow
evening and is to make the survey of the line from Sonning to
Streatley inclusive with Book of reference, etc., etc. Came to town
by Mail.

The second entry reveals his continuing difficulties with
Townsend who, as before, had been charged with the Bristol
end of the survey:

September 24th. – Arrived at Bath. (After an all-night coach
journey from London.) Met Townsend. Breakfasted and started
in his phaeton. Went as far as Keynsham; got out and walked
over line. Arrived at the valley at Brislington found the staffs up
– all to double the curve agreed on. Could not make him understand the theory or rationale.

After a long meeting with the Bristol Committee discussing proposed branch railways, we next find him busy on
the Bristol Dock works until a late hour.

While Brunel was thus engaged, Charles Saunders was occupied hardly less fully stumping the West of England in the
attempt to raise the necessary capital. Despite tireless efforts
on his part, however, the 50 per cent of the full working
capital which was necessary before the Bill could be brought
before Parliament was not forthcoming. To wait any longer
for money to come in would mean missing the current Parliamentary session. The Directors therefore decided to apply
for powers to construct the Bristol–Bath and the Lon-

don–Reading sections whose cost Brunel had estimated at £1,250,000 and to apply for additional powers to complete the remainder during the next session, by which time the additional capital should have been raised. This decision was reached at a meeting in Reading on 18 October whereat Brunel was instructed to stop work on the survey between Reading and Bath immediately.

The results of his labours, the plans which were deposited in November 1833, did not vary very widely from the Great Western Railway as it was eventually to be built except at the London end where, after passing through a short tunnel to the south of Ealing, the line was to be carried on a viaduct 20 ft high for four miles through South Acton, Hammersmith, Brompton and Pimlico to a terminus near Vauxhall Bridge. There was to be a tunnel under Sonning Hill and the line was shown crossing the Avon at four points, later reduced to two, between Bath and Bristol.

The second reading of the Bill was moved in the Commons on 10 March 1834 and after some hours' debate it was carried by 182 votes to 92. It was then referred to Committee where there ensued one of the epic battles of railway parliamentary history lasting fifty-seven days and remarkable for two things: the amount of fantastic nonsense delivered by opposition witnesses and the extraordinary forensic skill displayed by Brunel. The Committee at last approved the Bill, but when Lord Wharncliffe moved its second reading in the Lords it was thrown out by 47 to 30.

Undismayed by this defeat, the Great Western directors at once set about raising additional capital to enable them to present a fresh Bill for the whole railway to the Parliamentary session of 1835. Negotiations were entered into with the London & Birmingham Company as a result of which the prospectus of November 1834 showed the two lines uniting at a junction near Wormwood Scrubs and proceeding thence to a joint terminus 'near the New Road in the Parish of St Pancras'. When this new Bill came before the

Commons in March 1835 it soon became clear that the struggle of the previous year had not been altogether in vain. The opposition was much discomfited when Charles Russell, the Member for Reading and the chairman of the Commons Committee, ruled that the value of a railway from London to Bristol had already been proved and that no further evidence on this score was required. The Bill was passed by the Commons at the end of May and passed its second reading in the Lords on 10 June. It was then referred to a Committee of the Lords presided over by Lord Wharncliffe where the opposition waged another stubborn struggle which went on for forty days. At last, however, the Committee declared the Bill proved; it passed its third reading and received the Royal Assent on the last day of August. The victory had been won.

This great battle of wits cost the new Company no less than £88,710 in legal and Parliamentary expenses, but it unquestionably established Brunel's reputation and he emerged from the long ordeal as a major star in the engineering firmament. His cross-examination before the Commons Committee occupied eleven days, during the greater part of which counsel for the opposition endeavoured with all their subtle art to rattle him and to make him contradict himself or betray ignorance. Few witnesses have ever been subjected to such a protracted and exhausting intellectual duel. Yet he never flagged and never gave a point away. One of the crowd of landowners and engineers who packed the Committee room during his examination said of him: 'His knowledge of the country surveyed by him was marvellously great. . . . He was rapid in thought, clear in his language, and never said too much or lost his presence of mind. I do not remember ever having enjoyed so great an intellectual treat as that of listening to Brunel's examination.' Knowing Brunel's passionate temperament and the nonsensical and provocative arguments of opposition witnesses and counsel, we can but marvel at his self-control, at the impenetrability of that mask behind which he had learned to conceal his feelings. In that crowded room, only his most intimate friends can have known what lay behind the cool, untiring thrust and parry of his intellect.

Some of the eminent engineers, Locke, George Stephenson, Palmer, Price and Vignoles, who supported Brunel, showed themselves less patient under examination. One particularly inept and provocative remark by opposition counsel goaded George Stephenson past endurance. 'I wish you had a little engineering knowledge – you would not speak to me so,' he expostulated. 'I feel the disadvantage,' replied counsel suavely, and we can picture the slight, ironic bow. But his irony was lost upon that bluff north-countryman. 'I am sure you must,' he retorted. The 'Father of Railways' was warm in his praise for the results of Brunel's survey. 'I can imagine a better line,' he said, 'but I do not know of one.'

The opposition represented diverse and occasionally conflicting interests. Landowners, coach proprietors and the Kennet & Avon Canal Company opposed for obvious reasons. The Provost of Eton, one of the most formidable of opponents, declared that the proximity of the railway would certainly undermine the morals and discipline of the school, whereas the representatives of the town of Windsor opposed on the grounds that the railway did not come near enough. Maidenhead pleaded the certain loss of tolls on their Thames bridge, while the promoters of the little London & Windsor Railway joined in the fray out of sheer spite as they had previously decided to withdraw their own Bill. But the most formidable opposition of all came from the London & Southampton Railway Company. This Company had originally wooed the Great Western proprietors to join their route at Basingstoke. When this proposition was rejected they had grown bitterly hostile and had instructed their engineer, Francis Giles, to survey a rival line which became known as the Basing & Bath. From Newbury westwards this followed closely that southern route which had been surveyed in 1832 by Brunel's predecessor, William Brunton. That worthy was certainly persistent, for Brunel had scarcely completed his survey when he submitted to the Bristol Railway Committee a report in which he claimed to offer a better route. Brunel dismissed it as 'a well-written unmeaning, *uncommitting* thing' and nothing more was heard of it. This drove Brunton into the enemy's camp and he became one of the leading engineer witnesses for the opposition.

The first Great Western Bill for the two ends of the line only had offered the opposition an easy target and one of their counsel had dismissed it as 'neither "Great" nor "Western" nor even a "Railway" at all but a gross deception, a trick, and a fraud upon the public in name, in title, and in substance'. On the other hand the champions of the Basing & Bath did not distinguish themselves. In answer to the criticism that their line was much more heavily graded than that proposed in the Bill, they put forward the delightful argument that because the 'ups' practically cancelled out the 'downs' the line would, in effect, be level. The Chairman's reaction to this roller-coaster theory was to remark drily that if it was correct then the Highlands of Scotland should offer ideal opportunities for railway building. Undaunted by such heavy sarcasm, the Basing & Bath faction organized a public meeting in Bath during the interval before the second Great Western Bill came up. Several flowery speeches were made and the meeting appeared to be going splendidly but, alas for their hopes, the woodpile of that packed and attentive audience harboured the most formidable of niggers in the person of their arch-enemy himself. Despite his confessed hatred of public meetings, Brunel, having bided his time, rose to his feet and proceeded to demolish their arguments to such ruthless and deadly effect that a resolution in favour of the Great Western Railway was carried with acclamation, the meeting broke up in disorder and the opposition champions retired discomfited to lick their wounds.

By the time the stage was set for that last desperate battle in the Lords' Committee room many of the opposition's guns had already been spiked and they concentrated all their remaining fire upon that 'monstrous and extraordinary, most dangerous and impracticable' tunnel which Brunel proposed to drive under Box Hill at a gradient of 1 in 100. The construction of such a tunnel must inevitably lead to a wholesale destruction of human life which no conceivable care or foresight could prevent. 'No person would desire to be shut out from daylight with a consciousness that he had a superincumbent weight of earth sufficient to crush him in case of accident,' declared one witness. 'The noise of two trains passing in the tunnel would shake the nerves of this assembly,' another declaimed dramatically. 'I do not know such a noise,' he added. 'No passenger would be induced to go twice.' To these impassioned pleas the egregious Doctor Dionysius Lardner, one of the mainstays of the opposition, added all the ponderous weight of that pseudo-science of which he was one of the first and greatest masters. He had proved by elaborate calculation that if the brakes were to fail as a train entered the tunnel on the falling gradient it would emerge at the other end at a speed of 120 m.p.h., a speed, he added, at which no passenger would be able to breathe. At this Brunel pointed out drily that the factors of friction and air resistance must evidently have become lost in the Doctor's calculations because owing to their combined effect the speed would be 56 miles an hour and not 120. It would appear, however, that the eminent Doctor, who made a livelihood by writing popular scientific works on subjects ranging from steam power, hydrostatics and optics to astronomy and the habits of the white ant, floated through life upon an unpuncturable balloon of self-esteem. So far from accepting his defeat he returned to the charge on several subsequent occasions though invariably with the same result.

During these two crucial years of 1834 and 1835 when Brunel's future hung in the balance with that of the Great

Western Railway we are, unfortunately, left almost as ignorant of his hopes and fears as were those who flocked to committee room or public meeting to hear him speak. For the pressure of work became so great that there was no longer time for him to record his private thoughts or even his day-to-day doings. During the period of the survey the entries in his journal become more and more laconic and intermittent, their handwriting more hurried, until finally the entries close with the brief record of an abortive conciliatory visit to Eton with Saunders on 24 January 1834, and the book is not re-opened until the end of 1835. Such a blank is understandable. Indeed it is astounding that one man could have crammed so much work into the space of two years and at the same time, without the aid of railways, have covered so much ground in the West Country.

The interval between the rejection of the first and the presentation of the second Great Western Bill was no breathing space for Brunel. Like Charles Saunders he travelled the country untiringly, conciliating hostile landowners, addressing the public meetings held in town after town to rally support for the railway, or superintending his assistants who were already engaged upon fresh surveys. For his prophetic vision of the line to Bristol, not as the over-ambitious project which others saw, but as a mere sapling which would grow into a mighty trunk with wide-spreading branches, was already beginning to take shape. In the summer of 1835 his assistants were surveying the route of the Cheltenham & Great Western Union Railway, while one of his most trusted men, William Gravatt, an associate of Thames Tunnel days, was at work on the line of the Bristol & Exeter through Bridgwater and Taunton. Both railways had obtained their Acts in 1836, the latter having an easy passage with little opposition and the former a hard struggle. For the proprietors of the London & Birmingham had become alarmed by the precocious growth of Brunel's railway empire and made every effort to prevent it from spreading north of the London to Bristol line. They attempted to counter Brunel's advance from Swindon through Stroud and Gloucester with a proposal for a line from Tring through Oxford to Cheltenham, but their attack was repulsed. The Thames & Severn Canal Company and Squire Gordon of Kemble were also formidable opponents of the Cheltenham line who had to be bought off to the tune of £7,500 apiece. The Squire of Kemble had to be placated in other ways. A tunnel nearly a quarter of a mile long had to be built so that the hated railway should be hidden from Kemble House, while because of the stipulation that there should be no public station on his property, Kemble, the junction for Cirencester, could be used only for interchange traffic and did not appear in the timetable until 1872 when the ban was lifted.

Already Brunel was looking west of Exeter to Plymouth and beyond, while the Cheltenham line he saw as the key to South Wales. As early as 1834 he had staked a claim in South Wales through his friendship with Anthony Hill, the Taff Vale ironmaster, whom he had met in connexion with the ironwork for the Clifton Bridge. It was with Hill that Samuel Homfray of Penydaren ironworks had laid his famous 500-guinea wager that Trevithick's locomotive could haul 10 tons of iron from Penydaren to Abercynon. Trevithick had won that wager for Homfray two years before Brunel was born and now Hill was no longer sceptical of the new power. He consulted Brunel on the prospects of building a railway from Merthyr to Cardiff which would bring the coal and iron of the Taff Vale down to the coast for shipment and break the monopoly of the inadequate Glamorganshire Canal. Lady Charlotte Guest recorded in her diary Brunel's first flying visit to discuss this projected railway on 12 October 1834 when he got her husband, Sir Josiah Guest, soon to be first Chairman of the Taff Vale Railway, out of his bed at 6 a.m. for what she describes as 'a *very* early meeting'. The railway engineer had become the man of the hour and was no respecter of persons. Those on railway business bent must be early birds if they would catch Brunel on his headlong course to fame, for he seldom stayed long in one place. It was on this visit to South Wales that he was discovered one morning asleep in his chair with the ash of a complete cigar lying undisturbed upon his chest.

Throughout these hectic months the only intimate glimpses we get of Brunel are through correspondence or in the personal reminiscences of others. Thus just before the parliamentary battle was joined he sent a copy of his plans to his friend G. H. Wollaston, the chemist and physicist, and in the covering letter dropped for a moment the air of confidence to betray his doubts and fears. 'Will you accept a copy of my survey of the Gt West Railway?' he writes. 'The dress is somewhat more showy than I like but it was done during my absence and you must not therefore imagine it to be to my taste. However, perhaps we may as well mount gay colours while we can and before we enter that dreadful place the House of Commons from which no similar act ever came out alive at the first time. ... However, of course we *make sure of it* and it is only in confidence that I express my doubts.'

St George Burke, K.C., has left a personal memoir of his friendship with Brunel at this time which, for all its pompous style, has the stamp of authenticity and truth. Burke evidently found in him the same light-hearted companion who had so charmed Charles Macfarlane on his journey from France four years before.

'For a period of nearly three years,' writes Burke,

viz. during the contest for the Great Western Railway Bill, I think that seldom a day passed without our meeting. ... He could enter into the most boyish pranks and fun, without in the least distracting his attention from the matter of business ...

I believe that a more joyous nature, combined with the highest intellectual faculties, was never created, and I love to think of him in the character of the ever gay and kind-hearted friend of my   early years, rather than in the more serious professional aspect . . .

. . . [we] occupied chambers facing each other in Parliament Street. . . . To facilitate our intercourse it occurred to [him] to carry a string across Parliament Street, from his chambers to mine, to be there connected with a bell, by which he could either call me to the window to receive his telegraphic signals, or, more frequently, to wake me up in the morning when we had occasion to go into the country together, and great was the astonishment of the neighbours at this device, the object of which they were unable to comprehend.

I believe that at that time he scarcely ever went to bed, though I never remember to have seen him tired or out of spirits. He was a very constant smoker, and would take his nap in an armchair, very frequently with a cigar in his mouth; and if we were to start out of town at five or six o'clock in the morning, it was his frequent practice to rouse me out of bed about three, by means of the bell, when I would invariably find him up and dressed, and in great glee at the fun of having curtailed my slumbers by two or three hours more than necessary.

No one would have supposed that during the night he had been poring over plans and estimates, and engrossed in serious labours, which to most men would have proved destructive of their energies during the following day; but I never saw him otherwise than full of gaiety, and apparently as ready for work as though he had been sleeping through the night.

. . . [He] had a britszka, so arranged as to carry his plans and engineering instruments, besides some creature comforts, never forgetting the inevitable cigar-case[1] among them; and we would start by daybreak, or sometimes earlier, on our country excursions, which still live in my remembrance as some of the pleasantest I have ever enjoyed . . .

I have never known a man who, possessing courage which to many would appear almost like rashness, was less disposed to trust to chance or to throw away any opportunity of attaining his object. . . . In the character of a diplomatist . . . he was as wary and cautious as any man I ever knew.

I frequently accompanied him to the west of England, and into

1. An immense case which held fifty cigars.

Gloucestershire and South Wales, when public meetings were held in support of the measures in which he was engaged, and I had occasion to observe the enormous popularity which he everywhere enjoyed.

Reading this, and remembering the impression which he made before the Parliamentary Committees, it becomes clear that Brunel's meteoric success was due not only to his undoubted genius as an engineer and his unlimited capacity for hard work, but also to the magnetic power of his extraordinary personality. He had discovered that power at his first public meeting in Bristol and now had learnt to wield it with all the apparent artlessness of an accomplished actor.

It was not only Brunel's new railway schemes that prospered. Some of those older 'irons' of which he had so recently despaired began to grow warm and to show promise of fulfilment before 1835 was out. The Monkwearmouth Dock scheme was going forward at last, albeit on a more modest scale; the Clifton Bridge Committee had launched a fresh appeal for funds and the response had been such that a resumption of work on the bridge was imminent. Even the Thames Tunnel on which Brunel had written so gloomy an epitaph had now awakened from its long sleep, the Government having at last agreed to make a loan of £246,000 to the Tunnel Company. The first instalment of £30,000 had been paid in December 1834 and Marc Brunel had moved to Rotherhithe to be nearer the work. For how long had both father and son struggled on against bitter disappointments and adversities, but now at last, it seemed, the Brunel star was in the ascendant. Brunel could hardly bring himself to believe it; it seemed too good to be true.

On Boxing Night 1835 he sat late and alone by the fireside in his little office in Parliament Street meditating, perhaps for the first time, on the momentous events of the year just past. Taking down his long neglected diary from the shelf he opened it at random and wrote:

What a blank in my journal! And during the most eventful part of my life. When last I wrote in this book I was just emerging from obscurity. I had been toiling most unprofitably at numerous things – unprofitably at least at the moment. The Railway certainly was brightening but still very uncertain – what a change. *The Railway* now is in progress. I am their Engineer to the finest work in England – a handsome salary – £2,000 a year – on excellent terms with my Directors and all going smoothly, but what a fight we have had – and how near defeat – and what a ruinous defeat it would have been. It is like looking back upon a fearful pass – but we have succeeded. And it's not this alone but everything I have been engaged in has been successful.

Clifton Bridge – my first child, my darling, is actually going on – recommended work last Monday – Glorious!!

Sunderland Docks too going on well –

Bristol Docks. All Bristol is alive and turned bold and speculative with this Railway – we are to widen the entrances and the Lord knows what.

Merthyr & Cardiff Railway – This too I owe to the G.W.R. I care not however about it –

Cheltenham Railway. Of course this I owe to the Great Western – and I may say to myself. Do not feel much interested in this. None of the parties are my friends. I hold it only because they can't do without me – it's an awkward line and the estimate's too low. However, it's all in the way of business and it's a proud thing to monopolize all the west as I do. I must keep it as long as I can but I want *tools*.

Bristol & Exeter Railway – Another too!!

This survey was done in grand style – it's a good line too – and I feel an interest as connected with Bristol to which I really owe much – they have stuck well to me. I think we shall carry this bill – I shall become quite an oracle in Committees of the House. Gravatt served me well in this B. & E. Survey.

Newbury Branch a little go almost beneath *my* notice now – it will do as a branch.

Suspension Bridge across Thames – I have condescended to be engineer to this – but I shan't give myself much trouble about it. If done, however, it all adds to my stock of irons.

I think this forms a pretty list of real profitable, sound professional jobs – unsought for on my part, that is given to me fairly by the respective parties, all, except MD [Monkwearmouth Dock] resulting from the Clifton Bridge – which I fought hard for and gained only by persevering struggles and some manoeuvres (all fair and honest however). *Voyons.*

I forgot also Bristol & Gloster Railway.

Capital:

70,000	Clifton Bridge
20,000	Bristol Docks – to come – Portishead Pier
2,500,000	G. W. Railway – to come – Oxford Branch
750,000	Chelt^m Railway
1,250,000	Bristol & Exeter do. do. – perhaps Plymouth etc.
250,000	Merthyr & Cardiff do. Gloster & S. Wales
150,000	Newbury Branch
50,000	Sunderland Docks
100,000	Thames Suspension Bridge
450,000	Bristol & Gloster Railway

5,590,000

A pretty considerable capital likely to pass through my hands – and this at the age of 29 – faith not so young as I always fancy tho' I really can hardly believe it when I think of it.

I am just leaving 53 Parliament St where I may say I have made my fortune or rather the foundation of it and have taken Lord Devon's house, No. 18 Duke St – a fine house – I have a fine travelling carriage – I go sometimes with my 4 horses – I have a cab & horse, I have a secretary – in fact I am now somebody. Everything has prospered, everything at this moment is sunshine. I don't like it – it can't last – bad weather must surely come. Let me see the storm in time to gather in my sails.

Mrs B. – I foresee one thing – this time 12 months I shall be a married man. How will that be? Will it make me happier?

BOOK II

[6]

Ellen Hulme and
Mary Horsley

'SHALL I make a good husband? – Am doubtful – my ambition, or whatever it may be called (it is not the mere wish
to be rich) is rather extensive . . .' 'After all I shall most likely
remain a bachelor and that is I think best for me. My profession is after all my only fit wife. . . .' 'As long as health
continues, one's prospects tolerable, and present efforts,
whatever they may be, tolerably successful, then indeed a
bachelor's life is luxurious: fond as I am of society, "*selfish
comfort*" is delightful. I have always felt so. My *châteaux
d'Espagne* have mostly been founded on this feeling, what
independence! For one whose ambition is to distinguish himself in the eyes of the public, such freedom is almost indis-
pensable – but, on the other hand, in sickness or
disappointment, how delightful to have a companion whose
sympathy one is sure of possessing! . . . I have always wished
and intended to be married, but I have been very doubtful
on the subject of – children – it is a question whether they
are sources of most pleasure or pain.'

Thus, at the age of twenty-one, Brunel had confided his
musings on the problems of marriage to his secret journal as
he sat on night duty in the little cabin by the tunnel shaft at
Rotherhithe.

From the portraits that Macfarlane and Burke have left of
him as a young man, it is obvious that Brunel must have
been attractive to women, and, equally, that he would be
attracted by them. It is impossible to think of him playing
the part either of the gauche youth or the aloof mysogynist.
His was far too passionate and impulsive a temperament. 'I

have had, as I suppose most young men must have had, numerous *attachments*, if they deserve that name,' he wrote in
November 1827. 'Each in its turn has appeared to me *the
true one*. E.H. is the oldest and most constant, now however
gone by. During her reign (nearly 7 years!!!) several inferior
ones caught my attention. I need hardly remind myself of
Mlle D.C., O.S., and numerous others.'[1]

Brunel never divulges the full name of his seven years'
love, but from other references in his journals it is possible
with reasonable certainty to identify 'E.H.' as Ellen Hulme
of the Manchester family whom he knew well. She had not,
in fact, 'gone by' at that time, for a month later he was writing: '*qu'on retourne toujours aux premiers amours* – Ellen is
still it seems my real love. I have written her a long letter
yesterday – her answer shall decide, if she wavers, I *ought* to
break it off for I cannot hope to be in a condition to marry
her and to continue in this state of suspense is wronging her.
... Oh! Ellen! Ellen! if you have kept up your Musick and
can even only play tolerably we might be very happy yet.
And starve – it won't do – however we'll see.' Five days later
he writes: 'No answer yet from E—n, and I'm afraid when it
comes it will be a quizzing one without any decisive answer –
a shocking habit that of quizzing, it prevents a person thinking seriously. I'm almost afraid of an answer, however, for to
marry would be absurd and to remain for years engaged
would be painful. ... If I have anything like an answer it
would probably decide my state of life.'

We do not know what Ellen's answer was, but nine months
later he notes laconically: 'I have had long correspondence
with Ellen which I think I have managed well. I may now
consider myself independent.' Perhaps she displayed that
shocking habit of quizzing; perhaps, grown tired of waiting,
she had given her heart elsewhere; perhaps it was Brunel
who decided it were best to disentangle himself, but if this be

1. As he was only twenty-one when he wrote this he had evidently
not wasted any time.

so he certainly managed it adroitly for he continued to be on terms of friendship with the Hulmes. But he did not remain unattached for very long.

It was Benjamin and Sophia Hawes who first introduced Brunel to their friends the Horsleys of No. 1, High Row, Kensington Gravel Pits. This may have been in the spring of 1831 although the name does not appear in his journal until 13 Febraury 1832, when he notes briefly: 'Went to Mrs Horsley in the evening – much amused.' He was by no means alone in finding pleasure in the society of that talented and hospitable family.

William Horsley was an organist, music teacher and composer and had married Elizabeth, daughter of John Wall Calcott, who was likewise an organist and composer and had been William's tutor in vocal composition at Oxford. It is therefore not surprising that their five children displayed strong artistic and musical talent. John Horsley, destined to be a Royal Academician, was an art student who soon became Brunel's devoted friend. His younger brother Charles was studying music, while of the three girls Fanny was a romantic and talented artist, fated to die young, and Sophy, the youngest, was a pianist of true brilliance who was, alas, prevented by her family from practising her art professionally. Mary, the Horsleys' eldest child, was nineteen at this time and the family beauty, but she seems to have lacked both the artistic talent of the rest of the family and the warm spontaneity and vivacity which distinguished both her younger sisters. It was unkindly said of her that 'she had nothing to be proud of except her face'. Certainly her portraits confirm that she was a classic beauty, but it is a cold beauty strangely without charm. Her features seem to lack animation yet fail to convey either tranquillity or repose; they explain why she was nicknamed 'the Duchess of Kensington' and they suggest that her air of hauteur may have been the expression of self-will as much as vanity. Nevertheless, Mary's charms undoubtedly dazzled the talented young men who delighted to visit the Horsleys to dine, to make music in the spacious drawing room, or, on summer evenings, to stroll with the girls in the walled garden until, at an upper window, Mamma's lamp, 'the domestic moon', as they called it, gave them the tactful signal to be gone.

No. 1 High Row is now No. 128 Church Street and it is difficult to believe that when the Horsleys came to the house in 1823 London had not yet engulfed it. Its windows surveyed open fields which stretched away to the grounds of Notting Hill House, later known as Aubrey House, on Campden Hill, and on one sultry September evening Fanny could write: 'Howard's hayrick has smelt abominably for the last two nights and Mamma says that Uncle William fully expects it will burn ere long.'

There were very few figures in the musical world of London whom the Horsleys did not number amongst their friends or acquaintances and who did not at some time or other visit High Row. Sophy kept a tiny autograph album measuring only two inches by one and a half in which were inscribed the names of Brahms and Chopin, of Joseph Joachim and Nicolo Paganini the great violinists, of Ignaz Moscheles the pianist and composer of Prague, of Vincenzo Bellini the Sicilian opera composer and of Felix Mendelssohn-Bartholdy. On his first visit to England, Mendelssohn had been introduced to the Horsleys by his friends, Klingemann, a young attaché at the Hanoverian Legation, and Rosen, Professor of Oriental Studies at University College. He was charmed and thereafter whenever he was in London he became a regular visitor at High Row, where he would sit at the piano in the drawing room playing his latest compositions to a rapt little audience, the first in England to hear his music for 'A Midsummer Night's Dream'. He was obviously captivated by Mary's beauty, and on his return to Berlin in 1832 he wrote to Klingemann: 'Was it a chance that, during the night, somewhere near Boitzenberg, Mary's dear flowers which I was carrying in my buttonhole and which had kept so fresh during the sea voyage, suddenly smelt as sweet as if she were sitting near me?' Thus the romantic young composer, but Mary was not for him.

The Horsleys lived for the arts and many a young engineer might have felt himself to be rather a fish out of water amongst the company which frequented their drawing room at High Row. Not so Brunel, who could speak their own language. Although no executant himself, he was passionately fond of music, delighting to visit the opera when his busy life allowed, while he and John Horsley were naturally attracted to each other by their mutual love of drawing and painting. Amidst the troubles and disappointments which beset him during 1832 it was delightful of an evening to be able to put them aside for a while and to relax in such congenial surroundings and among such charming companions. However depressed he might be feeling about his future prospects, in the company of the young Horsleys his sense of fun quickly asserted itself and he would enter wholeheartedly into the amateur performances, the oratorios, the charades and the plays, frequently written by themselves, in which they delighted. John Horsley recalled in his old age how on one occasion Brunel stage-managed the drawing room production of a tragic dramatic poem entitled *King Death*. He evidently thought little of the script and decided that it needed livening up. So he reversed the leading roles, casting Sophy for the name part and himself playing a grief-laden widow with such comic effect that the audience, which included the author, became helplessly convulsed with laughter. Unfortunately, however, this burlesque proved too much for the author's wife, who interrupted the applause with some withering remarks and flounced out of the room greatly, no doubt, to Brunel's secret amusement.

Towards Mary he was always admiringly attentive and respectful, but her younger sisters found him a rather mysterious and enigmatic figure and for a while they could not make up their minds whether they wholly approved of him

or not. He was evidently fond of pulling their legs so that
they were never sure when to take him seriously. But as time
went by and he became an old friend their feelings altered.
'Isambard Brunel came on Sunday to call,' wrote Fanny in a
letter to her young aunt Fanny Calcott, 'Mamma asked him
to stop to dinner, which he did. Only Mary and I were at
home, but he was in a very *un*-satirical mood, so I did not
mind. Indeed I do not know whether I am not a wee bit
prejudiced about him. ... I must try and root out this weed
from the parterre of my bosom.'

It is from the letters which both Fanny and Sophy wrote to
their aunt that we learn most about this phase of Brunel's
life, for he had ceased to keep his private journal and his
diary records but briefly or not at all his visits to High Row
and the occasional excursions he made in their company.
Thus he dismisses in a few lines a visit to the Zoological
Gardens, which took place on an occasion when the Horsley
girls were staying at the Barge House, whereas Sophy gives
us a much fuller and more entertaining account of the proceedings:

On Sunday, after Church Mr and Mrs Hawes and ourselves
and Isambard Brunel walked to Trafalgar Square, with Thomas
Hawes riding by our side. He then rode round the Hackney
Coaches and at last found one that took us to the Zoological
Gardens, Isambard riding on the Groom's horse. We did not see
much of the animals, for the gentlemen were all so very much
fatigued that they sank down on some seats opposite the eleph-
ants, and there remained till near dinner-time. I was very glad
that they rested themselves in front of a decent animal, for some
of them are very indelicate; indeed, the monkeys are so very
nasty that I told Thomas Hawes I would rather not look (really,
with a gentleman, I think it quite indelicate) at them; he said he
quite agreed with me, so we went to the Ottar [*sic*], while the
others remained before the monkeys.

There were quiet evenings spent at High Row also, when
no distinguished visitors were present. 'Thursday it poured

all day,' wrote Sophy in August 1834. 'I.B. came and mended a pair of compasses all the evening while we marked; we spent a very pleasant evening.' After a dance, Sophy wrote: 'Isambard Brunel was there, and he and Mary were of course a good deal together,' yet neither she nor the rest of the family appear to have been aware of the serious nature of the romance which was going on in their midst. Yet if what Brunel said later is to be believed, he had been attracted to Mary soon after his first visit but had decided not to commit himself or her until his professional prospects improved. This attraction, it would seem, was mutual, but at what juncture they acknowledged their feelings for each other no one will ever know for they kept their secret well. He certainly had not proposed marriage by the end of 1835, yet he must have been very sure of her to write as he did in his journal on that December night in Parliament Street.

He did not open that journal again until 14 April 1836, when he writes:

When last I wrote I was in high spirits it seems but dreading a reverse. I dread it still yet everything has prospered since and is still going on well. Since that time I have added to my stock in trade the Plymouth Railway, the Oxford branch and today somewhat against my will the Worcester & Oxford. Here's another 2,500,000 of capital – I may say 8,000,000 and really all very likely to go on. And what is satisfactory all reflecting credit upon me and most of them almost forced upon me. . . . Really my business is something extraordinary.

In the light of these encouraging prospects he evidently decided that the long-awaited moment had come, for a month later he proposed marriage to Mary Horsley. She accepted him. Upon this momentous occasion Brunel himself is silent and the story is best told in the excited words of Fanny Horsley as she sends the news to her aunt:

It literally came on us all like a thunderbolt, though certainly one of a very pleasant description. I think he called once in March, and that was all till last Thursday week, when he called on his way to Hanwell, and said he would come back to tea at nine o'clock; which he did, and staid chatting very pleasantly till eleven. A long time ago he told us he was very fond of musk plants, and Sophy and I had often given him bunches of it out of our garden, and Mary promised that some day or other she would give him a pot of it. So that morning, she got one, and when he was going away she gave it to him, but he said he would rather leave it as he was going to walk home, and would send for it the next day. No one came, however, either Friday or Saturday, and I, as you may imagine, made many wise reflections on forgetfulness and so forth, when on Sunday afternoon he arrived in person. It was near dinner-time, and Mamma asked him to stay, which he agreed to with great alacrity. . . . At seven o'clock Mr Klingemann and Dr Rosen came to tea, and Isambard expressed a great wish to see Lord Holland's Lanes, so, by way of doing a very genteel thing, we all agreed to go. Isambard offered Mary his arm – Mamma went with John, I with Mr Klingemann, and Sophy with Dr Rosen. We walked all through the lanes to the house, and then back. Mr Brunel and Mary walked all the way very slowly, but when, on our return, we were quite at the bottom of Bedford Place, they were only just visible, and Mamma got quite vexed and annoyed, never thinking of the real reason. They were some minutes after us in finding their way up to the drawing-room, and when they did enter, Mamma said, 'Upon my word, Mr Brunel, I never knew anyone walk so slowly in my life.'

'Why indeed,' he said, 'I walk so seldom that when I do, I like to make the best of my time,' which as we have since discovered, was rather a witty answer. He almost immediately took leave, with his musk, and I certainly saw a look which ought to have flashed conviction on my mind, but it did not. Mary was silent and pale all the evening, but I thought nothing of it; Mr Klingemann staid late, and directly he had gone, Sophy and I went up to bed. In about half an hour we heard Mary come up and called her in.

'Well, what could you be doing lagging behind in that way?' said Sophy.

'Indeed, Mary,' I said – but quite in fun, without any idea of the truth – 'one would think he had been making you an offer.'

'And what would you say if he really had?' said Mary in an awful, hollow voice which I shall never forget. Sophy and I immediately fell into such tremendous fits of laughing, that Mary said she must go away. It certainly seemed very unfeeling – and she, poor thing, with tears in her eyes – but so it was, and I must confess I was much the worst of the two. However, we soon got composed, and listened with delight to the little she had to tell, I mean little in quantity for such happy and excellent facts are great, if anything is. He made her the offer as they were coming home, and told her he had liked her all the five years he had known her, but would never engage her till he was fully able to keep a wife in comfort – I do admire his conduct very much, so honourable and forebearing – not shackling her with an endless engagement, as so many men would have done, but leaving her free, with her mind clear to enjoy pleasure, and to gain improvement and experience during the years of her youth. I do not mean that her youth is over, poor thing; there everything is as it should be, she two and twenty and he thirty. . . . I always thought he admired and paid her more respect than anyone else, but never dreamt of its coming to this. . . . I often said and thought that Mary would have chosen him before anyone else in the world. Well, I hope we are all, as you say in your little note, not only joyful but thankful. Indeed we have very great cause. He came every day till Friday, and on Friday, melancholy to relate, he was obliged to go to the country, and does not return till Sunday. I think Mary has borne it very well with the constant aid of pen, ink and post, and sundry double letters. You know I must have a little laugh, but really, *Love* is such a very new character in our family, not to speak of *Marriage*, that I only wonder at my good behaviour on the occasion.

Now I must tell you how delightful it is that all his family approve it so much, and are so very kind. Nothing can equal Mr and Mrs Brunel's and Emma's kindness, and Mr and Mrs Benjamin Hawes are just the same. We all had our fears about this till their letters came, which were everything that could be desired. . . . On Monday, Mamma and I and Mary set off at half-past ten to town in a fly. I took a book and well it was I did, for the hours they spent at Turner's making endless substantial pur-

chases would have been unbearable. We got about four to Rotherhithe, and found the family at home. What a perfect old man
Mr Brunel is! I leave Mary to adore the son, but I really must be
allowed to adore the father.

Mamma and Mrs Brunel retired to a private conference after
some time, and then Miss Brunel and Mary, so he proposed
taking me to see the Tunnel, which is only six or seven yards
from their house. It was the first time I had seen it, and I cannot
tell you how much I was impressed with wonder and admiration
at it and at the mind of the man who could conceive it. There are
numbers of men at work at it now, and it is going on most
briskly. I believe it will be finished in about two years.

... Now I think I have pretty well told you the *How, When*,
and *Where. How*, perfectly delightful. *Where*, Sunday week in
Lord Holland's Lane. No day is, or indeed can be, fixed, for I
suppose it must depend entirely on his engagements, which, as
you know, are numberless and imperative. She is going to be
dressed completely in white, and married in Kensington Church,
and have the bells rung in the good old style; and there is to be a
breakfast, only a quiet family one, which is much the best, and
then to my particular delight, a gay dance in the evening. ...
They mean to go to N. Wales, and come back through Dev-
onshire. Really everything is charming ...

Once having resolved to take the plunge, Brunel was not
one to teeter long upon the brink of matrimony. There was
also, no doubt, the consideration that the longer he delayed
the more deeply would he become embroiled in railway affairs. So the wedding took place on 5 July 1836, after which
the couple left for their fortnight's honeymoon, making first
for Capel Curig and from thence southwards along the
Welsh Border and into the West Country. At Cheltenham,
the faithful Saunders met them bringing letters from London. In Mary's mail came news from High Row of the dance
which had followed their departure. From this it appears
that Brunel had invited most of his assistants to see him to
the altar and that they had acquitted themselves well with
the exception of poor William Gravatt, who might be a very

competent civil engineer but was evidently no lady's man. 'Pray tell Isambard', wrote Fanny, 'that we liked all his friends extremely except Mr Gravatt who I think partakes of the Wild Beast. Sophy and I, however, paid him every attention, so don't frighten yourself about it, and from what Papa says, I think he must have emptied half his snuff box.'

The honeymoon over, the couple returned to London where they settled at once into No. 18 Duke Street, Westminster, which was to remain their home for the duration of their married life. It was to prove a tranquil partnership undisturbed, it would seem, by any storms of conflicting temperament. Yet the marriage had little or no influence on the course of Brunel's life, and although the exact nature of such a partnership is a secret that can never be known to others, the suspicion remains that in this case matrimonial harmony was based on the mutual acceptance of a relationship of no very profound depth. If this be so it stands in complete contrast to his father's marriage. For there undoubtedly existed between Marc and Sophia Brunel a relationship of the most profound and mutually enriching kind, and Marc was revealing the sweet and simple truth of this when in his old age he wrote that touching tribute: 'To you my *dearest* Sophia I am indebted for all my successes.' We cannot imagine the younger Brunel writing in this strain to his Mary. 'My profession is after all my only fit wife' were probably some of the truest words he ever wrote. For it seems clear that the severance of his long relationship with Ellen Hulme marked that critical moment in his life which decided his future course. He determined then to make perfection of his work the supreme goal and from that resolve he never subsequently wavered.

He was not destined for very long to bask in the unclouded sunshine of success; as he feared, bad weather soon came, and when it did the courage with which he rode the storms only revealed more clearly his quality of greatness. But always he fought alone and it was fortunate for him that these storms never drove him to financial shipwreck. Had they done so, the lack of the more enduring riches which he had forsworn for ambition's sake might have been most tragically revealed to him. For while it is difficult enough to imagine him languishing in a debtors' prison, it is impossible to picture Mary uncomplainingly sharing her husband's adversity, sitting quietly beside him at her sewing as her mother-in-law had once done in similar circumstances.

Once fixed in his resolve, Brunel no longer required a wife who would share his life whether in triumph or in adversity, but one of whom he could be proud as the walking manifestation, as the intimately personal symbol of a success in which she played no part. He would have men turn their heads in admiration and say of her 'There goes the wife of Isambard Brunel!' There can be no doubt that in this regard he chose wisely. Mary may have been obstinate and self-willed, and she undoubtedly became very much the *grande dame* in her family circle, but she was too unemotional and far too level-headed to cross passionate swords with her husband. Her beauty and that quality which had won her as a girl the nickname 'Duchess of Kensington' enabled her to play to perfection the part that her husband assigned to her. No stage director ever lavished more loving care in the presentation of his leading lady to the public than did Brunel upon his Mary when his success was at the flood. No expense in silks and jewels was spared to enhance her remarkable beauty and she certainly turned heads when, exquisitely dressed, she strolled in St James's Park, followed respectfully by a liveried footman, or drove out in one of her silk-upholstered carriages. When she and her little niece Maria were presented at Court by Sophia Hawes, Mary cut so superb a figure in her billowing ivory silks and Honiton lace that poor little Maria Hawes was quite overshadowed and wrote sadly: 'The Queen never took her eyes off Aunt Mary, but followed her to the end of the room, and I had no chance of being noticed, coming behind her immense crinoline.'

Brunel delighted in his ability thus to enhance Mary's charms, not as a *nouveau riche* in the mere vulgar display of wealth which, as such, did not interest him, but as a crafts-man in his handiwork. It was to satisfy his self-confessed 'love of glory, or rather approbation', that he endowed his own wife with something of that same splendour which dis-tinguished his great engineering works.

He was equally at pains to give Mary a worthy setting and in 1848 he acquired the house adjoining 18 Duke Street. On the ground floor he extended his offices, but above he laid out a great dining room which became known as the 'Shake-speare Room' because the walls were adorned with scenes from the plays which he commissioned from eminent paint-ers of the day and which included Landseer's often repro-duced 'Titania'. It would seem that prevailing fashion had by this time blunted his good taste, at any rate so far as interior decoration was concerned, for imagination quails at the de-scription of this dining room with its over-elaborate pendant ceiling, its plaster-panelled walls grained to imitate oak, its Venetian mirrors and red velvet curtains, its massive dining table staggering like Atlas under the weight of monstrous silver-gilt centre and side pieces presented by the Great Western Railway Company. Perhaps this display of gran-deur à la mode appealed to Mary more than to her husband who was so seldom at home to see it.

For Mary's life was rather like that of the wife of a sea captain and she generally reigned at Duke Street alone, taking little or no part in her husband's daily life except on such ceremonial occasions as the opening of a new section of railway. When Brunel's fame spread to the Continent and he was asked to survey two Italian railway routes, the line from Florence to Pistoja and the trans-Apennine route between Genoa and Alessandria, he took Mary abroad with him on several occasions. But she proved an indifferent travelling companion, insisting that looking at mountains made her dizzy and suffering with very ill grace the petty misfortunes and vexations inseparable from Continental travel at that time. In 1842, when she was unable to accompany him across the Channel, he seems to have found in his brother-in-law John Horsley a more congenial companion.

Mary bore three children at Duke Street: Isambard, Henry Marc and Florence Mary. Of these Isambard was a delicate child who was born with a slight leg deformity. This could have been rectified in infancy by a minor surgical operation, but Mary had that horror of surgery which was so common amongst women of her generation and boasted, with characteristic obstinacy, that she would never allow the knife to touch her children. Consequently her eldest son remained partially crippled for life. Showing no bent for engineering and subsequently entering the legal profession, Brunel's first-born sadly disappointed him, although he never betrayed it and earned the boy's devotion by his kindness. Henry Marc, on the other hand, delighted his father by displaying the keenest interest in his activities. He followed the family profession and ultimately entered into partnership with Sir John Wolfe Barry, the designer of the Tower Bridge. Unfortunately, neither son had issue and with them the male line of the Brunels died out. It was Brunel's daughter Florence who, by marrying Arthur James, an Eton master, carried on the blood.

That Brunel was always extremely fond of children is revealed by the many references in his early journals to his young nephew, Benjamin Hawes, 'Little Ben', as he calls him to distinguish him from his father. Busy though he was, he would drive over to see Little Ben at his school or play the indulgent uncle by taking him out for treats in London during his holidays. No wonder that whenever he was away on his frequent business trips his children looked forward eagerly to his return. While Mary sat in formidable state in her elegant long drawing room they would listen expectantly in their nursery for the familiar steps which took the shallow flights of the wide eighteenth-century staircase two treads at a time. For they could be sure then that after a very few minutes in the drawing room he would be with them to organize some wonderful nursery game or to entertain them with conjuring tricks just as their Grandpapa had done when Papa was a little boy.

It was during one of these nursery entertainments that there occurred the only incident to ruffle seriously the ordered calm of Duke Street. In performing one of his tricks, Brunel accidentally swallowed a half-sovereign which lodged in his wind-pipe and placed him in imminent danger of choking to death. Sir Benjamin Brodie the eminent surgeon was called in and after anxious consultation in which the patient himself joined it was decided to perform a tracheotomy operation using a most horrific instrument nearly two feet long which became known in the profession as 'Brodie's Forceps' although Brunel himself designed it. The operation proved unsuccessful. When the forceps were inserted through the incision in the wind-pipe Brunel found himself unable to breathe and the attempt to locate the coin had to be abandoned, leaving him worse off to the extent of a throat wound. In this serious pass and with the best medical brains of the day defeated, Brunel summoned his own engineering skill to his aid in the form of centrifugal force. He rapidly sketched out a simple piece of apparatus consisting of a board, pivoted between two uprights, upon which he could be strapped down and then swung rapidly head over heels. This was quickly made and the experiment tried while Mary and the children's old Irish nurse waited white-faced outside the door of his room. The first trial brought on so violent a fit of coughing and choking that those present feared that his death was imminent and the frame was stopped. But his choking finally subsided and he then signalled them to try once more. As he was swung round he began to cough again and then suddenly felt the coin leave its place. A few seconds later it dropped from his mouth. That same evening he wrote to his friend Captain Claxton in

Bristol: 'At four ½, I was safely and comfortably delivered of my little coin; with hardly an effort it dropped out, as many another has, and I hope will, drop out of my fingers. I am perfectly well, and expect to be at Bristol by the end of the week.'

This was in 1843 when Brunel, at the age of thirty-seven, had reached the height of his fame. The amount of interest and concern aroused by his predicament may be gauged from the fact that when Macaulay, having made inquiry at Duke Street and been given the glad tidings, ran pell-mell through the Athenaeum Club shouting excitedly, 'It's out! It's out!' no one thought he had lost his senses and everyone knew without question what he meant. The real drama of Brunel's life is to be found in his engineering career and it is a little ironical that the only dramatic episode in his home life should have become for some people his best known exploit. It was even immortalized in the *Ingoldsby Legends* by the lines:

> All conjuring's bad! They may get in a scrape
> Before they're aware, and, whatever its shape,
> They may find it no easy affair to escape.
> It's not everybody that comes off so well
> From 'leger de main' tricks as Mr Brunel.

The luxury and order of Duke Street over which Mary presided so efficiently and with such beauty and grace undoubtedly gave Brunel intense satisfaction. It was not only the symbol of his success but the one stable thing in his restless, hectic life. Yet to the question which he had asked himself that Christmas night in his old rooms in Parliament Street: 'Will marriage make me happier?' it is difficult to make an answer. It is doubtful. To the relentless pursuit of perfection in his work, to the realization of his lofty ambitions, his castles in Spain, it would seem that he deliberately sacrificed the quest for a relationship which might have changed the direction of his life and brought to it fresh meaning and purpose. But if this be true, his loss is our gain.

In 1841, Brunel personally escorted Lady Holland on her first journey by railway from London to Bowood. He was captivated by the brilliance and the wit of this most remarkable old lady, the great hostess of Holland House who had made it the very heart of the political, literary and artistic life of London. She, it appears, was no less attracted by the restless genius of the young engineer, for from that journey sprang a warm friendship, which ended only with her death. In her delightful book,[2] Brunel's grand-daughter, Lady Noble, writes: 'He who worked at home till midnight, and never went out, unless into the family circle, or to join the meetings of a learned society, used to dine with her and cherish her little notes pressing him to come "any day that may suit you, at seven o'clock to eat your *Soupe*", or reproaching him that she had not seen him for so long; for, with the sure touch of the great hostess, she poured "kind attention" upon Mary, but secured Isambard for her dinner table.' It is obvious that Brunel found in this friendship something that Mary was never able to give him, and through his admiration for Lady Holland he may perhaps have realized the price he had paid for his lonely greatness, seeing, perhaps, in her ageing face the ghost of the might-have-been.

2. *The Brunels, Father and Son* (1938).

7

The Battle for the
Broad Gauge

THE first important event to take place after Brunel's marriage was the laying, on 27 August 1836, of the foundation stone of the Leigh abutment of the Clifton Bridge by the Marquis of Northampton, then President of the British Association. As the Bridge Company had by now raised a sum which was deemed sufficient to complete the work, this was a ceremony of more practical moment than the absurd little affair that had been held on the Clifton side five years earlier. When the stone was lowered into place trumpeters sounded a fanfare which, as it echoed down the gorge, was answered by a storm of cheering from the vast crowds assembled on both sides of the river and from the beflagged decks of the many ships on the Avon. Beneath it were buried a number of current coins, a china plate decorated with a picture of the bridge, a copy of the Act of Parliament and a plaque inscribed with details of the work and its dimensions, including the name of the engineer. Who knows what natural or man-made cataclysm may ultimately lay bare this treasure trove; what wondering eye may first light upon it?

To provide a temporary means of transport for men and materials a wrought-iron bar 1,000 ft long and 1½ ins. diameter had been slung across the gorge just before the stone-laying ceremony took place. It was welded up section by section in Leigh Woods and hauled across by a cable. Unfortunately, when it was almost home the cable parted and the bar fell, happily without injuring anyone. When it had been recovered and successfully anchored it was found to have acquired a pronounced bend in the middle. It was intended to effect the crossing by means of a basket suspended from the bar on a roller and hauled to and fro by light ropes, but when some foolhardy individual tried it on the last Saturday in August the roller stuck fast when it reached the kink in the bar and the adventurer narrowly escaped with his life. This escapade was unauthorized and Brunel was furious when he heard the news of it.

The defective bar was immediately replaced by another and it was Brunel himself who, on 27 September, made the first crossing. It had apparently been suggested that Mary should accompany him on this vertiginous journey, but she understandably declined the honour and her place was taken by an unnamed youth. This was just as well. Having run by gravity down to the lowest point of the arc, the roller once again got stuck so that the basket could not be drawn up the other side. Fortunately, however, Brunel possessed, as well as courage, a good head for heights. While the swaying basket dangled nearly 200 ft above the river the workmen on both sides of the gorge held their breath when they saw their engineer climb out of the basket, swarm up one of its suspension ropes and free the roller.

We often tend to look upon our Victorian ancestors as a somewhat lily-livered lot too staid and too well satisfied with easy living to risk their necks in the hazard of high adventure. This is a fallacy based partly on the head-shakings of those who opposed railways on the score of their speed and danger (often with ulterior motives) and partly on their appearance. When we look at the sombre, sober and heavily bewhiskered figures in early photographs, at the tall hats, the high collars and the frock coats, we find it hard to credit that the men behind such a façade could possibly be young in heart and of a spirit capable, had they lived today, of handling a racing car with the best of us. Yet so it was, as Brunel's own life story reveals. And lest it be argued that in this respect he was exceptional, it is worth noting that when work on the Clifton Bridge came to a standstill for lack of funds in 1854, the Trustees collected £125 in fares from people anxious to cross the gorge by basket in this perilous fashion. The bar has vanished long since, but its anchorage can still be seen near the base of the Clifton abutment; likewise the landing place on the Leigh Woods side.

Although the Clifton Bridge was Brunel's first brain-child, he could not afford to devote a great deal of time to it. He had matters of much greater moment on his hands, for the construction of the Great Western Railway was by now under way. As early as September 1835 he had written from London to Osborne and to Townsend in Bristol ordering them to get the undergrowth cut down in the Avon Valley at Brislington so that he could set out the exact course of the line and decide where the shafts for the tunnels should be sunk. 'We shall have our flags flying over the Bristol Valley tomorrow,' he wrote. 'I should not wish that Bristol should fancy itself left behind.' The first contract – for the Wharncliffe Viaduct over the Brent – was let in November and by the autumn following the works were well in hand between London and Maidenhead and also at certain points on the far more difficult section between Bath and Bristol.

It is hard for us today to appreciate the immensity of the task which confronted Brunel in the construction of the Great Western Railway. History holds no previous record of engineering adventure upon so heroic a scale. When he had written, before scarcely a sod had been turned, that he was engineer to the finest work in England he was making no idle boast. He knew that it would be so because, as any artist or craftsman must, he had already conceived the completed work in his imagination. In the course of the survey he had covered every yard of the way and had seen with the mind's eye his iron road lying wide and true; had seen those ample curves through the Thames Valley, those long level miles across the White Horse Vale and the sudden stoop to Bath through the great tunnel under Box Hill. He was resolved that the execution should be worthy of the conception, 'but', as he wrote in 1836, 'I want *tools!*'

Apart from the fact that he was but thirty years of age, he could call to his aid no experts or elaborate machines and there were but few precedents to guide him; not that he was of those who timidly accept precedents in any circumstances. The art of railway construction and operation was still in its infancy and its definitive form, unrecognized as yet, was only just beginning to emerge from the mists of trial and error. The best men in the country with practical experience of railways were of the school of Stephenson and most of them were already fully engaged elsewhere, notably on the Grand Junction and the London & Birmingham Railways. So Brunel had to find and train his own assistants. To gather about him men who could satisfy his exacting standards and match his tireless energy was not easy and he was often disappointed. To Saunders from the Bear Inn at Wantage he wrote: 'Stephenson is himself a much better man of business than I am for certainly his tools are not quick. I have now three of them and he said they were some of his best but they are all desperate slow-coaches and have no great notion of hard work.' But gradually the chaff was winnowed from the wheat, 'Hudson has left me,' he wrote in April 1836, 'Stokes is an imbecile and I am rather in a mess, but S. Clark is a godsend and Hammond is a good fellow and very useful and has come up and endeavoured to put my accounts in order.' While Hudson and Stokes are thus consigned to limbo, Seymour Clark was to become the Great Western Railway's first Traffic Superintendent, while J. W. Hammond was appointed resident engineer in the London area. He was also Brunel's chief assistant until his death, when R. P. Brereton succeeded him. Another assistant, G. E. Frere, was resident engineer, Bristol division.

However efficient these assistants might prove themselves, and Brunel was ever quick to recognize merit and to reward it, they were left in no doubt as to who was master.

For it was an inviolable rule of Brunel's that he would never, under any circumstances, accept an appointment which involved divided responsibility. In any work upon which he engaged there could be only one engineer and he must have the full responsibility for the work and for the conduct of his staff. He would never accept any position as Joint Engineer or Consulting Engineer and once, in later life, when he was offered such a post, he set down his views in decisive terms: 'The term "Consulting Engineer" is a very vague one,' he wrote,

and in practice has been too much used to mean a man who for a consideration sells his name but nothing more. Now I never connect myself with an engineering work except as the Directing Engineer, who, under the Directors, has the sole responsibility and control of the engineering, and is therefore 'The Engineer'. ... In a railway the only works to be constructed are engineering works, and there can really be only one engineer.

So far as the Great Western Railway was concerned, Brunel regarded himself as commander-in-chief of the engineering staff and would brook no interference with them from any man, directors included. The following letter, written to Charles Saunders in 1842, is a good example of his reaction to such interference:

It was lately ... intimated that a pair of boxing gloves had been seen in one of the Company's offices, and that the *Directors had observed it*. Now I really do not know why a gentlemanly and industrious young man like — should be subject to have his trifling actions remarked upon more than I myself, unless the observer gave him credit for a much more gentle temper than I possess; because I confess, if any man had taken upon himself to remark upon my having gone to the pantomime, which I always do at Christmas, no respect for Directors or any other officer would have restrained me. I will do my best to keep my team in order; but I cannot do it if the master sits by me, and amuses himself by touching them up with the whip.

But while he was thus ready to come to the defence of his

assistants whenever the occasion warranted and would allow
no one to take the reins from him, woe betide the defaulter,
for he could himself wield the whip with most deadly effect.
The erring assistant who received the following missive must
have been left feeling decidedly weak in the knees:

Plain, gentlemanly language seems to have no effect upon you.
I must try stronger language and stronger measures. You are a
cursed, lazy, inattentive, apathetic vagabond, and if you continue to neglect my instructions, and to show such infernal lazi-
ness, I shall send you about your business. I have frequently told
you, amongst other absurd, untidy habits, that that of making
drawings on the back of others was inconvenient; by your cursed
neglect of that you have again wasted more of my time than
your whole life is worth, in looking for the altered drawings you
were to make of the Station – they won't do. I must see you again
on Wednesday.

After so devastating a blast of invective one is left wondering whether the victim ever plucked up sufficient courage to
keep that Wednesday appointment or whether he did not
take to his heels forthwith to seek refuge with some less
exacting master.

In such a fashion did Brunel forge and temper his human
tools. Mechanical tools he virtually had none to command. It
was the pith and thews of the navvy gangs aided only by
horse strength that built the Great Western. But what magnificent men they were, these railway navvies, working up to
sixteen hours a day and shifting as much as four hundredweight of spoil at a time on their huge barrows! It was
thanks to their herculean efforts in foul weather as in fair no
less than to the indomitable energy of their engineer that
the first train ran from London to Bristol only five and a
half years after the first contract had been let. The only
work of modern times with which this achievement may
fairly be compared is the construction of the new Woodhead
Tunnel, whose length is roughly equivalent to that of Box
plus the tunnels between Bath and Bristol. Here, with the

aid of every modern mechanical device and the advantage of the longitudinal sections prepared by Locke's resident engineer W. A. Purdon when the old tunnel was built, the work occupied no less than four and a half years.

As is well known, the so-called 'standard' gauge of 4 ft 8½ in. was arrived at quite arbitrarily because it happened to be the width of the early coal-waggon ways of Tyneside. It was natural that George Stephenson should have accepted this precedent at Killingworth and later on the Stockton & Darlington. The moment for him to have considered whether so narrow a gauge was in fact the most desirable for trunk lines of public railway came when he was appointed engineer to the Liverpool & Manchester Railway. Here there was no reason whatever for following the Tyneside precedent, yet Stephenson did so without, apparently, pausing to consider the question on its merits at all. For when Robert Stephenson was asked by the Gauge Commissioners whether his father had actually advocated the gauge of 4 ft 8½ in. for the Liverpool & Manchester he replied: 'No. It was not proposed by my father. It was the original gauge of the railways about Newcastle-on-Tyne, and therefore he adopted that gauge.' This classic *non sequitur* betrays the conservative, rule-of-thumb method which the conscientious and careful but unimaginative craftsman so often adopts when he ventures into strange fields. He proceeds slowly and cautiously by the light of precedent and his own practical experience, mistrusting bold experiment as a leap into the dark. On the contrary, experiment was the breath of life to Brunel and for him precedents only existed to be questioned. To say this is not to belittle the Stephensons and exalt Brunel but simply to state that they represent two perennially opposed schools of thought, each an essential foil to the other. Brunel's temperament was such that conflict became inevitable so soon as his star rose in the railway firmament, and the issue was first joined on this gauge question.

Having surveyed a peerlessly straight and level road, and

aimed to achieve upon it travel of a speed and smoothness unparalleled, that Brunel should reject what his supporters would later refer to contemptuously as 'the coal-waggon gauge' was only to be expected. Precisely at what juncture he determined to use a gauge of seven feet he could not himself say. 'Looking to the speeds which I contemplated would be adopted on railways and the masses to be moved,' he told the Gauge Commissioners, 'it seemed to me that the whole machine was too small for the work to be done, and that it required that the parts should be on a scale more commensurate with the mass and the velocity to be attained.'

'I think the impression grew upon me gradually, so that it is difficult to fix the time when I first thought a wide gauge desirable; but I daresay there were stages between wishing that it could be so and determining to try and do it.'

Whenever it was that he reached this momentous determination, Brunel kept it to himself. Had the first Great Western Railway Bill passed into law, nothing more might have been heard of the broad gauge for although no copy of this Bill survives it is said to have contained a clause limiting the gauge to 4 ft 8½ in. When the second Bill came up early in 1835 Brunel was able to persuade Lord Shaftesbury, the Chairman of Committees in the Lords, to agree to the omission of the gauge clause. He had noticed that for some reason, probably a mere oversight, the clause did not appear in the London & Southampton Railway Bill, so that he was able to cite this fact as a precedent. It was only when the Great Western Railway Company was safely incorporated that he sprung his scheme upon the Directors in a letter dated from 53 Parliament Street on 15 September 1835. In this the chief plank of his advocacy was a curious one. He aimed, he said, to reduce both the frictional resistance and the centre of gravity of the rolling stock by using wheels of very large diameter and mounting the coach and waggon bodies within instead of above them. The inconvenience and danger of this arrangement, particularly in the case of passenger carriages, scarcely needs stressing and it was an argument which he and his supporters very speedily dropped. More powerful engines, increased accommodation, and higher speed with greater safety and stability soon became the battle-cry of the broad-gauge champions. Brunel concluded his letter by saying that the only valid argument against the adoption of the broad gauge was the proposed junction with the London & Birmingham Railway and the use of a joint London terminus. This objection was automatically disposed of when negotiations between the two companies broke down over the question of land lease and an amending Act was passed authorizing the extension of the Great Western from Acton to its own terminus at Paddington. The possible inconvenience of the break of gauge elsewhere does not seem to have occurred to Brunel at this time, or if it did he made light of it. Although the London & Southampton already threatened his southern flank, he foresaw his broad-gauge metals monopolizing the traffic of the West Country and the question of interchange traffic at junctions with narrow-gauge lines seemed to him unimportant. Such a view strikes us today as extraordinarily short-sighted. Brunel's temperament was such, however, that once he had determined upon a scheme, enthusiasm brushed aside difficulties and disadvantages. It seems clear, too, that he believed that the advantage of his broad gauge would prove so overwhelming that other railways would soon follow the example of his Great Western. Had he been earlier in the field this might have proved true, but, alas, he was five years too late. The narrow-gauge empire had already grown too powerful. Brunel's broad-gauge proposal came before the Great Western Board on 29 October 1835 and was adopted by a large majority.

Not only in the matter of gauge but in the whole design of his permanent way, Brunel rejected precedent and proceeded from first principles to design what he confidently believed would prove to be the perfect railroad. The unusual construction of his permanent way and his reasons for adopting it have often been misrepresented although he explained matters fully in a letter to his Directors. In the first place it should be appreciated that at this date the aim of all railway engineers was an absolutely rigid and unyielding road. That a certain resilience in the road was essential to smooth high-speed running was a lesson which was only learnt by practical experience. So far the Stephensons and the engineers of their school had laid their rails in chairs mounted on massive stone blocks. There were apparently no cross ties, the engineers relying on the sheer weight of the blocks to hold the rails to gauge. Brunel maintained that with this type of track, or for that matter the later orthodox type using cross sleepers, unequal settlement of the blocks or sleepers under traffic was inevitable, with the result that true levels could only be maintained by constantly packing or tamping the ballast under them. He proposed to eliminate this difficulty by laying his rails upon continuous longitudinal baulks of timber. Another factor which influenced him in this decision was the very inferior quality of the first wrought-iron rails owing to faulty rolling.

At this period Brunel kept what might be called a commonplace book in which he collected any material which was likely to be of use to him in building the railway. This has survived as an example of his thoroughness and of his astonishing mastery of all the infinite detail involved in railway construction, including as it does a list of the species of grasses most suitable for growing in different soils for the purpose of consolidating new earthworks and tables of rainfall figures and local times[1] for places along the route. Here, amongst much fascinating information of the most diverse description we find a note to the effect that whereas the original cast-iron rails laid on the Hetton Colliery Railway had a life of from ten to twelve years, the malleable iron rails laid

1. e.g. Bristol local time was 10 mins 19 secs and Exeter 14 mins 18 secs after Greenwich Mean.

by the manager, Luke Dunn, in November 1831 between Hunter's Lane Engine House and the fourth incline had a life of only four years. Trouble with 'laminations and partings' kept a blacksmith fully occupied. Another note states that: 'Mr Stephenson, a relation of Mr Hugh Taylor who has had experience in laying Railways for 10 or 12 years states that he would recommend cast-iron in preference to malleable.' This note goes on to say that the cast-iron rails had proved themselves the more durable by 50 per cent on the Stockton & Darlington Railway, where all the malleable rails laid had had to be replaced while the original cast-iron rails of 1825 were still giving good service.

Despite these gloomy reports it was as obvious to Brunel as to his fellow engineers that the malleable rail had come to stay, but he decided to counteract its imperfection and prolong its life as much as possible by giving it continuous support. Yet another consideration was that by doing so he could safely use a rail of lighter section and so effect a considerable economy. As he himself pointed out, this method of laying railway was not original but had been employed before and although he does not say so it seems practically certain that Brunel's permanent way was based on that laid by his father to convey logs to the sawmill at Chatham Dockyard. It is true that his friend Charles Vignoles had already evolved the type of flat-bottomed rail which still bears his name and had unsuccessfully advocated its use, laid on continuous timbers, on the Midland Counties Railway. But whereas Vignoles's rail was of solid 'I' section, Brunel evolved his bridge rail which was of inverted 'U' section. In doing so he most probably had in mind the manufacturing problems of the time. In iron rolling great difficulty was experienced owing to the formation of scale due to cooling during the process and it was this which caused the lamination and splitting which gave such trouble at Hetton. The heavier the section the greater the difficulty in rolling and the risk of this trouble which Brunel therefore sought to minimize by designing a rail which was light in section relative to the strength which it derived from its hollow arch form.

The four longitudinals carrying both sets of metals were united at intervals of 15 ft 6 in. by substantial cross ties, or 'transoms' as Brunel called them, so that the whole formed a solid timber frame. Each of these cross ties was spiked to two 10 in. piles driven at 15 ft centres. Thus the piles were not situated directly beneath the longitudinals because their object was not, as is sometimes supposed, to support the permanent way. On the contrary the purpose of the piles was to hold the framework down so that ballast could be thoroughly tamped under the longitudinals without forcing them up and thus distorting their levels. Brunel decided, probably as a temporary measure only, that there should be no piling on embankments where the new formation was liable to settlement, and that in cuttings the outer piles should be driven diagonally the better to resist the thrust of the cutting sides and possible deformation of the road-bed.[2] Pine was used for the longitudinals and, to prevent the rails cutting into them, after the timbers had been laid a strip of hardwood was nailed to their top surface and then planed true but slightly wedge-shaped to give the rails an inwards inclination. Finally, after the formation under the longitudinals had been thoroughly consolidated by packing, the new road was tested by passing over it a trolley fitted with what Brunel calls in his notebook 'loaded wheels'. In effect, these were rollers, each wheel weighing just under one ton.

It will be apparent that the labour involved in laying this 'baulk road', as it was sometimes called, was prodigious, but, as Sir John Wolfe Barry was to point out in after years in his book *Railway Appliances* (1876), it showed an economy in timber as well as in iron. Barry proved that with a gauge so wide as 7 ft the use of cross sleepers would have consumed

2. These two provisions appear in Brunel's own notes on the permanent way but I have discovered no other references and have been unable to determine whether they were put into effect or not.

more timber than the longitudinals. Barry also maintained that a 62 lb rail on longitudinals was the equivalent of 75/85 lb rail on cross sleepers.[8]

Such were the technical and economic considerations which led Brunel to design his unusual form of permanent way. To his critics, both during his lifetime and subsequently, it appeared to be an example of a perverse striving for originality for originality's sake, of a desire to be different which was a form of exhibitionism. To be dubbed eccentric or exhibitionist by critics who, as in this case, have little or no grasp of the technical considerations involved or of the aims of the designer, is almost invariably the fate of original genius.

On this method of laying the permanent way, as on the question of gauge, Brunel was able to carry his Directors with him and his baulk road duly went down on the first section of the railway to be completed, that between Paddington and Taplow (called Maidenhead) which was opened to the public on 4 June 1838. The only major engineering work upon it was the Wharncliffe Viaduct over the Brent valley where, in the design of the piers, Brunel again used the Egyptian style with splendid effect.

In constructing this first section the bricks, timber and other materials used were delivered by lighter and barge to Bulls Bridge or West Drayton, where the Grand Junction

3. According to contemporary statistics the broad-gauge baulk road also cost slightly less to maintain than narrow-gauge cross-sleepered permanent way. From the maintenance figures for the half year ending 31 December 1868, A. W. Gooch of Engineer's Office, Oxford worked out the following cost per mile figures:

> Broad: £165–16–4
> Narrow: £176–1–1
> Mixed: £251–18–5

It is worth remarking here that in the early years of the Great Western Railway permanent-way maintenance was carried out by contract. The first contractors were often those who had built the section in question.

Canal adjoined the new line, and to Thames Wharf, Maidenhead. Thence they were distributed at first by carts and later by waggons on temporary ways. We find Joseph Winter of Brentford quoting Brunel four shillings a ton for lightering timber from West India Dock to West Drayton and ten shillings a ton to Maidenhead, while John Nicholl of Adams Mews, Edgeware, agrees to distribute timber by cart over the first eight miles of line at the rate of tenpence per ton per mile. At the landing wharves tanks were installed where the timber was pickled or 'kyanized' in a preservative solution of corrosive sublimate according to the method invented by Doctor Kyan.[4] For supplying 'Earth Wagons' of two cubic yards capacity with American elm and oak sides for use on the temporary way, Brunel was quoted ten guineas each.

It is difficult for us today to conceive the strain and the pressure to which Brunel was subjected as engineer of an undertaking so stupendous and so unprecedented. The construction of the London & Birmingham Railway was a prodigious engineering feat but it was not, as was the Great Western, the original conception of a single man who must, by reason of that originality, keep a finger on every detail and wrestle practically unaided with all the unexpected difficulties which inevitably arose. Although some of his directors, notably George Henry Gibbs and Charles Russell, supported him loyally, Brunel possessed at this time only one intimate friend in the enterprise – Charles Saunders. He alone was privileged to glimpse the real Brunel, the very human, distracted, troubled but undaunted spirit that lay concealed behind a cold impassive exterior. 'My Dear Saunders,' he wrote from Duke Street on 3 December 1837,

A hint or two from the other end is useful now and then to remind me of what, however, I am fully sensible of and always thinking of, your exceeding kindness in relieving me of every-

4. Bulls Bridge remained until recently the main sleeper depot for the railway, the timber being brought from the Port of London by canal.

thing you possibly can, and still more strongly shown in your silence, and the absence of complaints.

In my endeavour to introduce a few – really but a few – improvements in the principal part of the work, I have involved myself in a mass of novelties.

I can compare it ·to nothing but the sudden adoption of a language, familiar enough to the speaker, and, in itself, simple enough but, unfortunately, understood by nobody about him; every word has to be translated. And so it is with my work – one alteration has involved another, and no one part can be copied from what others have done.

I have thus cut myself off from the help usually received from assistants. No one can fill up the details. I am obliged to do all myself, and the quantity of writing, in instructions alone, takes four or five hours a day, and an invention is something like a spring of water – limited. I fear I sometimes pump myself dry and remain for an hour or two utterly stupid.

As regards the Company, I never regretted, one instant, the course I have taken. .... And, as regards myself, if I get through it with my head clear at all, I shall not regret it, but I certainly never should but for your kindness, and the corresponding forbearance and kindness of our Directors.

I have spun this long yarn, partly as a recreation after working all the night, principally to have the pleasure of telling a real friend that I am sensible of his kindness, although he hardly allows me to see it, and partly because I wish you to know that if I appear to take things coldly it is because I am obliged to harden myself a little to be able to bear the thought of it . . .

If ever I go mad, I shall have the ghost of the opening of the railway walking before me, or rather standing in front of me, holding out its hand, and when it steps forward, a little swarm of devils in the shape of leaky pickle-tanks, uncut timber, half-finished station houses, sinking embankments, broken screws, absent guard plates, unfinished drawings and sketches, will, quietly and quite as a matter of course and as if I ought to have expected it, lift up my ghost and put him a little further off than before.

Nevertheless, despite the activities of Brunel's devils and the exasperating delays so caused, the Directors were able formally to open the line as far as Taplow on 31 May 1838, travelling behind the locomotive *North Star*. 'A very pretty sight it was,' wrote George Gibbs in his diary.

At 11.30 we entered the carriages of the first train and, proceeding at a moderate pace, reached Maidenhead Station in 49 minutes, or at about 28 miles an hour. After visiting the works we returned to Salt Hill where a cold luncheon for about 300 was laid under a tent. After the usual complement of toasts we returned to the line and reached Paddington (19 miles) in 34 minutes, or 33½ miles an hour.

On this return journey Thomas Guppy demonstrated his *joie de vivre* by walking along the roofs of the carriages while the train was travelling, a feat from which we may infer the convivial nature of the festivities at Salt Hill.

But for Brunel the storm clouds were already gathering. Only a fortnight after the opening, Gibbs was writing in his diary:

Numerous reports have been spread here and in Liverpool in the course of the week injurious to the railway, and they appear to have originated partly in the vile manoeuvres of the Stock Exchange and of other parties speculating in shares and partly in the ill-will of those who are connected with the old system.

By the latter, Gibbs refers to the considerable body of north of England shareholders who later became known as the Liverpool Party. They believed that George Stephenson was the only man who knew how to build a railway and from the outset they had been bitterly opposed to Brunel and his innovations. In the troubles which beset him – teething troubles as we would now call them – they found a heaven-sent opportunity to assert themselves and without allowing him time to surmount his difficulties they attacked him mercilessly. For the fact was that the new railway fell very far short of expectations. Instead of the smooth, high-speed travel which had so confidently been predicted the quality of the riding was very rough and uneasy, while the much vaunted 'giant' locomotives of which so much had been ex-

pected were not only unreliable but so lacking in power that
they were incapable of attaining a high speed with any considerable load even if the permanent way had permitted it.
There were derailments, too. On 7 July Gibbs wrote:

> Went at 4 o'clock to Paddington and soon after news was
> brought us that the *Vulcan* was off the line and had sunk up
> to the axle. This led to an accumulation of trains and people, and
> in the attempt to correct the evil another engine got off the line
> and sank in the same way. The consequence was that many hun-
> dreds of people were disappointed, and the 4 o'clock train did not
> reach Maidenhead till past 10. I was so sick of the scene that I
> made off . . .

Later in the same month, Gibbs, Casson, another director
and Saunders made a journey over the rival concern from
Euston to Denbigh Hall and were somewhat relieved to find
that the travelling was no smoother or faster, but this was
cold comfort.

Brunel attributed the bad riding partly to the springing of
the coaches and partly to inadequate ballasting under the
longitudinal timbers which allowed them to sag in between
the piles. The introduction of six-wheeled coaches with
modified springs and the repacking of the permanent way
with coarser ballast effected a considerable improvement,
but the opposition was not to be so easily appeased.

Although, as will presently appear, considerable modifications were made to the permanent way, it eventually became clear to Brunel in the light of experience that a great
deal of this early trouble was due, not to his permanent way,
but to the crudity, faulty construction and defective maintenance of the coaching stock. This in turn was due to the
understandable but mistaken notion that the men best
qualified to build and maintain railway coaches were not engineers but craftsmen trained in the old road coach tradition. As a result, the first coaches not only rode badly but
seriously damaged the permanent way in the process. Four
years later, Brunel would write to Saunders as follows:

One very great defect which has annoyed us so long namely that dreadful thumping of the wheels when going fast I think would not have occurred had an ordinary millwright or engineer examined and received the wheels. After a great deal of trouble and very close examination I discovered the cause – an inequality in the thickness of the tyre which threw the wheel out of balance.

The dreadful side motion [he continues] which has shaken our carriages to pieces should never have arisen. I discovered quite by accident that the bearings of the axles having worn longer and the pattern of the brasses perhaps shorter, they were in the habit of using brasses half an inch or more too short. No *mechanic* would have done that.

There are fifty other points I could refer to but they all prove the same thing – that a railway carriage requires engineering superintendence just as much as the locomotives and that it is quite impossible that Lea and Clarke can conduct such a department.

If, as seems very probable, Brunel here refers to Seymour Clark or his brother, it would indicate that carriage maintenance was originally the responsibility of the Traffic Department. It was as a result of this letter that responsibility for carriages and waggons was transferred to the Engineering Department and to new repair shops at Swindon.

For the motive power troubles which beset the Great Western in its earliest days Brunel was far more culpable; indeed it is safe to say that these first locomotives represent the greatest and most inexplicable blunder in his whole engineering career. One of his arguments in favour of the broad gauge, it will be remembered, was that it would facilitate the construction of locomotives of greater power and speed. That he gave considerable thought to the question of locomotives and that he did not fail to acquaint himself with existing locomotive practice is revealed by his own sketches and by the appearance in his commonplace book of a table of performance figures taken in a series of tests with the locomotive *Star* on the Liverpool & Manchester Railway.

Yet, when orders were placed for the first engines, he imposed upon their builders conditions which not only failed to take advantage of the unique opportunity which the broad gauge offered but ensured that their performance would be inferior to that of many locomotives already in service on the narrow gauge. He stipulated that a speed of 30 miles an hour should be considered 'the standard velocity' and that at this rate the piston speed must not exceed 280 ft per minute. His second crippling condition was that the weight of the engine, less tender but with fuel and water, must not exceed 10½ tons if mounted on six wheels or 8 tons on four.

It is impossible to fathom Brunel's motives in imposing conditions which completely hamstrung the unfortunate builders and threw away one of the greatest advantages of his broad gauge. Locomotives of considerably greater weight were already in service on the narrow gauge, where piston speeds exceeding 500 ft per minute were not considered abnormal. Whereas the largest locomotive driving wheels hitherto seen were 5 ft 6 in. in diameter, Brunel's limitation of piston speed forced the builders to produce wheels of seven, eight and even ten feet in diameter and to adopt a short stroke into the bargain. Moreover such was the weight of these huge wheels that in the attempt to keep within the weight limitation the engines were, with two exceptions, grossly under-boilered. The two exceptions were *Thunderer* and *Hurricane*, which were built by Hawthorns of Newcastle to the designs of T. E. Harrison, later to be Chief Engineer of the North Eastern Railway. Two more extraordinary locomotives had never run on a British railway. In order to keep the axle loading down to the figure set by Brunel, Harrison mounted the boiler on a separate six-wheeled carriage and connected it to the motive unit by flexible steam and exhaust pipes. In the case of the *Hurricane* this motive unit was carried on six wheels, the single driving wheels being 10 ft in diameter. *Thunderer* had four 6 ft coupled wheels but in this case the crankshaft was geared up to the driving axle in the ratio 27/10 which meant that the effective diameter of the driving wheels was no less than 16 ft. There were two other geared locomotives, *Snake* and *Viper*, built by the Haigh Foundry, Wigan, but they proved such lamentable failures that they quickly disappeared from the scene and no mechanical details survive.

The first two locomotives to arrive were Mather Dixon's *Premier* and the *Vulcan* from the Vulcan Foundry. They were delivered by sea and canal barge from Liverpool to West Drayton where they were slung ashore with the aid of an adjacent elm tree early in November 1837. *Thunderer* was delivered in March 1838 and in May George Gibbs was given a trial run on her over four miles of line near West Drayton. 'Along the greatest part of the four miles,' he wrote, 'the engine ran beautifully smooth and for some of the way we cleared sixty miles an hour.' No wonder Gibbs was impressed; with the addition of a four-wheeled tender, *Thunderer* was almost a train in itself and the spectacle of this fantastic equipage travelling at a mile a minute must indeed have been awe-inspiring.

With such a collection of oddities and freaks, the motive power position of the new Company at the time of the opening to Maidenhead was parlous. It would have been downright impossible had it not been for the efforts of one man – Daniel Gooch – and one locomotive – *North Star*. After serving his time under Homphry at Tredegar, and at the Vulcan and Dundee Foundries, Gooch became an experienced mechanical engineer before he was out of his teens. In 1836, when he was nineteen, he was working for Robert Stephenson & Company of Newcastle, one of his tasks being to design two 5 ft 6 in. gauge locomotives intended for the New Orleans Railway. It was this job which first made Gooch aware of the advantages of a wider gauge. 'I was much delighted', he wrote in his reminiscences, 'in having so much room to arrange the engine.' In October of this year he left Stephenson to join Robert Hawkes, who proposed setting up a locomotive works at Gateshead with Gooch as manager. Gooch was fascinated by the idea of the broad gauge and soon travelled south to Bristol in the hope of seeing Brunel and canvassing an order for some locomotives. He was disappointed; the author of the broad gauge was not in Bristol at the time. The Gateshead scheme miscarried and soon after his return to the north he took a job under his brother, Tom Gooch, who was engineer to the Manchester & Leeds Railway. It was no use; he had become a broad-gauge enthusiast and on 18 July 1837 he wrote to Brunel applying for a situation. On 9 August the two men met for the first time in the Manchester office of the M. & L. R., and Brunel engaged him on the spot as his Chief Locomotive Assistant.

Meanwhile, owing to some financial difficulty, the two locomotives which Gooch had designed for Stephenson had never been exported to America but were still lying at Newcastle. Although he does not say so in so many words it is safe to infer that Gooch's first step in his new post was to persuade Brunel that the Company should acquire these two engines, their gauge being first altered by the builders to seven feet. His advice was followed and at the end of November one of them, now named the *North Star* and fitted with 7 ft single driving wheels, was delivered by barge to Maidenhead where she lay forlorn under tarpaulin until the rails reached her in the following May. The reason she could not be delivered to West Drayton like the others was that in concentrated weight and size she exceeded the load limits of the Grand Junction Canal, being the first Great Western locomotive which did not conform to the crippling conditions which Brunel had imposed. For this reason, until the arrival of her sister *Morning Star*, *North Star* was the only truly reliable locomotive which the Company possessed. She was also the ancestor of a great line of broad-gauge flyers. The derivation of these later and greater locomotives from this little prototype is unmistakable. Nevertheless, Daniel Gooch has come to be regarded as a conservative locomotive designer of no great originality, the fact that he was responsible for *North Star*, although clearly stated by Gooch himself, having been strangely overlooked or ignored by subsequent writers. Had he been such an innate conservative he would never have rallied to Brunel's banner as he did at a time when the majority of his contemporaries held fast to the narrow gauge. Events were to prove that Brunel could not have made a wiser choice for the post of Locomotive Assistant, for the name and fame of Daniel Gooch would soon rank second only to his own, while he would find in him a great engineer, an indefatigable champion of the broad gauge, and a staunch friend whose unswerving loyalty extended beyond death. Once he had evolved a locomotive design of proved merit, Gooch's policy was to develop it to the limit rather than to innovate. This was certainly a conservative policy, but it yielded results which astounded the railway world.

That a business association should so soon have ripened into a close and lifelong friendship was remarkable because in character and in way of life Brunel and Gooch were poles apart. In Brunel's temperament, more Gallic than English, there was no streak of puritanism. Behind his vast capacity for hard work there was no masochistic notion of duty. His career was to him a tremendous adventure, but when he allowed himself to forget it for a while he was capable of indulging in the fleshpots with the same gusto and with a delight which was quite uninhibited. Daniel Gooch, on the other hand, was a supreme example of the dour north-country Puritan to whom civilized pleasures appeared a sinful waste of God's good time. That only a few months after his appointment, in January 1838, Brunel should have invited his young Locomotive Assistant to a party at High Row shows how cordial their relations so soon became. The sequel also underlines the difference between them: 'Went to my first London party on January 29th,' wrote Gooch in his diary.

It was at Mrs Horsley's, Mrs Brunel's mother. I believe I did succeed in getting as far as the staircase, and left it disgusted with London parties, making a note in my memo. book never to go to another. I had to drive all the way from West Drayton to get there, and to walk from a public house, where I put up and got a bed, in silk stockings and thin shoes.

Evidently Gooch felt himself to be even more of a fish out of water in that elegant and witty company than had William Gravatt, and we may safely assume that Brunel forthwith gave up the attempt to launch his friend in polite society.

We may imagine Gooch's feeelings during that winter when his new charges began to arrive from the manufacturers. With their huge driving wheels they may have looked vastly impressive to the layman, but to the shrewd eye of the locomotive man it was apparent at a glance that they were grossly under-boilered. In some cases, indeed, the steam space was so small that when working the engines primed[5] continually and it was difficult to keep the crown of the firebox covered. As for *Thunderer* and *Hurricane,* Gooch wrote that on seeing them he came to the conclusion that they would have enough to do to pull themselves along. He was right. Though they possessed adequate boilers they naturally lacked adhesive weight, while the flexible steam joints were a constant source of trouble.

Poor Gooch! While his days and nights were spent in the Engine House in frantic efforts to keep this brood of lame ducks in commission he was quite unjustly blamed by the Directors for the repeated engine failures which inevitably occurred. 'Our engines are in very bad order,' wrote Gibbs in December 1838, 'and Gooch seems very unfit for the superintendence of that department.' The young man was in an extraordinarily difficult position. He knew very well the source of his troubles and though he must have been sorely

5. 'Priming' is said to occur when water from the boiler is carried over with the steam into the engine cylinders.

tempted to state the facts for the sake of his own reputation, he could not do so without disloyalty to Brunel. Eventually a direct request from the Board for a report on the locomotives forced him to tell the truth and provoked an acid letter from his Chief. But the breach was no sooner made than it was healed for Brunel had an immense respect for Gooch's abilities and knew in his heart that his assistant was in the right.

As Gooch wrote afterwards, there was at the outset only one engine upon which he could depend and that was *North Star*. She was extravagant on fuel but at least she was reliable and it was for this reason that she was chosen to draw the Directors' special on the opening day. In appearance, *North Star* easily surpassed all her sisters and she was greatly admired. In a letter to T. E. Harrison, Brunel wrote:

We have a splendid engine of Stephenson's, it would be a beautiful ornament in the most elegant drawing-room and we have another of Quaker-like simplicity carried even to shabbiness but very possibly as good an engine, but the difference in the care bestowed by the engineman, the favour in which it is held by others and even oneself, not to mention the public, is striking. A plain young lady, however amiable, is apt to be neglected.

Unfortunately, this solitary star could not lighten the darkness that was gathering about Brunel. The Liverpool Party were becoming increasingly fractious. The broad gauge was a complete failure, they said; they demanded representation on the Board and it seemed that short of receiving the Engineer's head upon a charger, nothing would satisfy them. 'Poor fellow,' wrote Gibbs in July, 'I pity him exceedingly, and I know not how he will get through the storm which awaits him. With all his talent he has shown himself deficient, I confess, in general arrangement. ... But I cannot help asking myself whether it is fair to decide on a work of this kind within a few weeks of its opening; and is not the present outcry created in a great measure by Brunel's enemies? I hear that at the meeting Brunel's dismissal is to be moved.' Gibbs's suspicions grew, for three days later he wrote: 'The present outcry ... has much less to do with the deficiencies of our rails than with the machinations of the party which has long been trying to crush Brunel and to get a share in the management of our line.'

The reason for this animosity towards Brunel is easy to understand. As sparrows will mob a bird of bright plumage, so the champions of orthodoxy will never miss an opportunity of attacking original genius. It was an unequal conflict for we must remember that whereas the Liverpool Party had nothing to do but make trouble, the railway was not a *fait accompli*. Westwards of Maidenhead and between Bristol and Bath the railway works were going forward as actively as ever which meant that Brunel's immense load of work and responsibility was in no way lightened. The need to defend himself against these adversaries was thus an additional burden which weighed upon him the more heavily because he believed that he had lost the confidence of his Directors and even of Charles Saunders. Nevertheless, even if he must fight alone he was determined to give battle and his opponents underestimated his strength. They had set themselves to bring down a falcon, not some exotic cage bird.

The crisis had certainly divided the Great Western Board, some of whom were in favour of conciliating the Liverpool Party by agreeing to the nomination of another engineer to act with Brunel and to giving them representation on the Board. Gibbs, however, believed that this would simply prove to be the thin end of a wedge aimed to wrest control of the Great Western away from London and Bristol and he was opposed to conciliation.

... We should have a nest of hornets about our ears [he wrote]. My own decided feeling is that we ought to firmly resist everything of this kind and that if we do not do so I had better retire at once from the Direction; but, on the other hand, I do not like to abandon Brunel and Saunders . . .

It was a pity that on the resignation of Benjamin Shaw, the first Chairman of the Company, Gibbs had declined the chairmanship on the score of ill-health. Had he not done so the issue might have been decided more speedily, for William Unwin Sims, though a man of considerable charm, was a weak Chairman like his predecessor.

The first clash occurred at the half-yearly meeting at the Merchants' Hall, Bristol on 15 August 1838 which went on for no less than seven hours before it was adjourned until 10 October. The Liverpool Party was present in great strength and the broad gauge had a narrow escape. They put forward their resolutions calling for representation and a second engineer but unaccountably failed to force them to a division. They would have done so, said Gibbs afterwards, 'if they had known, as I did, that they had a majority'. Notwithstanding their failure to press home their advantage, Brunel's opponents continued their clamour with the consequence that at the end of the month the Directors yielded to them to the extent of agreeing to invite an independent engineer to examine and report upon the railway. In fact, Brunel had already agreed to such a course in the previous July and the Directors had then invited James Walker, President of the Institution of Civil Engineers, Robert Stephenson and Nicholas Wood to examine and report. Only Wood had accepted but he was not able to begin his inspection until September when he brought along none other than our old friend Doctor Dionysius Lardner to assist him. He was now joined by John Hawkshaw, the twenty-seven-year-old engineer of the Manchester & Leeds Railway, who was called in to placate the Liverpudlians.

Hawkshaw's was the first report to appear. It was brief, forthright and critical, but not very constructive, nor was it calculated to settle the controversy one way or the other. Hawkshaw rightly and acutely pointed out the evils which were likely to arise in the future owing to the use of different gauges, but whereas the Directors had hoped for some constructive guidance from him on the question of the permanent way, he dismissed the whole subject in three lines by saying: 'The mode adopted in laying the rails is, I think, attempting to do that in a difficult and expensive manner, which may be done at least as well on a simple and more economical manner.' He also criticized the locomotives as too heavy, a view with which Daniel Gooch, for one, must have vehemently disagreed. Gibbs summed up this report as 'a very ill-natured production from beginning to end, the greater part of which might have been written without coming near the line'.

As Nicholas Wood had not completed his report, Brunel drafted a reply to Hawkshaw and when the adjourned meeting was reopened on 10 October the opposition was once again defeated. 'The Liverpool men,' wrote Gibbs, '... brought forward their points very feebly. Brunel defended himself from their charges with coolness and great effect.' As a result the opposition were obliged to withdraw the amendments which they had put forward in August and the motion that the Directors' Report be adopted was carried amidst great applause. Once again Brunel had exercised the magic of his personality to good purpose, but the final trial of strength at a Special General Meeting was still to come. So was Nicholas Wood's report.

When this long-awaited document at last appeared on 12 December it consisted of no less than eighty-two closely printed octavo pages. To this there was subsequently added Doctor Dionysius Lardner's characteristic contribution in the form of an even longer appendix in which he recounted in page after page of scientific jabberwocky, the details of his elaborate experiments. This last, however, did not appear until after the great controversy was over, not that it was likely to have influenced the issue in any case. Whereas Hawkshaw had said too little, Wood said too much. His long rambling statements, in which he scarcely ventured any opinion without immediately proceeding to qualify it, were worthy of a permanent civil servant and merely wrapped the simple issues at stake in an impenetrable fog of verbiage. The question of the piles was practically the only one on which Wood dared to commit himself. He decided that they did not contribute to the firmness of the base of the railway but that they 'seemed to prevent the contact of the timbers with the ground'. He recommended using heavier timbers and doing away with the piles. If this was done he thought the road would be both better and cheaper than if laid with stone blocks or cross sleepers. There was nothing very startling about this conclusion, however, because Brunel had himself decided to abandon piling as long ago as July when he had written in a report to the directors:

I find that the system of piling involves considerable expense in the first construction, and requires perhaps too great a perfection in the whole work, and that if the whole or a part of this cost were expended in increasing the scantling of timber and the weight of metal, a very solid continuous rail would be formed; for this as a principle, as for the width of gauge, I am prepared to contend and to stand or fall by it, believing it to be the most essential improvement where high speeds are to be obtained.

Despite the verbose and inconclusive character of Wood's report its general tenor was unfavourable to Brunel, who now faced, almost alone, the greatest crisis of his career. Even those who had hitherto staunchly supported him began to waver. The Bristol directors were weakly prepared to capitulate to a man, while even Saunders and Gibbs now favoured the appointment of a second or consulting engineer, the names of Locke and Stephenson being suggested. Such a course not only implied the abandonment of the broad gauge but Brunel's dismissal. For he had made it quite clear that he would never compromise and would brook no divided responsibility.

If it was proposed to connect another engineer with him [wrote Gibbs] he could not see how such a scheme could possibly work ... nor could he understand the meaning of a consulting engineer. He gave us clearly to understand that he could not and would not submit to either of these alternatives, but that he would resign his situation as engineer whenever we pleased.

Such a courageous stand was entirely in character, but it took his directors aback and threw them into a fever of doubt and indecision. Of the whole board, only one man, Charles Russell, still stood firm in his allegiance to Brunel. For the rest there were many comings and goings, strained board meetings and private heart-searchings as the date of the crucial Special General Meeting, already postponed once and now fixed for 9 January 1839, drew near.

The thing which perturbed the directors most about Wood's report was the miserable result of Lardner's experiments with *North Star* which they had looked upon with such pride as their crack locomotive. He had found that while the engine was capable of hauling 82 tons at 33 m.p.h., her load had to be reduced to 33 tons before she would achieve 37 m.p.h. and to a mere 16 tons in order to reach a maximum of 41 m.p.h. Moreover, in pressing her to this highest velocity the consumption of coke went up from 1.25 to 2.76 lb per ton per mile. These wretched figures appeared to refute all the boasted advantages of the broad gauge and in accounting for them Wood accepted the learned Doctor's conclusion that the great falling off in performance at high speed was due to wind resistance caused by the increased frontal area of the broad gauge engines. This was somewhat ironical in the light of Lardner's classic Box Tunnel diatribe in which he had left the factor of wind resistance out of his calculations altogether. If anything was needed to rouse Brunel to action it was such a pontification by the Doctor and the day after the report was published Gibbs wrote: 'He [Brunel] is perfectly convinced that a great fallacy pervades it, as may be shown and proved by experiment, he proposes to devote all his mind and energies to show this in the next three weeks.'

Brunel and Daniel Gooch at once began their own experiments with *North Star*. They very soon found that the orifice of the blast pipe was much too small, with the result that the engine was throttled at high speeds. More than this, the blast pipe itself was wrongly placed in relation to the chimney so that the blast could not exert its proper effect upon the fire. By opening up the blast pipe and altering its position in the smoke box the improvement both in the performance and the economy of *North Star* was extraordinary. We can well imagine what a tonic effect this must have had on Brunel and his sorely tried Locomotive Assistant in so dark an hour. They lost no time in arranging a trial trip for Gibbs and some of his fellow-directors. On 29 December, with a train weighing 43 tons, *North Star* took them from Paddington to Maidenhead at an average speed of 38 m.p.h., start to stop, and a coke consumption of only 0.95 lb per ton mile. Later, as his correspondence with Saunders reveals, Brunel fitted a new blast pipe of different shape which was cast to his design and which brought about an even better performance. But the first improvement was striking enough and the directors were deeply impressed. Later, it would enable Brunel to produce a crushing reply to Wood's report and to dispose of Doctor Lardner with the remark that it was a pity that Wood as a practical engineer had not conducted the locomotive experiments himself instead of relying upon another who possessed only theoretical knowledge. But for the present the results of the tests were divulged only to the Board. 'We kept our trials in this matter very quiet,' wrote Gooch, 'intending to spring it as a mine against our opponents.'

Thus, at the eleventh hour, Brunel vindicated himself and was able to face the final trial of strength on 9 January with increased confidence. This crucial meeting which was to decide once and for all the future of Brunel and his Great

Western Railway opened at noon at the London Tavern, Bishopsgate Street. Gibbs and his fellows had been so heartened by *North Star*'s demonstration that their air of confidence had a great psychological effect upon the meeting. 'In a few minutes we found we had the whole room with us,' wrote Gibbs enthusiastically. This was not strictly true, for the Liverpool Party had mustered all their forces and the outcome was in fact very much in doubt. A show of hands made it appear that the majority were in favour of adopting the Directors' Report and rejecting the opposition amendment, but the latter demanded a poll. Balloting then began and continued until the meeting was adjourned at 6 p.m. After a night of suspense the meeting was re-opened the following morning and the voting continued. Then came the counting and at one o'clock the result was announced. It was as follows:

Liverpool Party, for the amendment	6,145
London, for the adoption	7,790
majority	1,645[6]

'Being thus completely beaten,' commented Gibbs, 'our opponents declared in conversation with different members of our board that our dissensions would now cease; that being beaten, they had nothing to do now but to cooperate with us heartily for the common good, etc. I hope these sentiments may be acted upon with honesty, but it is evident that they are a mere rope of sand and that they have amongst them some very ill-conditioned fellows.' Ill-conditioned or no, after this defeat nothing more was heard from them. Brunel had triumphed and under his sole command the broad gauge metals went forward into the west.

6. Those attending the meeting held proxy votes in considerable numbers.

8

'The Finest Work in
England'

WHEN the site of the Thames crossing at Maidenhead had
been determined upon, the Thames Commissioners had
stipulated that the projected bridge must in no way obstruct
the towpath and the broad navigation channel beside it. This
limited Brunel to the use of only one river pier in a breadth
of 100 yds and as he had planned to cross at no great height
he was thus faced with a nice problem. He solved it by designing a bridge with two of the largest and flattest arches
that have ever been built in brickwork. Each had a span of
128 ft with a rise of only 24 ft 3 in. to the crown.

This typically bold conception was naturally hailed by
Brunel's enemies as yet another example of his extravagant
folly and its fall before ever a train passed over it was confidently predicted. On 1 May 1838 the contractor, Chadwick,
eased the centerings and the critics howled with delight
when the eastern arch showed signs of distortion. For 13 ft
on either side of the crown the bottom three courses of
bricks separated to the extent of half an inch. Much to their
annoyance, however, the western span stood firm and they
were even more disgruntled when, in July, Chadwick admitted that he was to blame for easing the centres too soon
before the Roman cement had properly set, and agreed to
make good the damage at his own expense. Because a temporary way had already been laid over the bridge and Old-
ham, another contractor, was using it to convey spoil to form
the western approach embankment, Brunel ordered the repairs to be deferred until this work had been completed. The
job was then speedily tackled by Chadwick and the centres

were successfully eased again on 8 October. Brunel's critics
still refused to be convinced and now maintained that when
the time came to remove the centering altogether the bridge
would surely collapse. The engineer himself had no doubts
whatever about his bridge but he ruled that the centres
should not be removed finally until it had stood through
another winter. The suspicion that this was due not so much
to excessive caution as to an impish sense of humour is hard
to resist. Certainly the fact that the bridge was standing entirely free for nine months while his jealous opponents supposed that the centering was still helping to support it was a
joke that Brunel must have relished keenly. Its point was
revealed and his critics confounded by a violent storm one
autumn night in 1839 which blew all the useless centering
down.

By this time, trains were running over the bridge, for the
railway had been opened as far as Twyford in July 1839, and
it was still predicted that the bridge would give way beneath
their paltry weight. Thus one night in the stormy November
of this year when the Thames was in high flood a rumour
spread abroad that the bridge was unsafe.

On Sunday morning [wrote Brunel to Saunders], about 2
o'clock, Mr Hammond was called out of his bed by Low from
Maidenhead with a message from Bell that the Maidenhead
bridge was reported by Ld Orkney [?] as in a dangerous state and
that the 6 o'clock train must not go over it. Mr H^d immediately
got a post-chaise, went to Twyford, got an engine and went to the
bridge. As well as he could by lanterns he examined everything –
could find of course nothing wrong and nobody on the line knew
of anything wrong. He then went for Bell who could only say
that he heard of this but, it seems, never took the trouble to go
and see for himself, as if anything had been wrong it might have
been necessary to send for me, but went quietly to bed – leaving
it to chance, I suppose, whether Hammond might be at home
and whether the train might tumble in the river or not – in fact
taking no more trouble in the matter than to forward this cock-
and-bull story on to H—d who was ten miles off while Bell was

within a few hundred yards. I shall have my own quarrel to settle with Bell ...

November 1839 was a dark month in the history of the Great Western.

We have had some tragical events this week which have made it a very unhappy one [wrote Gibbs on the 16th]. On Monday evening I went to Paddington. Sims was in the Chair; his brother came afterwards to see the Electro-Magnetic Telegraph. We went afterwards to dine with Mills and spent a quiet evening. Sims walked home with me and talked cheerfully all the way. On Tuesday I attended the usual Committee – Sims was in the chair. On Thursday Sims was in Princes Street and at the Bank. The next morning he was found dead in his bed, undressed and with a pistol in his hand, with which he had lolged a ball in his head. We were all dreadfully shocked with this most unexpected tragedy, as there was nothing in his manner or conduct or circumstances to create the slightest suspicion of such an event.

This tragedy, however, had a sequel which was to prove very favourable both to the Railway and to Brunel when Sims was succeeded in the chairmanship by Charles Russell, the only man on the Board whose confidence in Brunel had never faltered. For the first time the young Company now had a strong man at the helm, for Russell was not only, like Saunders and Gooch, a doughty champion of the broad gauge, he was also one of the greatest railway chairmen of the Victorian era.

Meanwhile Brunel, having vanquished his mortal enemies, was now contending against the elements, for his Chairman's untimely death, like the murder of Duncan, was accompanied by weather so tempestuous that many a wiseacre must have shaken his head and echoed the words of Shakespeare's old man:

> Three score and ten I can remember well:
> Within the volume of which time I have seen
> Hours dreadful and things strange; but this sore night
> Hath trifled former knowings.

It was indeed as if, man having failed, nature herself now sought to humble Brunel; to confound the man whose pride had pitted an army against her. The gales which blew down the centerings of the Maidenhead bridge were followed by torrential rains which drowned the valleys of Thames and Avon and reduced the unfinished works to a quagmire. Nowhere were conditions worse than in the deep defile of the mighty cutting, 60 ft deep and nearly two miles long, which was being hewn through Sonning Hill. Two contractors, Ranger and Knowles, had already been defeated here so Brunel had himself assumed control, throwing into this one sector alone a force of 1,220 navvies and 196 horses. They had been shifting no less than 24,500 cubic yards of spoil a week, but now men and horses floundered helplessly in a morass of mud.

> I don't know how the weather has been with you [wrote Brunel to Saunders from Bristol]. We *have* had two tolerable days but otherwise it is dreadful -- out of doors work -- cuttings and culverts are all at a stand -- all flooded. Unless very fine weather comes soon I don't know what's to become of us.

Such appalling conditions defy imagination, but there was no waiting for the spring; as soon as it was humanly possible the work went forward. By the end of the year, Sonning cutting was completed, and on 17 March 1840 the first train, a special carrying a party of directors and their friends, steamed through to Reading.

In the previous autumn Robert Stephenson had sent two additional locomotives to the aid of Daniel Gooch's sorely tried locomotive department, *Evening Star* and *Dog Star*. By this time Brunel had obviously become fully alive to the importance of free steaming for we find him writing to the directors advising them not to accept *Evening Star* because he had found that her exhaust ports were smaller than those of *North Star*. Evidently the trouble was soon overcome for in January 1840 he wrote to Newcastle:

My Dear Stephenson – We have given out that we shall open
our line 30 miles further by 1 May. Our line will be ready, but we
shall be short of steam power. Another 'Star' would make us
comparatively easy, particularly the Directors, who consider the
'Stars' double 'Stars'; I suppose as they always reckon them for
two. Now can you by any extra exertions deliver us one in
March?

Despite the success of the 'Stars', Daniel Gooch believed
he could do better and Brunel, who had complete confidence
in him, agreed that he should draw up a design and specifications. The result was a 7 ft single which was a development of *North Star*, having a larger boiler with a high
'Gothic' firebox that provided ample steam space. It was
Firefly, the prototype of this new class, soon to be so numerous, which drew the Directors' special to Reading and back,
having been delivered from her builders only five days earlier. On the return journey this new locomotive covered the
30¾ miles from Twyford to Paddington in 37 minutes, start
to stop, an average speed of 50 miles an hour. Such sustained
high speed running was without parallel in 1840 and was a
tribute, not only to Gooch but to the improvements which
had been effected in the permanent way.

E. T. MacDermot, in his *History of the Great Western
Railway*, states that the heavier longitudinals and rails (62 lb
per yard) which Brunel had decided to use were first laid west
of Maidenhead and that the original line between Paddington and Maidenhead was not relaid until 1843, when an
even heavier rail weighing 75 lb to the yard was used. Yet it
seems improbable that *Firefly's* high-speed run would have
been possible on the old road which had proved so unsatisfactory and in fact Brunel's letters to Saunders indicate
that during 1839 the whole of the Paddington–Maidenhead
length was relaid with the 75 lb rail. As early as February
1839 he reports that the work of reinstating the line was proceeding well 'while 16, 18 and even latterly 28 trains a day
have been running without interruption'. On 29 May he

writes to say that he is considering using a lighter rail for the
new line. 'Each pound weight per yard represents £40 per
mile,' he pointed out. 'Our present rail is 18 to 19 lb more
than the original and I suggest 10 or 12 might do and offset
the cost of the proposed improved joint. I intend to lay half a
mile beyond Boyne Hill Bridge with the old rails as an experiment. If 55 lb is suitable it would save £24,000 on the
whole line.' It seems clear from this that Brunel decided that,
thanks to the more substantial timbers, 75 lb rail was need-
lessly heavy and that as a result of his experiments 62 lb was
decided upon as the future standard for the line.

With the exceptions of the Purley Park cutting and the
bridges over the Thames at Basildon and Moulsford, the
works west of Reading and through the White Horse Vale
were comparatively light, so that in spite of the bad weather
of the previous winter, the railway was opened for traffic as
far as Farringdon Road (now called Challow) on 20 July 1840.
This lonely place, lost in the wide levels under the White
Horse Hill, became the terminus of the Great Western Railway until the following December, when services were extended to Hay Lane near Wootton Bassett.

When he rode through these fields on his survey, did
Brunel ever pause to glance up at that haunting shape, glimmering white on the green shoulder of the down, to be reminded that he was by no means the first road maker to pass
that way? For above the white horse and just beyond the
high skyline runs the road of the Stone Age men, the Ridge-
way which linked their capital of Avebury with the flint
mines of Grimes Graves in East Anglia. Below, skirting the
steeper contours, runs the later Bronze Age road, the so-
called Port Way or Icknield Way, while at a still lower level
there winds through the fields on its course from Abingdon
to Semington William Whitworth's Wilts & Berks Canal.
This last was of great value in the construction of the railway, but having used it, the new road killed it, and its light-
reflecting waters slowly seeped away to leave a dry ditch as

lonely and as lost as the Stone Age road on the chalk height above. Thus on this one narrow strip of English soil all the ages of man have left their mark and the great driving wheels of Gooch's broad gauge flyers pounded in the wake of uncountable generations of slower and humbler travellers.

It was while Farringdon Road was a temporary terminus that Brunel witnessed his first railway accident. With two of his assistants he was waiting in the chilly darkness of an early morning in late October for an engine to take him to Paddington when he heard the sound of the night goods train approaching at unusual speed. To his consternation it was soon obvious that it was not going to stop. As the train swept past the driver could be seen standing motionless on his open footplate while the guard and four third-class passengers who were travelling in an open truck next to the engine were heard shouting frantically in their efforts to attract his attention. The locomotive, *Fire King*, another of Gooch's new 7 ft singles, crashed through the closed doors of the temporary engine house beyond the station and the unfortunate driver was killed. It was assumed that he had fallen asleep at his post.

When it was opened, the only intermediate station on the Farringdon Road–Hay Lane section was Shrivenham. This marked the boundary of the London Division. Westwards, the railway was the responsibility of the Bristol Directors. Although work in both divisions had been started simultaneously, progress in the Bristol Division so far as track mileage was concerned was very much slower because the works involved were far heavier. With the exceptions of the Wharncliffe viaduct, the Thames bridges and the cuttings at Sonning and Purley there were no major engineering works in the London Division whereas between Wootton Bassett and Bristol the route bristled with them. It was here, then, that Brunel was given the fullest opportunity to display his powers. The entire terminus at Bristol Temple Meads with its splendid timber roof was constructed of arches 15 ft above ground level. Bath station was similarly elevated and was approached by a viaduct of 73 arches. Between these two stations there was another viaduct of 28 arches at Twerton, four major bridges carrying the line over the floating harbour at Bristol, the feeder from the Netham dam and the Avon, and no less than seven tunnels,[1] the longest, known as No. 3, being 1,017 yards in length. The first contract on this section, for the three tunnels at the Bristol end in the neighbourhood of Brislington, was let in March 1836, but owing to trouble with contractors and bad weather it was not until 31 August 1840 that the line was opened for traffic between Bristol and Bath. The line was laid with the same type of bridge rails which had been used west of Maidenhead. Rolled at Rhymney, Dowlais and Merthyr they were conveyed to Newport and thence over the Severn Sea to Bristol. Because this western section was still cut off from Paddington, before it could be opened locomotives and rolling stock had either to be built locally or brought by sea. *Arrow* and *Dart*, two of Gooch's seven-footers, were built by Stothert & Slaughter of Bristol, while three more, *Fireball*, *Spitfire* and *Lynx*, were shipped from Liverpool along with *Meridian* of the 'Sun' class.

The next section of the Bristol Division to be opened was that between Hay Lane and Chippenham. Here the deep cuttings and the long embankments between Wootton Bassett and the bridge over the Avon near Christian Malford gave Brunel infinite trouble and anxiety, particularly during the terrible winter of 1839, due to the constant slipping of the heavy clays. One embankment in particular slipped so persistently that in the spring of 1841, in order to get the line finished, he was forced to drive piles down each side, lashing their heads together with chain cables carried through the bank. The section was finally opened on the last day of May when, to celebrate the occasion, Brunel, Saunders, Russell

1. The number is now five. The two short tunnels at Fox's Wood were opened out in 1894.

and a number of other directors and officers were entertained to a public breakfast by the Mayor of Chippenham. But the clay still gave a great deal of trouble. A slip during the night of 7 September caused the derailment of the up mail train drawn by *Rising Star* and *Tiger*, but although two carriages were badly damaged there was no injury more serious than a broken leg.

It would be interesting to know how many thousands of miles Brunel must have covered in the country between London and Bristol, not to mention excursions farther afield to Exeter and South Wales, between the conception and completion of the Great Western main line. As the railway extended westwards his long black britzka would be loaded on a special truck at Paddington and he would then travel in it to the temporary terminus where a team of four post-horses waited to whirl him away to Chippenham, to Bath or to Bristol as the case might be. His grand-daughter, Lady Noble, quotes a letter which he wrote to Mary Brunel at Duke Street from Wootton Bassett at the time when the difficult Hay Lane–Chippenham section was under construction. It is so delightful and so revealing that it must be quoted in full. Illuminated with graphic little sketches in the margin it shows that even so immense a burden of responsibilities had not succeeded in extinguishing that sense of humour which Macfarlane had remarked in him years before. It also reveals that the country inn of the stage coach era was by no means the snug abode of olde worlde good cheer that several generations of industrious Christmas card artists have induced us to believe.

My Dearest Mary,

I have become quite a walker. I have walked today from Bath-ford Bridge to here – all but about one mile, which makes eighteen miles walking along the line – and I really am not very tired. I am, however, going to sleep here – if I had been half an hour earlier, I think I could not have withstood the temptation of coming up by the six ½ train, and returning by the morning goods train, just to see you; however, I will write you a long letter instead. It is a blowy evening, pouring with rain, my last two miles were wet. I arrived of course rather wet, and found the *Hotel*, which is the best of a set of deplorable public houses, full – and here I am at the 'Cow and Candlesnuffers' or some such sign – a large room or cave, for it seems open to the wind everywhere, old-fashioned with a large chimney in one corner; but unfortunately it has one of these horrible little stoves, just nine inches across. [*A sketch shows a diminutive hob grate.*] I have piled a fire on both hobs, but to little use, there are four doors and two windows. What's the use of the doors I can't conceive, for you might crawl under them if they happened to be locked, and they seem too crooked to open, the two ones with not a bad looking bit of glass between seem particularly friendlily disposed. [*The doors are portrayed leaning towards each other with a fine eighteenth-century mirror between.*]

The window curtains very wisely are not drawn, as they would be blown right across the room and probably over the two extra greasy muttons which are on the table, giving just light enough to see the results of their evident attempts to outvie each other, trying which can make the biggest snuff. One of them is quite a splendid fellow, a sort of black colliflower . . .

There is a horrible harp, upon which really and honestly somebody has every few minutes for the last three hours been strumming these chords, always the same:

Good-bye my dearest love, Yours, I. K. Brunel.

By the end of May 1841 there remained only one missing link in Brunel's iron chain from London to Bristol and it was the hardest of all to forge. This was the section between Chippenham and Bath. It included much deep cutting and embankment, yet another crossing of the Avon at Bath, viaducts at Chippenham and Bath, and the diversion of the Kennet & Avon Canal, but, over and above all, there was that 'monstrous and extraordinary, most dangerous and impracticable tunnel at Box'. Nearly two miles long, it was by far the greatest railway tunnel so far attempted, exceeding the length of Kilsby on the London & Birmingham Railway by 786 yards. Apart from its length, the fact that Brunel had planned it upon an incline of 1 in 100 provoked, as we have seen, much headshaking amongst the pundits.

Work on the tunnel had begun as early as September 1836 with the sinking of six permanent and two temporary shafts 28 ft in diameter from the hill top down to rail level. One of these, sunk near the road from Bradford to Colerne, was 300 ft deep. All had been completed by the autumn of 1837 and contracts for the tunnel were then advertised. George Burge of Herne Bay, one of the first of the great railway contractors, undertook to build over three-quarters of the length of the tunnel. His section ran through clay, blue marl and inferior oolite and was to be lined with brick throughout. The remaining half mile at the eastern end was undertaken by two local men, Brewer of Box and Lewis of Bath. It was to be driven through the great oolite and Brunel planned to cut it in the form of a Gothic arch and to leave it unlined. One of his personal assistants, William Glennie, remained in charge of the whole work until its completion.

It was an immense undertaking and the more staggering when we remember that, apart from the steam pumps which kept the workings clear of water and the power of gunpowder which was used to blast away the rock, it was accomplished entirely by the strength of men and horses working by candlelight. For two and a half years the work consumed a ton of gunpowder and a ton of candles every week. The lining bricks were burnt by another local man, Hunt, at a yard in the meadows to the west of Chippenham and for three years he employed one hundred horses and carts to convey a total of 30,000,000 bricks to the site.

The task undertaken by Lewis and Brewer in driving the great Gothic arch through half a mile of solid Bath stone was almost superhuman and was made the more difficult by the inrush of water through fissures, particularly during the wet winters. In November 1837 there was such an overwhelming influx that the water overpowered the pumps, filled the tunnel and rose 56 ft up the shafts. The situation was mastered by the installation of a second pumping engine of 50 h.p., but another inundation in the following November held up work for ten days.

The spoil excavated, 247,000 cubic yards in all, was hauled up the shafts in buckets by means of horse gins at the surface. It is entirely typical of Brunel that when he was accompanied by some of his directors on one of his many visits to Box he should have insisted that they join him on his descent of one of the shafts. Experience of this kind was the breath of life to him, but we may imagine the feelings of those staid and prosperous gentlemen as they huddled together in the dirty spoil bucket to be swung down by the creaking gin into a pitch-dark abyss reeking with the fumes of spent gunpowder and reverberating with the muffled thunder of the blasting. No doubt they secretly envied one member of the party who dutifully stayed on the surface, having confessed that his wife had forbidden him to descend.

Lewis and Brewer had commenced cutting their section from opposite ends and Brunel was present in the workings on the historic occasion when the two gangs at last met. Whether or not a true line had been kept was a question which had been the subject of much anxiety, and when it was found that the junction was perfectly true Brunel was so delighted that he took a ring from his finger and gave it to the foreman in charge of the gang. This ring is treasured by his descendants in Bristol to this day.

August 1840 was the date which had originally been stipulated for the completion of the tunnel contract, but despite every effort the work fell far behind schedule until, in December 1840, Brunel decided to throw all his reserves into the last battle. Henceforward a force of 4,000 men and 300 horses laboured night and day to complete the tunnel. We can imagine how eagerly that completion was awaited by the local inhabitants, whose only link with the outside world had hitherto been the London turnpike where the stagecoaches and the smart chariots of the quality came spanking down Box Hill on their way to Bath. For upon this quiet countryside which had not changed greatly since the days of Shakespeare the railway builders had descended like the roystering legions of an invading army. In Corsham, in Box and in all the neighbouring villages and hamlets every available bed was occupied and not one was ever allowed to grow cold. As soon as a day-shift worker left his bed of a morning, a night worker was waiting to move into it. There was much drunkenness and fighting, and because there was no police force, on Sundays the foremen were allotted the invidious task of trying to preserve order in the different villages.

The railway navvy certainly played as violently as he worked, but too much has been said of his pugnacity and his capacity for strong drink and too little of his sterling qualities; his amazing powers of physical endurance, his fundamental good humour, and his unswerving loyalty to his foreman, to the mates in his gang and to his own simple code of honour. No strikes and labour disputes marred the building of the Great Western. Every man from Brunel down to the navvy with shovel and pick was endowed with an astounding capacity for hard work and seems to have been inspired with a determination to see the job through which enabled the work to continue even under the most appalling weather conditions. That there was another and gentler side to the character of the railway navvy was revealed by the Reverend H. W. Lloyd, Rector of Cholsey cum Moulsford, who described in a little pamphlet how he conducted the burial service for one of them, John Stanley of Brandon in Norfolk, who was known to his mates as 'Happy Jack'. Clad in white smocks and wearing white ribbon bows in their hats, two by two all his old companions silently accompanied

'Happy Jack' on his last journey to Cholsey churchyard. He
was not the only stalwart to lose his life in the building of
the Great Western; in the cutting of Box Tunnel alone about
one hundred men perished. Such prodigious feats are never
accomplished without risk and sacrifice.

In June 1841, the great tunnel was at last completed and
the way opened throughout from London to Bristol. The
whole work had cost £6,500,000, considerably more than
double Brunel's original estimate, but the dream of the 'Bristol Railway' which had seemed to many so impossible of
realization only eight years before was now a magnificent
reality. On the last day of the month a decorated train pulled
out of Paddington and arrived in Bristol four hours later. But
the fears of Box Tunnel which the sceptics and the prophets
of doom had aroused were not easily at rest. A suggestion
that the tunnel should be illuminated throughout was
scouted by Brunel as impracticable, and for some years apprehensive travellers suffered the inconvenience of leaving
the trains at Corsham or Box and posting over the hill rather
than brave its dark and sulphurous depths. Then, although
our old friend Dr Lardner seems to have held his peace,
another of his kind, an eminent geologist named the Reverend Doctor William Buckland of Oxford, started a fresh
scare. Although he had never ventured into the tunnel himself, he declared that the unlined portion was highly dangerous and would certainly fall owing to 'the concussion of the
atmosphere and the vibration caused by the trains'. To this
Brunel replied with forgivable sarcasm that while he regretted his lack of scientific knowledge of geology, he had
had some recent practical experience of excavating the rock
in question and considered that it was absolutely safe. Nevertheless, to settle the question, the Board of Trade Inspector-
General of Railways, Major-General C. W. Pasley, was
invited to inspect the tunnel. The General did so and reported
on its safety in such forthright terms that Brunel, who was
anxious to dispose of the scare without raising a public con-

troversy, wrote to Saunders: 'Thank you for General Pasley's
report. As far as we are concerned it is all we could wish ...
but I regret very much that it is going to be published as
there is a decided *hit* at Buckland which will in all probability bring him out and undo all the quiet good that we might
have derived from the report.' Nevertheless, any controversy
that the report may have provoked seems to have soon died
down. Subsequently, as a result of occasional small falls due
to frost action, a brick arch, originally protected by a mattress of faggots, was turned under part of the stone vault, but
a portion of the original unlined section remains open to this
day.

As early as June 1839 the question of where to establish
the chief locomotive depot and repair shop for the railway
was being debated between Brunel, Gooch and Saunders.
Reading was at first mooted, and then Didcot, the junction
for the branch to Oxford. It was Gooch who finally expressed
himself in favour of siting it in the fields below the little
market town of Swindon where the Cheltenham Railway
joined the main line, and his choice was governed by the
gradients of the railway.

Just as his great predecessor of the canal age, James Brindley, had made a practice of concentrating his locks together
and thus securing long unbroken levels, so Brunel in his survey had concentrated the gradients on his main line. For no
less than 71 out of a total of 118 miles the line is either level
or of a gradient no steeper than 1 in 1,000. From Paddington
to Didcot the ruling gradient is 1 in 1,320 and from thence to
the summit level at Swindon there is only one short length of
1 in 660, a gradient which is not exceeded on a further 43
route miles of line. This means that the western fall from
Swindon is mainly concentrated in the remaining four miles
of line which is represented by the Box Tunnel incline and
by the 1 mile 550 yards gradient in the cutting west of Wootton Bassett, which is also inclined at 1 in 100. Because of
these two gradients, Daniel Gooch proposed to provide

locomotives with smaller driving wheels to work the traffic between Bristol and Swindon and then to use his fast 7 ft singles on the long level galloping ground between Swindon and Paddington. Although Swindon was farther from London than from Bristol it was on this account an obvious engine changing point. Moreover, as Gooch pointed out in a letter to Brunel, owing to the use of locomotives with smaller driving wheels in the Bristol Division, if the distance was measured in terms of engine revolutions instead of miles, Swindon became the halfway point. Brunel and his directors accepted Gooch's argument and thus, on purely mechanical grounds, the site of a new town was determined. Before the line was opened throughout, the workshops which would give the name of Swindon a new meaning were already building, while over the green fields the streets of terrace houses began to extend their geometrical pattern. Swindon works were opened on 2 January 1843.

It was at Swindon station that the railway refreshment room first acquired that unsavoury reputation which it has never since succeeded in living down. With a singular disregard for the nocturnal peace of potential guests, a hotel and dining-room, connected by a covered overbridge reminiscent of the Bridge of Sighs, was incorporated in the Swindon station buildings. Most unwisely, the Company let the management of this hotel on a long-term contract in which they agreed that all regular trains should stop at Swindon 'for a reasonable period of about ten minutes' so that passengers might refresh themselves. This enforced stop at Swindon became a bane to the Company, the more so when successive proprietors of the hotel grossly abused their monopoly. The first, and perhaps the worst, S. Y. Griffiths of the Queens Hotel, Cheltenham, provoked Brunel into writing what is, perhaps, his most frequently quoted letter:

Dear Sir,

I assure you Mr Player was wrong in supposing that I thought you purchased inferior coffee. I thought I said to him I was surprised you should buy such bad roasted corn. I did not believe you had such a thing as coffee in the place; I am certain I never tasted any. I have long ceased to make complaints at Swindon. I avoid taking anything there when I can help it.

Yours faithfully,

I. K. Brunel.

The urn in which Mr Griffiths made his execrable coffee was the work of the still extant firm of Martins of Cheltenham. Worthy of a better brew, it was made in the form of a broad gauge locomotive and is now preserved in the Board Room at Paddington.

Swindon was not the only curious station on the Great Western, for in station design as in all else Brunel displayed his originality. At Reading, and a little later at Slough, he built the first of his celebrated 'one sided' stations. This design really consisted of two separate stations for 'Up' and 'Down' passengers, both situated on the same side of the running lines. Brunel claimed that this not only obviated the necessity of crossing the lines, but was of great convenience to passengers. Judging from the total lack of a sense of direction from which many railway travellers appear to suffer, there was probably some truth in this and the one-sided station may have saved many a passenger from boarding a train going in the wrong direction. The type survived for a number of years despite the fact that, as any railwayman will appreciate, it possessed chronic operational disadvantages. It is no exaggeration to say that not only the stations but every yard of the way between London and Bristol was impressed with the stamp of Brunel's original thought. Even the signals, the tall discs and crossbars and 'fantails' which seemed so suggestive of broad gauge speed, were of his unique design. He was thus responsible for creating a complete and unmistakable railway landscape which was entirely his own.

When the Great Western Railway was built, the London and Bristol Committees into which the Board was divided enjoyed considerable autonomy and in the execution of the works in the Bristol Division the latter exercised a far less stringent control over expenditure than did their colleagues in London. For this liberality we, their posterity, are deeply indebted for it enabled Brunel to display not merely his engineering skill but his architectural gifts to the full. Even if it had been carried out in the most utilitarian and unimaginative manner the line from Wootton Bassett down to Bath and Bristol would still have been a great engineering achievement. But the magnitude of that achievement would soon have been eclipsed and forgotten did not every detail remind us of it, by reflecting Brunel's infallible eye for proportion and his sense of grandeur. His exquisite sketches of the architectural detail of tunnel mouth, bridge, or viaduct, of pediment or balustrade remain to reveal, not Brunel the engineer, but Brunel the artist at work. He had set himself to build 'the finest work in England' and he was bringing this splendid highway to Bath, one of the most beautiful cities in Europe. Like his predecessor John Rennie, the engineer of the Kennet & Avon Canal, Brunel determined to be worthy of such an occasion and to pay due tribute, as architect, to those forerunners who had handled the honey-coloured stone to such urbane and noble purpose. The day when the new machine age would grow arrogant, ignorant and barbarous and scrawl its crude graffiti across the face of stone was not yet. To Brunel, his great tunnel at Box must be something more than a mere hole in the ground, it must be a triumphal gateway to the Roman city. It was for this reason that he crowned Box with that huge Classic portico which towered high above Gooch's flying locomotives, *Lord of the Isles*, *Tornado* or *Typhoon*, as they shot from shadow into sunlight and swept down on their imperial way to Bath.

[9]

The Gauge War

EIGHTEEN hundred and forty-one was a year of triumph for both the Brunels. On 12 August, just two months after the completion of his son's great tunnel at Box, Marc Brunel was able to write in his diary at Rotherhithe: 'Just returned from Wapping by land at 2 p.m. These are the first lines I have laid on paper of this event, and to my dearest Sophia I owe this triumph.' 'Sans toi, ma chère Sophie, point de Tonnelle.'

To detail in full the story of the completion of the Thames Tunnel would be out of place here because Isambard Brunel was unable to play any part in it and it was his erstwhile assistant Richard Beamish who succeeded to his old post of resident engineer on 22 January 1835. The first task was the replacement of the original tunnelling shield by a new one, an operation of such extraordinary difficulty and delicacy that it was not completed until 1 March 1836. The slow advance under the river then began once again. This time, with the object of avoiding a repetition of previous disasters, much clay was flung down in the river bed ahead of the advancing shield, but not withstanding this precaution the river broke into the tunnel on no less than five occasions. At one stage, so soft and treacherous was the ground that to remove a 'poling board' was quite impossible without certain disaster and the shield advanced simply by the use of the poling screws. The atmosphere became even more lethal than before; men collapsed in the frames; sewer gas would suddenly ignite in flashes of flame that flickered twenty feet across the shield and Marc's diary is a tragic record of sickness and death. Yet still the shield went forward, although throughout the whole of June 1837 it only advanced one foot.

With no son by his side and age telling against him, the
strain and anxiety which fell upon Marc may be well imagined. Night and day, samples of the soil ahead of the
shield were brought to his house at Rotherhithe, a cord,
pulley and bell being rigged for this purpose outside his bed-
room window so that the sample bucket could be hauled up
to him. Even when low water mark on the Wapping side had
been reached, danger was not passed, for on 4 April 1840
there was suddenly a sound in the tunnel which the men
described as 'like the roaring of thunder' as a part of the
foreshore thirty feet in diameter suddenly subsided, leaving a
cavity thirteen feet deep which had to be stopped with clay
before the work could go forward.

In October of this year the sinking of the shaft at Wapping was begun, and after thirteen months' work it had
reached its full depth, by which time the shield was only 60 ft
away. The better to drain the tunnel a driftway was then cut
from beneath the tunnel invert into the shaft, and in the
spring of 1841 Marc Brunel's three-year-old grandson, Isambard III, was handed through this newly completed driftway
to become the first person in history to pass beneath the
Thames. At the same time his indomitable grandfather received at the hands of his Queen the richly deserved honour
of a knighthood.

On 15 December, the top staves of No. 1 Frame of the
shield touched the brickwork of the Wapping shaft and the
long battle against almost overwhelming odds was over. It
left the old engineer seriously weakened in health, and in
November 1842 he suffered a paralytic stroke. He recovered
sufficiently, however, to attend with the rest of his family the
official opening celebrations on 25 March 1843 when he received from the great crowd an ovation which moved him
deeply.

Thus Marc Brunel was able to falsify his son's gloomy
prophecy that he would not live to see the completion of his
Thames Tunnel. His great work finished at last, he and

Sophia removed from Rotherhithe to a little house in Park Street, Westminster, which overlooked St James's Park and where he became a near neighbour to his son whose fame was a great source of pride to him and in whose activities he took the liveliest interest. The only shadow on his declining years was cast by the fact that, characteristically, he never received the £5,000 which had originally been promised to him on the completion of the tunnel, a circumstance which made him dependent on his son's generous financial aid. In 1845 a second stroke left him partly paralysed so that until his peaceful death on 12 December 1849 in his eighty-first year he became a benign and white-haired invalid in a wheeled chair. He was buried in Kensal Green Cemetery where, after five years spent in seclusion at Duke Street, his beloved Sophia was laid by his side. So passed Sir Marc Isambard Brunel. He left, wrote Richard Beamish in his memoir, 'a name to be cherished so long as mechanical science shall be honoured'.

*

The Brunels were both men of short stature and no great physical strength. That they appear to us as giants was due to the immense reserves of nervous energy which they possessed. Marc Brunel never drove himself so hard as did his son, but both, so soon as these reserves had been unleashed upon a particular project, could perform feats of endurance and sheer hard work which were out of all proportion to their physical powers. Their bodies, in other words, were the grossly overworked slaves of their minds and when a project was completed and the lavish expenditure of energy no longer required there was apt to be a dangerous physical reaction.

Just as his father had suffered a stroke when the Thames Tunnel was finished, so there is evidence that Brunel's constitution, which had never once failed him since his Thames Tunnel accident and which he had abused so unmercifully

during the surveying and building of the Great Western
Railway, began to show signs of strain after this first great
task was over. The evidence is slight, but in the light of after
events it is none the less ominous. We have grown so accustomed to reading of his hurried departures in the small hours
after nights spent on drawings, calculations or reports that
we begin to credit him with super-human powers and it is
with something of a shock that we are reminded of his mor-
tality in this letter, so unlike his usual self, which he wrote to
Saunders in the summer of 1842 in apology for not attending
a board meeting: 'I cannot get out early in the mornings and
this evening I feel that it would have been impossible for me
to have been at Steventon[1] tomorrow and doubt my attending even the arbitration. My state of health indeed renders
me very anxious to get away entirely for a week or ten days
or I see no prospect of my getting well.'

Evidently he had his rest and soon recovered, for his next
letter to Saunders, written exactly ten days later from the
Engineer's Office of the Bristol & Exeter Railway, reveals
him pursuing his railway business at the same old hectic
pace: 'Farr says you want to know my probable movements
this week. I go tomorrow down the Exeter line. On Wednesday I shall be in Bristol and Bath; on Thursday at Steventon if there is anything connected with engineering coming
on, if not I should be in London – Friday at Cirencester to
meet the Cheltenham directors about the contract. . . .' Saunders knew his Brunel and the most he ever expected to learn
was his 'probable movements'. Not so Thomas Osler, one-
time secretary of the Bristol Committee. 'I really am teased
out of my life by the man, he expects to know every day
where I am to be every next day,' wrote the exasperated
engineer.

Brunel's references to Exeter and Cheltenham reveal how

1. From July 1842 to January 1843 the Superintendent's house at
Steventon was selected as a convenient half-way house for Great
Western Board meetings.

rapidly his broad gauge was extending its tentacles through the West Country. To follow that expansion in detail and to explore the intricacies of inter-company politics which it involved would be to write, not a biography of Brunel so much as a history of the Great Western Railway, a task which the late E. T. MacDermot so ably carried out in two monumental volumes. The construction of the original main line has been dealt with in some detail because it was Brunel's first major work by which, as he himself said, he was prepared to stand or fall. At the time he undertook the first flying survey for the Bristol Committee he was practically unknown. He had acquired a certain local reputation but beyond that his name carried only the lustre acquired by his father's achievements. That when he had completed the line to Bristol at the age of thirty-five he enjoyed a reputation second only to that of George Stephenson himself gives us the measure of his success and shows us how decisively he had triumphed over those adversaries who had striven so hard to bring about his downfall. Henceforth we need notice only those of his railway works which affected that reputation for good or ill and the events chiefly responsible for repelling the advance of his seven-foot gauge.

The first train to reach Bristol from Paddington in June 1841 did not terminate at Temple Meads but was able to continue over the newly opened metals of the Bristol & Exeter Railway as far as Bridgwater. The works on this first section of the satellite company presented few difficulties and had been proceeding rapidly under the charge of Brunel's assistant William Gravatt while the last links of the Great Western main line were being forged. There were two deep cuttings at Pylle Hill and Uphill, the latter crossed by a flying arch road bridge of singular grace and delicacy, having a span of 110 ft. But from Uphill cutting, a little distance from the junction of the short branch to Weston-super-Mare, the line ran level and straight as a ruler with scarcely an earthwork of any kind across the flat lands be-

tween Axe and Parret to Bridgwater. The course from Bridgwater to Taunton was also easy and the first train reached
Taunton on 1 July 1842. It is worth noting, however, that
in the design of his bridge over the Parret just south of
Bridgwater, Brunel over-reached himself. A masonry span of
100 ft with a rise to the crown of only 12 ft, or nearly twice as
flat as the controversial Maidenhead bridge, it evidently proved a failure. It was begun in 1838 and completed in 1841, but
two years later it was replaced by a timber structure which
stood until 1904, when it was rebuilt in steel. The reconstruction was carried out without interrupting traffic and
with such celerity and lack of fuss that the news of this failure never seems to have reached the ears of his critics.

Construction of the remaining section from Taunton to
Exeter proceeded more slowly. This was mainly due to the
cutting of the White Ball tunnel through the spine of the
Black Down Hills which here form the Devon–Somerset
border. While the tunnel was building, train services were
extended to a temporary terminus at Beam Bridge near the
Exeter turnpike but on 1 May 1844 the line was opened
throughout to Exeter. Of this opening to Exeter Daniel
Gooch wrote:

We had a special train with a large party from London to go
down to the opening. A great dinner was given in the Goods
Shed at Exeter Station. I worked the train with the *Actaeon* engine, one of our 7 ft class, with six carriages. We left London at
7.30 a.m. and arrived at Exeter at 12.30, having had some deten-
tion over the hour fixed. On the return journey we left Exeter at
5.20 p.m. and stopped at Paddington platform at 10. Sir Thomas
Acland, who was with us, went at once to the House of Commons, and by 10.30 got up and told the House he had been in
Exeter at 5.20. It was a very hard day's work for me, as, apart
from driving the engine a distance of 387 miles, I had to be out
early in the morning to see that all was right for our trip, and
while at Exeter was busy with matters connected with the opening, so that my only chance of sitting down was for the hour we
were at dinner. Next day my back ached so that I could hardly

walk. Mr Brunel wrote me a very handsome letter thanking me for what I had done, and all were very much pleased.

Brunel's delight in such an achievement was fully justified. A time of 4 hours 40 minutes for the 194 miles between Exeter and Paddington represented a performance unparalleled for sustained high-speed running, especially when we remember that the time included the frequent stops for water which the locomotives of the day, with their small tenders, had to make. It must have given Brunel's narrow gauge rivals furiously to think. It was also a remarkable tribute to Gooch's courage and power of endurance. Having thus so convincingly demonstrated the potentialities of Gooch's locomotives and Brunel's broad gauge road, in the very next year the best trains were scheduled to reach Exeter in five hours from Paddington inclusive of stops. They were the first expresses, easily the fastest trains in the world, and soon they would travel faster still.

A month after the opening to Exeter, the Oxford branch was completed and a year later, following the completion of the long Sapperton tunnel through the ridge of the Cotswolds and of the heavy works in the Stroud valley, the first train ran through to Gloucester from Swindon over what was now a branch of the Great Western, the Cheltenham & Great Western Union Company having ended its purely nominal existence in 1844. This opening to Gloucester coincided with the passing of the Act for the South Wales Railway from Fishguard and Pembroke Dock through Carmarthen, Swansea, Neath, Cardiff and Newport to Chepstow. In the original prospectus of this railway Brunel had planned to bridge the Severn at Hock Cliff between Fretherne and Awre and to form a junction with the Gloucester line at Standish. But although this treacherous estuary had already been by-passed by the Gloucester & Berkeley Ship Canal which Brunel proposed to cross by swing bridge, the Admiralty objected to the Severn bridge and even the engineer's offer to make a new navigable cut across the Arlingham promontory would not placate them. He therefore had to fall back upon the detour via Gloucester and the South Wales Railway finally terminated in a junction with the Monmouth & Hereford Railway at Grange Court. The South Wales Act received strong Irish support and across St George's Channel the Waterford, Wexford, Wicklow & Dublin and the Cork & Waterford Railways were promoted by Great Western interests with Brunel as their engineer. But when the great potato famine struck Ireland these dreams of profitable Anglo-Irish traffic faded. The Fishguard extension was abandoned, the Dublin line terminated at Wicklow, the Cork at Youghal and none of them would be completed until long after Brunel's death.

Worked by Gooch's locomotives, the Bristol & Exeter at its inception was virtually an extension of the Great Western main line, but in this turbulent period of railway history the honeymoon of the two companies was brief. The Government of the day had stressed that railway promotion schemes in Dorset and east Somerset should be made with a view to providing a more direct route to Exeter than that via Bristol. This state of affairs placed the Great Western in a dilemma. To promote such a route themselves would be to threaten the Bristol & Exeter with a serious loss of traffic, while failure to do so would be to leave the door open to the rival London & South Western Railway. The solution adopted was to proceed with the direct route scheme and at the same time to offer to purchase the Bristol & Exeter. Although the very favourable terms proposed were acceptable to the Bristol & Exeter directors they were decisively rejected by the shareholders and for a time relations between the two companies became distinctly strained. This is revealed in a letter which Brunel wrote to Saunders in which he discusses the proposed direct route into Exeter.

The only way of entering [Exeter] independently [he writes] is to run *alongside* the B. & E. from near Stoke Cannon, miss their station and join the S. Devon. Now I question the policy of such a demonstration – it's decidedly like in appearance, tho' not in reality, a parallel line from Gloster to Stonehouse or some contrivance hereafter to be projected at Bristol. I think it on the whole wiser to go on as if with the best understanding with the Exeter. A threat is very short lived and we should keep up the character of never barking unless we mean to bite! What say you?

It is unnecessary to enter into the ramifications of the dispute and the details of the war with the London & South Western Railway over the territory west of Salisbury which ended in the defeat of the broad gauge forces. The broad gauge direct line to Exeter was never built, but it is worth remarking that the route which Brunel ultimately proposed for it was almost identical with that built many years later which the West of England expresses now follow.

The rift between the Great Western and the Bristol & Exeter made Brunel's position as engineer to both companies most invidious and it was for this reason that he resigned from the latter in 1846. But he remained engineer to the South Devon, Cornwall and West Cornwall Companies which were formed to carry the broad gauge metals to the farthest west where, as we shall see, he carried out both his most ill-fated railway experiment and his greatest work of railway engineering.

On the South Wales Railway and, despite the friction with the Bristol & Exeter, on this long line from Bristol to Penzance the advance of the broad gauge was not seriously impeded, nor was its future immediately threatened. But whereas South Wales, Devon and Cornwall were still virgin territory, narrow gauge interests were ready to repel any attempt to advance into the southern counties or the Midlands and the celebrated battle of the gauges became inevitable. Because we tend to think of this 'battle' in terms of our own apathetic day when, having made his mark on a ballot paper, the man in the street tends to leave the conduct of affairs to bureaucrat or business man, we wrongly suppose that it was waged solely in the board room and on the floor of the House of Commons. True, these were the centres of conflict, but the sympathies of the whole country were most passionately engaged upon one side or the other and each bombarded the other with books, pamphlets and articles on the gauge issue.

It was at Gloucester that the rival forces first came into serious collision as the narrow gauge Birmingham & Gloucester line advanced upon the city from the north to meet the broad gauge line from Swindon and the Bristol & Gloucester Railway. The latter was first projected as a narrow gauge line, but Brunel was their engineer and his arguments, backed by the imminence of the Swindon line which would join theirs at Standish, induced the directors of the Bristol & Gloucester to change their minds. So the broad gauge metals were laid, straddling, at the Bristol end, Townsend's old Bristol & Gloucestershire tramway. When the Company joined forces with the Birmingham & Gloucester and the new Bristol & Birmingham Railway Company opened negotiations with the Great Western a victory for the broad gauge seemed certain. Amalgamation with the Great Western was discussed at a meeting in Bristol on 24 January 1845 but the two parties failed to come to terms and it was adjourned until the 27th, when there was still no agreement. Next day the Bristol & Birmingham was leased by John Ellis, Deputy Chairman of the Midland Railway, acting solely on his own responsibility. It was one of the boldest strokes in railway diplomacy and a bitter blow to the broad gauge champions.

Between Cheltenham and Gloucester lay the first section of mixed gauge permanent way, the rival factions, after much argument, having each laid half the line. Hence it was at Gloucester that the evils of the break of gauge were first felt and where they continued unabated until June 1854, when the Midland opened their narrow gauge route to Bristol.

The next battle, fought over the Oxford & Rugby and the

Oxford, Worcester & Wolverhampton Railways, brought the Great Western into conflict with that most ruthless and unscrupulous of all railway buccaneers – Captain Mark Huish. As Manager of the Grand Junction Railway, Huish had appeared favourable to the broad gauge but only, as was soon apparent, to further his own ends. He was then at war with the London & Birmingham Railway and his flirtation with the Great Western was designed solely to bring that company to its knees. It had the desired result – the alliance of the two companies as the London & North Western Railway with Huish in command. His object achieved, Huish then became a bitter enemy but despite all his efforts he could not prevent the broad gauge from extending north of Oxford. This time the spoils went to the Great Western although, in the case of the Oxford, Worcester & Wolverhampton, it was to prove a hollow victory.

It was while this conflict was raging that the gauge question became a national issue and as a result of a motion by Cobden a Royal Commission was appointed on 9 July 1845 to investigate the matter. One of the three Commissioners was Sir Frederick Smith, late Inspector General of Railways, but the other two seem somewhat oddly chosen. They were George Biddell Airey, 'Astronomical Observator in Our Observatory at Greenwich', and Peter Barlow, 'Professor of Mathematics in Our Military Academy at Woolwich'. But we must remember that it would have been impossible to find a distinguished engineer who was not an interested party. Knight, astronomer and mathematician, set to work with a will, devoting thirty days to the protracted examination of 48 witnesses, a total which included such seemingly irrelevant characters as Her Majesty's Inspector General of Fortifications. Brunel was naturally examined at great length. 'Having seen the working of other railways and of the Great Western since its entire opening,' the Commissioners asked, 'are you at all inclined to think that it was an injudicious arrangement to alter the gauge to seven feet or that less difference would have been better?' To this Brunel's retort was brief and typical: 'I should rather be above than under seven feet now if I had to reconstruct the lines,' he replied. The Commissioners then asked why, in that case, he had not adopted the seven-feet gauge on the lines he had engineered abroad and to this implication of inconsistency Brunel replied as follows:

I did not think that either the quantities or the speeds likely to be demanded for many years to come in that country [Italy] required the same principles to be carried out that I thought was required here, and I thought it very important that they should secure the goodwill of certain other interests which would lead into or out of the railway and as a question of policy as much as of engineering I advised them to adopt the gauge [1.43 metres]. I thought it was wise to conciliate the interest of the Milan & Venice Railway and others which are likely to be connected with us.

Brunel, who, as always, preferred deeds to words and practice to theory, concluded the proceedings by suggesting that the merits of the two gauges so far as locomotive performance was concerned should be put to practical test. Evidently this challenge appealed to the sporting instincts of the Commissioners for they agreed and deputed Brunel and Stephenson's friend G. P. Bidder to decide upon the terms of the contest. In view of the fact that new narrow gauge locomotives of greatly improved design had recently been put into service whereas no locomotives had been built for the broad gauge since 1842, Brunel's offer was certainly a sporting one. He proposed that his challenger should run between Paddington and Exeter but as the narrow gauge party would not agree to such a long distance trial of strength it was eventually decided that the rival courses should be Paddington to Didcot, 53 miles, and Darlington to York, 44 miles. Bidder would not accept his rival's suggestion that the tests should be conducted simultaneously but insisted that the broad gauge should perform first. It was Brunel's intention that the Commissioners should observe the tests from the footplate, but in the event the knight and the mathematician displayed a marked reluctance to avail themselves of this opportunity and waived the honour in favour of the intrepid astronomer.

As broad-gauge champion, Daniel Gooch picked from daily service *Ixion*, one of his numerous seven-foot singles of the 'Firefly' class which had been built in 1841 by Fenton, Murray & Jackson. In mid-December *Ixion* made three round trips between Paddington and Didcot with loads of 80, 70 and 60 tons. She attained a maximum speed of 60 m.p.h. with the 80-ton train, while with the 60-ton load she averaged 50 m.p.h., start to stop, on the down journey and 53.9 m.p.h. on the return. As no rule had been made to preclude it, the feed water in the tender was pre-heated.

In reply, the narrow-gauge party produced two locomotives, a brand new Stephenson long boiler locomotive with 6 ft 6 in. driving wheels and 4–2–0 wheel arrangements which was referred to as 'Engine A', and a 2–2–2 North Midland engine named *Stephenson*. Not only was 'Engine A' given a flying start from York with hot water in the tender, but Bidder had artfully installed a portable boiler at Darlington to provide an artificial blast to liven up the fire before making the return journey, a ruse which infuriated Gooch. Nevertheless, a maximum speed of 53¾ m.p.h. was the most that 'Engine A' could achieve with a load of only 50 tons. She also made one run from York to Darlington with 80 tons at an unrecorded speed, but the return trip was abandoned on the pretext of bad weather. As for the *Stephenson*, her effort came to a most ignominious end after 22 miles, when she ran off the road and turned over. The astronomer was not on board at the time or he might, as the late E. L. Ahrons suggested, have seen some constellations such as he had never observed from Greenwich.

After such a decisive victory, the report of the Gauge Commissioners came as a bitter disappointment to Brunel and his supporters, for although broad-gauge superiority in speed was acknowledged, on every other count the verdict was for the narrow gauge. The report recommended that the adoption of the narrow gauge as the standard gauge of the country should be enforced by law and that existing broad-gauge lines should be narrowed. In coming to this conclusion the Commissioners took into account two damning facts: first, that it is much easier and cheaper to narrow the gauge of a railway than it is to widen it, and secondly that whereas the broad gauge totalled only 274 miles of line in work the narrow gauge could already boast 1,901 route miles.

Another thing which undoubtedly influenced the Commission was the chaos which prevailed at Gloucester due to the break of gauge, and here we come to the greatest mystery surrounding the gauge war and one which seems to have been insufficiently stressed by previous writers on the subject. The break of gauge was an evil which could not be remedied, but there were various ways in which it might have been mitigated. Brunel himself made a number of suggestions on this score, some impracticable but others which were quite feasible, such as the use of transporter wagons or the introduction of containers in which goods could be transferred from one wagon to another. For example, in his sketch books we find, dated 10 July 1845, a drawing of a simple and highly practical form of hydraulic hoist designed for transferring one-ton box containers loading 4 tons 10 cwt each. Yet there is no record that any such device was ever actually built or tried, all the transhipment being carried out laboriously by hand portage. Thus J. D. Payne, the narrow gauge Goods Manager at Gloucester, was able, as he openly confessed afterwards, to present to a visiting Parliamentary Committee a scene of shocking confusion which he had deliberately contrived. This inexplicable failure of the broad-gauge party to make any attempt to mitigate an inconvenience which, more than any other factor, prejudiced the public against them is not to be explained by a refusal on the part of the narrow gauge to cooperate. For later, when the Great Western themselves owned narrow-gauge lines, they still made no improvements at their own transhipment points.

The Commissioners' Report by no means put an end to hostilities. On the contrary the controversy blazed up all the more fiercely. Brunel and Saunders published a reply in which they stated their case so ably and threw such doubts on the reliability of the report that the Government and the Board of Trade seem to have been thoroughly bewildered. At any rate when the Gauge Act of 1846 passed into law its sting had been drawn. Its first section sternly forbade the future construction of any railway in Great Britain to a gauge other than 4 ft 8½ in., or 5 ft 3 in. in Ireland. Yet in the second section there was inserted a clause which exempted from this prohibition: 'Any railway constructed or to be constructed under the provisions of any present or future Act containing any special enactment defining the gauge or gauges of such railway or any part thereof.' So far as the broad-gauge party was concerned this was a saving clause which almost completely nullified the earnest and protracted efforts of the Royal Commission. The fight was still on.

While all these weighty and fruitless deliberations had been going on, the pamphlet war continued unabated. Because we are apt to think that the power of propaganda or 'psychological warfare' is a modern discovery it is interesting to note in this connexion that the London & North Western Railway employed an official, Sir Henry Cole, solely 'to create a public opinion' in favour of the narrow gauge, and that to this end he became an indefatigable pamphleteer. As the sole author of the broad gauge, Brunel was naturally the chief target of these broadsides, whose titles were as long as their texts were short. Perhaps Sir Henry Cole, under the pseudonym of 'Vigil', was the author of a little work entitled: *Inconsistencies of Men of Genius exemplified in the Practice and Precept of I. K. Brunel, Esq., and in the Theoretical*

Opinions of Charles Alexander Saunders. This bore on its title page a quotation from a certain Bishop Hall: 'How worthy are they to swart that marre the harmony of our peace by the discordous jars of their new paradoxicall conceits.' Describing Brunel as 'an eccentric genius', 'Vigil' declared that: 'He has not been able to find a single independent *railway* engineer to back his eccentricities, for we cannot regard Mr D. Gooch of the G. W. as but Sancho to Don Quixote.' Another writer (or was it the industrious Sir Henry again?) under the soubriquet '£ s. d.' produced *The Broad Gauge, the Bane of the G. W. R. Co. An Account of the Present and Prospective Liabilities saddled on the Proprietors by the Promoters of that Peculiar Crotchet.* Here the following quotation is: ' "A barbe de fol, on apprend a vaire" (which being translated for the benefit of Country Gents means) "Mr B. has learnt to shave on the chin of the Great Western Proprietors".' But '£ s. d.' very soon drops this note of ponderous whimsy to thunder: 'The state of the affairs of the G.W.R. is such as imperatively to demand inquiry. The Board Gauge', he continues, 'is I. K. Brunel's and Saunders' child and hobby and Shareholders may well ask "That boy will be the death of us".' Then, working himself into a fine frenzy of righteous indignation he demands: 'Is it fair, is it honourable thus to sport with the confidence reposed in him?' The Company's affairs, he maintains, 'have always been conducted more in the style of a small political state than in that of a great mercantile concern. The political system of managing a railway is the reverse of the mode pursued by the most successful railway leader of the day, Mr Hudson. ...' In view of the imminent and spectacular fall of George Hudson from the throne of his Midland empire this last sally was unfortunate.

It is unlikely that such tirades affected the issue materially, and a formidable body of public opinion, especially in the west of England, remained favourable to the broad gauge. This may seem surprising in view of the inequality of a contest in which Brunel, Saunders, Gooch and Charles Russell were pitted against the combined forces of all the other railway companies in the country. But the Englishman has a traditional respect for the plucky fighter, for minority opinion and the losing cause. Moreover, Brunel continued to act upon the principle that deeds spoke louder than shoals of pamphlets.

With another round of the gauge war imminent in the Parliamentary session of 1846, Brunel determined upon a still more convincing demonstration of broad-gauge superiority, and Gooch was ordered to design and build a new and more powerful express locomotive. She would be the first engine to be built by the Company in their new shops at Swindon, and speed, both in time of construction and in performance was vital. Daniel Gooch himself, his works manager Archibald Sturrock and the men in the shops worked night and day at drawing-board and bench. There was no time to prepare detail drawings, so 'Lightning', as the men on the job nicknamed her, was built from simple centre-line drawings and dimensioned sketches. Rumour spread abroad to the disquiet of the narrow-gauge party that a colossus was building at Swindon and when, only thirteen weeks from the date of the order, the *Great Western* steamed out of the shops she was indeed, by the standards of the day, colossal. Her single driving wheels were 8 ft in diameter and she had a boiler of prodigious size with a huge haystack firebox of hand-wrought iron which towered high above her driver's head. As in all Gooch's locomotives ample boiler power was the simple secret which ensured success, for the boiler was not only large, it carried steam at 100 lb to the square inch, an unusually high pressure at that time.

On 1 June 1846, only a month after leaving the shops, *Great Western* was given her first run with the Exeter express. Her running time for the 194 miles was 208 minutes down and 211 minutes up, representing average speeds of over 55 m.p.h., a fabulous achievement at that time. But the new broad-gauge champion capped this on the 13th of the month by hauling a train of 100 tons from Paddington to Swindon, start to stop, in 78 minutes, an average of 59 m.p.h. 'Mr Russell called this "a great fact", and it *was* a great fact,' wrote Gooch proudly in his diary.

By 1847, when the battle over the broad-gauge route to Birmingham was raging, six more magnificent 8 ft singles of improved design had rolled out of Swindon shops. They were the *Iron Duke*,[2] *Great Britain, Lightning, Emperor, Pasha* and *Sultan*. With such power at command the overall timing of the fastest booked trains to Exeter was reduced to 4 hours 25 minutes, a time which included a new 8 minute stop at Bridgwater and, of course, the agreed 10 minute stop at Swindon which the Company had unsuccessfully attempted to evade. On this new schedule 55 minutes were allowed for the $52\frac{7}{8}$ miles from Paddington to Didcot, while Bristol was reached in exactly $2\frac{1}{2}$ hours. There was a long-standing legend that *Great Britain* once reached Didcot from Paddington in only $47\frac{1}{2}$ minutes, but whether this story is true or not, such locomotive performances, achieved only 17 years after the birth of railways, were truly astounding and more than vindicated all that Brunel had ever claimed for his broad gauge. Admittedly the trains were featherweight by modern standards, but their roomy six-wheeled coaches carried second- as well as first-class passengers, whereas the best trains on the narrow gauge, though far slower, were even lighter and offered only limited first-class accommodation. No wonder the West Country rallied to Brunel's banner; no wonder Sir Henry Cole had to busy himself about his pamphlets.

Yet despite feats which would not be surpassed for many years to come, despite the victory of a broad-gauge route to Birmingham and the war-cry of 'the broad gauge to the

2. This locomotive gave its name to the class, being the first to appear with an improved type of boiler which dispensed with the towering firebox.

Mersey', it is clear to us when we survey that bygone battlefield that the ultimate issue was never in doubt, and one cannot but suspect that even Brunel's most loyal supporters realized in their heart of hearts that they were championing a lost cause. Happily, Brunel did not live to see that decline and eclipse, nor did he witness that advance of the mixed gauge over the system which brought the narrow-gauge metals into the very citadel of Paddington. The agent responsible for this impudent invasion was that cuckoo in the Great Western nest, the Oxford, Worcester & Wolverhampton Railway.

After being so stubbornly and expensively fought for, the O. W. & W. R. proved a constant source of anxiety to Brunel and a thorn in the flesh to the Great Western directors. The cost of the works far exceeded Brunel's estimate, money ran out, the Chairman, a Stourbridge banker, went bankrupt and there was a bitter wrangle over the extent of the G.W.R.'s financial guarantee. There was also the extraordinary affair of Mickleton Tunnel in which Brunel figured prominently.

There was trouble and difficulty from the outset in driving this tunnel, which was to carry the line under the escarpment of the north Cotswolds near Chipping Campden, and in June 1852 the contractor, Marchant, had a dispute with the Company and stopped work. Determined to resist the Company's intention to take over the works, he posted guards and a kind of guerrilla warfare went on until the end of July when, resolved to put a stop to such nonsense, Brunel himself appeared on the scene with his resident assistant, Robert Varden and a large body of men. Marchant, however, had been forewarned and Brunel found himself confronting, not only the contractor's private army, but the local magistrates backed by a posse of police armed with cutlasses. These apprehensive guardians of law and order hastily gabbled the Riot Act twice over, whereupon the contending forces withdrew. But if Marchant supposed that by this manoeuvre he had outwitted Brunel he had sadly underestimated his opponent. Unknown to him, throughout the week-end an army of navvies was recruited, not only from other points on the same line, but from the works of the Birmingham & Oxford at Warwick and elsewhere, and on Sunday night they began to march on Mickleton. With Brunel as Commander-in-Chief and his assistant engineers as colonels of regiments it was a highly organized military operation perfectly timed and executed. A regiment of 200 from Evesham reached the western mouth of the tunnel at 3 a.m. on the Monday morning, taking Marchant's men com-pletely by surprise. They fought back fiercely, but fresh regiments of Brunel's army kept arriving from different points of the compass until they found themselves sur-rounded by a force 2,000 strong. Seeing that further resist-ance was hopeless, Marchant called for a truce and the two generals retired from the stricken field (where there were many broken heads and limbs but happily no fatal casu-alties) to consider terms of truce. By the time the police, sup-ported by a detachment of troops from Coventry, appeared on the scene their services were not required. Brunel was in complete control of the works and order was already restored. So much for the fantastic 'Battle of Mickleton', an episode that was surely without parallel in England since the days of the private armies of medieval Lords Marcher.

It was not long after this[3] that Brunel resigned his post as engineer to the Oxford, Worcester & Wolverhampton, the reason being that the Company had gone over to the enemy camp. The Company's Act had stipulated that the mixed gauge should be laid on the section north of the junction with the Bristol & Birmingham Railway, but the Great Western party was shocked to discover that the Company had not only neglected to lay the broad-gauge rails on this section but were also actually laying narrow-gauge lines on

3. According to MacDermot, Brunel had resigned in the previous March, but if this is correct, Mickleton Tunnel would no longer have been his concern when the trouble with Marchant broke out.

the southern section through Evesham, which was to have been broad gauge only. As a result of strong pressure backed by legal action the O. W. & W. R. were eventually compelled to lay the broad-gauge metals, but that in doing so they acted in the letter rather than the spirit of the law by providing no broad-gauge crossing or siding accommodation we may gather from this letter from Brunel to Saunders regarding the Board of Trade inspection:

It cannot surely be consistent with the *public safety* and a line cannot be said to be fit to open with safety to the public that involves the necessity of the engine running backwards for *40 miles*. Is not this a simple way of putting the O.W.W. to business, because if Capt. Galton (the inspector) tries it with an engine & carriages he will so find it – for broad gauge trains.

Alas, neither the Great Western nor the Board of Trade succeeded in putting the O.W. & W.R. to business and it was as a narrow-gauge line that the railway was finally absorbed by the Great Western. It brought the parent company a great new territory in the West Midlands and South Wales but it also brought the narrow-gauge metals to Paddington.

Had it not been for the mixed gauge the broad gauge would not have survived so long as it did, but it was a costly compromise. It involved laying one narrow-gauge rail inside the broad gauge, the outer broad-gauge rail becoming common to both. On straight track this sounds simple enough, but at junctions and stations it was quite another story. In the plans room at Paddington there are preserved some of Brunel's original drawings of station layouts which reveal the complexity which the mixed gauge involved, a complexity which makes it surprising that the G.W.R. preserved such a good accident record. The installation of the mixed gauge was made all the more complicated owing to the frequent use which was made of wagon turntables instead of siding switches at this period. A wagon had to stand centrally on these turntables or it would throw them out of balance. Therefore at the approach to each turntable the three-rail layout had to give place to gauntleted track, that is to say to two independent narrow-gauge metals laid within the broad gauge.

Looking at such complexities today one can but admire and marvel at the lengths to which the magic of one man's personality could carry others with him in the single-minded pursuit of an ideal. It may have been commercial folly, but then like oil and water, ideals and commerce do not mix. Brunel spent his life trying to mix them and succeeded better than most men.

It is commonly supposed that once having conceived the idea of his seven-foot gauge Brunel became so stubbornly wedded to it that he would admit no other. That this was not so has already been shown in his reply to the Gauge Commissioners concerning his Italian railways. It is revealed even more clearly in the report which he made to the directors of the Taff Vale Railway in April 1839. He had previously advised them that on this sharply curved and graded line between Merthyr and Cardiff with its mineral line connexions with their inclined planes, the seven-foot gauge was not suitable. The directors then suggested a gauge of 5 ft. Brunel replied to this that he 'most decidedly' recommended the standard gauge of 4 ft 8½ in. The small increase to 5 ft was not justified, he wrote, while it would be easier to acquire engines and carriages, the standardization of parts being a great advantage. In this context he quoted the Bury inside frame locomotive and recommended it for their purpose. He went on to stress the importance of being able to connect with the probable trunk line along the coast and wrote: 'This main line will not be a 5 ft gauge; it will either be a 4 ft 8½ in. or a 7 ft – the latter you cannot have and therefore the former offers the only chance.' He recommended it in view of the curves, the inclined planes and because great economy and not speed was the object.

Had this report fallen into the hands of 'Vigil' he would have made good use of it in his pamphlet to support his charge of inconsistency. Indeed it might have caused even the Great Western directors some surprise. Yet from an engineering point of view there was nothing inconsistent about Brunel's approach to the gauge question, however short-sighted it may have been commercially. Had he possessed a commercial mind he would have recognized the prime importance of a unified railway system throughout the country and subordinated engineering considerations to that end. As it was he considered every project with which he was associated in isolation as an engineering perfectionist. Thus the Taff Vale, a small mineral line in the hills of South Wales about which, as he himself admitted, he cared little, was cut out, in his view, to be a standard gauge line. His Great Western, on the other hand, was something entirely different. It was the pride of his heart, a great trunk route which he had planned from the outset with high speeds in mind, and therefore he chose the gauge best calculated to achieve that object. In this, from the point of view of the Great Western shareholders, Brunel was wrong and his choice of the broad gauge was a costly mistake. He underestimated the speed with which the railway system would grow and how rapidly, with the establishment of the Railway Clearing House, through traffic between company and company would develop. In the fierce struggle for new territories between rival companies the broad gauge was checked in the south, in the West Midlands and in the Black Country. The Great Western directors were therefore faced with the alternatives of stagnation or further expansion upon the narrow gauge. They chose the latter, and as soon as the Company had acquired a considerable mileage of narrow-gauge route the ultimate extinction of the broad gauge became a foregone conclusion. Judged solely as an engineering achievement, however, the broad gauge was magnificent. Moreover the incomparable results achieved by Isambard Brunel and

Daniel Gooch acted as a tremendous stimulus to technical development on the narrow gauge. Without the broad gauge and the gauge war to which it led, locomotive design and performance would never have improved so rapidly and Britain's railway system might never have become, as it so soon did, a model for the world to copy.

Failure and Triumph
in the West

WHEN Brunel was appointed engineer to the South Devon Railway it presented him with new problems which he approached in his habitually original way. He had first surveyed a route from Exeter to Plymouth in 1836 when he proposed to bridge the estuary of the Teign, pass near Torquay and from thence carry the line across the Dart and through the South Hams country. The levels were favourable but the works would have been extremely heavy. So on economic grounds the plan was abandoned in favour of the route which the West of England main line now takes, followering the east bank of the Teign to Newton Abbot, crossing the Dart at Totnes and then skirting the southern fringes of Dartmoor through South Brent, Ivybridge and Plympton. It was in this form that the South Devon Railway was authorized in July 1844 with capital subscribed jointly by the Great Western, Bristol & Exeter and Bristol & Gloucester companies. The South Devon would be an extension of the broad-gauge main line to the west and therefore high speed running was once again the aim. But here conditions were very different. The economy of the new survey had been achieved only at the price of very severe gradients at Dainton, Rattery and Hemerdon on the section west of Newton Abbot, and Brunel doubted the capacity of the steam locomotives of the day to work such a railway either swiftly or economically. Hence the possibility occurred to him of using some other form of motive power. How logical such reasoning was has been demonstrated only recently by the advantages derived from the electrification of the heavily graded Woodhead route over the Pennines, but unfortunately Brunel's imagination was too far ahead of the means available. Electricity was not then recognized as a motive power so he decided to give the so-called 'atmospheric' system a trial.

The idea of propelling trains by compressed air, or by the pressure of the atmosphere acting upon a piston travelling in an exhausted tube, was no new one. An optimistic enthusiast named George Medhurst had started the hare in 1810. His patent proposed that the entire train should constitute a piston which would be forced by compressed air through a tube 30 ft in diameter. What fun this would have been for the passengers! In the wild simplicity of this first brain-wave Medhurst avoided the difficulty which confronted him and other inventors when they pondered the more practicable idea of attaching a carriage to a piston travelling in a continuous tube laid between the rails. The attachment of the piston to the carriage was the problem, for this must involve a continuous slot in the top of the tube. How, then, to keep this slot air-tight? Medhurst adopted Vallance's idea of using atmospheric pressure instead of compressed air, but the problem remained, although the former suggested an impracticable form of water seal. Then in 1834 another inventor named Pinkus came along with what he called his 'Valvular Cord' system in which a cord was laid in a groove over the slot, a device on the travelling carriage picking up the cord in advance of the piston and relaying it behind. Pinkus was the first to give a practical demonstration of his invention on a site near the Kensington Canal in 1835. Evidently the results were not encouraging, for we hear no more of Mr Pinkus.

The next development in atmospherics took place five years later not very far from the scene of the previous abortive experiment. That strange little railway with the grandiose title, the Birmingham, Bristol & Thames Junction, later more modestly known as the West London, agreed to lay an experimental length of 'Mr Clegg's Pneumatic Railway'. As the Company was in financial difficulties at the time, its own rails had not been laid and the vacant road-bed provided a most convenient playground for aspiring inventors. Before this 'Pneumatic Railway' was actually laid, Clegg's idea attracted the attention of the brothers Jacob and Joseph Samuda, who suggested various practical improvements which seemed so promising that a patent was taken out in their joint names in 1839.

The hopeful inventors had little capital to spare for experimental work, and the first atmospheric railway of 1840 consisted of 1¼ miles of badly laid contractor's rail which they had bought second-hand from the Liverpool & Manchester Railway. Midway between these rails there ran a pipe 9 in. in diameter which could be exhausted ahead of the travelling carriage by a small steam engine. Along the slotted top of this pipe lay the key feature of the invention – a continuous hinged flap valve of leather backed with iron which could be opened and re-sealed to allow the passage of the piston arm by means of rollers mounted on the carriage. With this first crude layout the inventors succeeded in hauling 5 tons at 30 m.p.h. up a gradient of 1 in 120 and achieved 22 m.p.h. with 11 tons. At this the railway world sat up and took notice. The most eminent engineers of the day including Brunel visited the little test track and opinions became sharply divided. George Stephenson dismissed the idea as 'a great humbug' and both his son Robert and Joseph Locke followed his lead. Brunel, Cubitt and Vignoles, on the other hand, believed that there might be a future in the atmospheric and that it deserved a wider trial.

Among the many interested parties who saw the experimental line and were suitably impressed were the directors and officers of the Dublin & Kingstown Railway. Their treasurer, James Pim, in particular, became a rabid enthusiast and it was largely due to his advocacy that the Company decided to give the atmospheric system a full-scale trial on a short branch line between Kingstown and Dalkey. The line, which was laid in cutting beside the road-bed of a pre-existing horse tramway (thus saving the expense of seeking fresh Parliamentary powers), was built by the Company's contractor, while the Samuda brothers supplied and installed the atmospheric equipment including the pumping engine. They evidently decided to make sure of things, for the latter was so large that it was never called upon to work at more than half power. It was a Fairbairn crank-overhead engine of 100 h.p. with a 36 ft flywheel. Fed by three Cornish boilers, it was installed at the Dalkey end of the line. The railway climbed from Kingstown at an average gradient of 1 in 128, some sections being much steeper, and it was sharply curved. The pipe, which was 15 in. in diameter, was 2,490 yards long and stopped 100 yards short of the Dalkey terminus, the idea being that the trains would cover this by their own momentum.

The method of working was as follows. Five minutes before the train was due to depart from Kingstown the engineman at Dalkey began pumping to exhaust the pipe. The train was then pushed forward until the piston on the motive carriage entered the pipe through an ingeniously designed treadle-operated entrance valve and was there held on the brakes. When the brakes were released it was drawn swiftly and smoothly up the incline. The return journey was made by gravity.

As soon as the line was opened on 19 August 1843 crowds flocked from Dublin to witness the new mechanical marvel and to enjoy the experience of being whirled along without the smoke, the smuts or any of the visible and audible evidence of effort associated with steam locomotion. Thirty miles an hour was the average speed of the regular trains up the incline, but enthusiastic experimentalists achieved some much more spectacular performances. The record ascent was made accidentally by an engineering student named Frank Ebrington, the son of a Regius Professor of Dublin University and one of Ireland's unsung heroes. Ebrington was about to make a test run one day when the motive carriage on which he was seated was suddenly sucked away without its train, someone having forgotten to couple it up. For the only time in its brief and chequered history the atmospheric achieved gale-force speed as the big pump drew the terrified Ebrington up to Dalkey in the astounding time of 1¼ minutes, representing an average speed of 84 m.p.h. This story was related to the Committee which was formed to investigate the atmospheric railway by the Rev. T. R. Robinson, who confidently predicted speeds of 100 m.p.h. As the curves on the line were laid out with considerable super-elevation, while the piston of the motive carriage would hold it down and resist the force of 'G', it is not incredible. Even allowing for some exaggeration it seems certain that Ebrington could safely have claimed the title of 'fastest man on earth'. At any rate he was given a most convincing demonstration of one of the inherent faults of the system. This was that those in charge of the train were entirely at the mercy of the enginemen in a pumping station anything up to three miles away, ineffective brakes being their only form of self-defence. One can only compare their situation to that of the driver of a car who finds that the throttle has stuck full open, that there is no means of switching off the ignition or disengaging the drive and that his brakes have faded. Owing to this almost total lack of control the run in to the Dalkey terminus with trains of varying weight was a hazardous proceeding where no amount of judgement on the brakesman's part could guarantee success. The piston of the motive carriage having burst out of the tube like a cork out of a champagne bottle – so violently that a leather sack stuffed with straw had to be laid between the rails to cushion the exit valve – the progress of the train for the last 100 yards was in the lap of the gods. Sometimes it stopped short and had to be pushed ignominiously into the station with the help of the passengers; sometimes its momentum was such that it overshot the platform and ran off the rails at the end, terminal buffer stops being very wisely not provided.

These drawbacks apart, the little railway worked well for a time and to the ingenious inventors the future must have appeared very rosy. Railway experts, not only from England but from the Continent, made elaborate experiments and recorded performances, while the promoters of many new railway schemes considered abandoning locomotive traction. Monsieur C. F. Mallet of the French Public Works Department was so impressed by his visit that he recommended the widespread introduction of the atmospheric on French railways. William Cubitt saw and was conquered with the result that the atmospheric was installed on the London & Croydon Railway. After he had paid two visits to Kingstown, Brunel recommended the adoption of atmospheric traction, not only on the South Devon Railway, but on the Newcastle & Berwick in the heart of the Stephenson country. A Board of Trade inquiry conducted by Sir Frederick Smith and Professor Barlow reported favourably on the possibilities of the system. Only the Stephensons remained sceptical.

Robert Stephenson was instructed to report upon the line by the directors of the Chester & Holyhead Railway, of which he was engineer. His conclusions were decidedly unfavourable. He doubted the great economy over locomotive working which the Samuda brothers claimed and he also refused to accept the contention that as much traffic could be worked with perfect safety over a single line of atmospheric railway as over a double line of steam railway. If, as he believed, this was untrue, then the cost of installing the atmospheric equipment to work a double line would be prohibitive. Two of his assistants, George Berkley and William Marshall, had made comparative experiments at Dalkey and on the cable-operated Camden incline out of Euston with results slightly in favour of the former as regards mechanical efficiency but, Stephenson went on, 'The wear and tear of the longitudinal valve and the degree of attention which it will constantly require are points upon which we yet have no information. At Kingstown about two men per mile are appropriated to the application of composition for the purpose of maintaining the tightness of the valve.' By contrast, Brunel, in his report to the South Devon Board, expressed the opinion 'that the mere mechanical difficulties can be overcome' and that, 'as a mechanical contrivance, the Atmospheric apparatus has succeeded perfectly as an effective means of working trains by stationary power, whether on long or short lines, at higher velocity and with less chance of interruption than is now effected by locomotives.' Yet even if all that Brunel said was true, Robert Stephenson's chief objection still remained. This was the fallibility of the system. Supposing, he said, the atmospheric was to be installed on the London & Birmingham Railway, it would involve a chain of no less than 38 pumping stations. A mechanical failure in any one of these stations would stop traffic completely, for not a wheel could turn on the affected section until the pump could resume work. To this objection no advocate of the atmospheric appears to have suggested any answer.

When Brunel visited Newcastle in connexion with the Newcastle & Berwick atmospheric scheme he met George Stephenson, who playfully seized him by the collar and asked him how he dared venture north of the Tyne. Certainly if we require evidence to show the nation-wide reputation which Brunel had acquired it is that this north country scheme of his was only defeated after a most protracted struggle. In the west of England his sway was undisputed and the installation of the atmospheric system on the South Devon Railway proceeded apace.

On 11 April 1846 the *Railway Chronicle* gave its expectant readers the following progress report:

The first portion from Exeter to Teignmouth is to be opened on the 1st of May. The rails have been laid to Teignmouth, also between seven and eight miles of atmospheric pipes laid and

caulked. Sixteen miles of pipe have been delivered from Bridgwater and Bristol foundries. Material for the valve is at the
station but none has been fixed yet. The pipe begins on the
bridge a short way southwards of the first engine house. It then
runs to the next engine house at Countess Weir, where there is a
blank for a short distance to allow of the turn off. Beyond this
the pipe is caulked to Turf Station. Masons and engineers are
proceeding with the engine houses, which are substantial looking
buildings. The House at the Exeter end is nearly completed, the
boilers are all enclosed and buildings erected. The cylinders of
the forty horse engine are set and the twelve horse engine are
commenced. The Engine House is not so far advanced as that
at Exeter. The engine boilers are covered in and finished and
the engine itself is on the ground and in course of erection. The
engine for the station at Turf is also on the ground and the
boilers are in. There are eight engines between Exeter and Newton inclusive . . .[1]

From this account we may picture very clearly the
intensive activity which was going on along the South
Devon coast and imagine the stir which it must have created
in the neighbourhood. In his recommendation to the South
Devon directors, Brunel had argued that the cost of constructing and installing the atmospheric apparatus would be
more than offset by the saving in other directions. For all its
faults the atmospheric carriage was the lightest form of motive power that has ever moved on rails because it carried no
power unit and adhesive weight was of no account. Therefore the weight of rail and the size of the longitudinal timbers could be substantially reduced. Again, owing to the
anticipated hill-climbing powers of the atmospheric, gradients could be stiffened at a great saving in the cost of earthworks. The original survey of the line west of Newton Abbot
was in fact modified in this way with the consequence that

1. These pumping stations, which were placed at intervals of about
three miles, were: Exeter, Countess Weir, Turf, Starcross, Dawlish,
Teignmouth, Summer House (Bishop's Teignton) and Newton Abbot.

its formidable inclines have severely taxed Great Western motive power resources from that day to this. Then there was the saving represented by laying single instead of double line and finally the saving in the cost of steam locomotives. Under this last head, Brunel argued that the average life of a steam locomotive was only ten years, an example of special pleading which must have caused Daniel Gooch to raise his eyebrows.

In the event it was Gooch's maligned locomotives which worked the first trains to Teignmouth in May and to Newton Abbot in December 1846, for the installation of the atmospheric apparatus proved a far more costly and protracted business than Brunel had ever anticipated. In the first place, as any iron founder will appreciate, the 15 in. pipes with their longitudinal slot were found difficult to cast with sufficient accuracy. This problem was eventually solved by T. R. Guppy, who evolved special equipment and tools for the job which made possible a very creditable delivery rate of a mile a week. Each 10 ft section of pipe (as against 9 ft on the Kingstown & Dalkey) had to be carefully fitted to its fellow, the joint being caulked with yarn steeped in putty and linseed oil. The continuous valve of best oxhide leather riveted to its iron reinforcement was another costly item which demanded skilled and laborious fitting. Then there were the numerous valves: exit and entrance valves at the stations, and, at each pumping engine house, separating valves to isolate one section from the next. At each passenger station a short branch pipe was installed. This was fitted with a pilot piston to which the train could be hitched and thus drawn forward until the travelling piston entered the main tube. Yet another gadget had to be provided at each level crossing. Here a hinged iron plate connected to a piston in a cylinder covered the atmospheric pipe and so allowed carts to cross the line. When the pipe was exhausted for the passage of a train the piston was sucked down, thus raising the iron plate like a drawbridge. This ingenious device threatened the apprehensive waggoner like a kind of inverted sword of Damocles. What was to prevent it springing up when a cart was in the act of crossing? someone had asked at the atmospheric inquiry. It was an awkward question to which there was no satisfactory reply.

Despite all these complications the greatest source of trouble and delay at the outset were the pumping engines. Brunel had anticipated difficulties, but not in this quarter, and we may share his surprise. The engines were the one orthodox engineering feature of the whole system and they were supplied by Boulton & Watt, Maudslay, Son & Field and the Rennie Brothers, three firms of the highest reputation who had been building stationary steam engines of proved reliability for many years. What the troubles were is not related, but the suspicion that they were not unconnected with the specifications which Brunel drew up for these builders is hard to avoid. Once again he may have asked too much from the manufacturing resources of the day. Certainly his stipulated working pressure of 40 lb per sq. in. was high for large engines of this type and date.

An experimental train worked from Exeter to Turf and back in February 1847, but owing to this trouble with the engines it was not until the following September that a regular service of four atmospheric trains a day each way began to operate between Exeter and Teignmouth. In January 1848 this service was extended to Newton Abbot. For a time they ran with reasonable regularity. In his *Life* of his father Isambard III writes:

The Atmospheric System was vaguely credited with every delay which a train had experienced in any part of its journey; though, in point of fact, a large proportion of these delays was really chargeable to that part of the journey which was performed with locomotives. It often happened that time thus lost was made up on the Atmospheric part of the line, as is shown by a record of the working which is still extant. In the week 20–25 September 1847, it appears that the Atmospheric trains are chargeable with a delay of 28 minutes in all; while delays due to the late arrival of the locomotive trains, amounting in all to 62 minutes, were made up by the extra speed attainable on the Atmospheric part of the line.

The highest speed recorded with these trains was an average of 64 m.p.h. over four miles (maximum 68 m.p.h.) with a load of 28 tons, while 35 m.p.h. was averaged over the same section with 100 tons. This was with a little over 16 in. of vacuum in the pipe, this being considered the desirable working figure.

In February 1848 Brunel submitted a report on the Atmospheric to the South Devon directors in which he wrote:

Notwithstanding numerous difficulties, I think we are in a fair way of shortly overcoming the mechanical defects, and bringing the whole apparatus into regular and efficient practical working, and as soon as we can obtain good and efficient telegraphic communication between the Engine Houses and thus ensure proper regularity in the working of the Engines, we shall be enabled to test the economy of working. At present this is impossible, owing to the want of the telegraph compelling us to keep the Engines almost constantly at work, for which the boiler power is insufficient, and the consequence is that we are not only working the Engines nearly double the time that is required, but the boilers being insufficient for such a supply of steam, the fires are obliged to be forced, and the consumption of fuel is irregular and excessive. There is every prospect of this evil being speedily removed, and as the working of the Atmospheric will then become the subject of actual experiment, and its value be practically tested, I shall refrain from offering at present any further observations upon it.

The Engine House at the Summit at Dainton, between Newton and Totnes, is completed; preparations are making for erecting the Engines, and the pipes are being laid upon the line.

For obvious reasons Brunel's reports to directors seldom

erred on the pessimistic side so that, reading between the lines, we can gather from this that things were very far from well. The atmospheric services were, indeed, only being maintained at this time by an unremitting struggle against a sea of troubles that would have daunted any lesser man. There were accidents in starting the trains with the auxiliary piston device; the numerous valves in the pipe frequently damaged or destroyed the cup leathers of the travelling pistons; the accumulation of water in the pipe due to condensation caused trouble; in the pumping stations there were more breakdowns, nor were these breakdowns and the overloads and uneconomical working methods which caused them due solely, as Brunel implied, to the lack of a telegraph. To a far greater extent they were the result of the inefficiency of the continuous valve. This was the most fatal defect of all. If the valve became unseated by only one-thousandth of an inch on one mile of pipe the opening was equivalent to that of a pipe 15 in. in diameter. It will therefore be obvious that the working efficiency of the system depended utterly on this valve maintaining a perfectly air-tight seal.

The valve seating consisted of a shallow trough filled with a special sealing compound. On the Croydon line the substance used was a mixture of beeswax and tallow. A copper blade on the travelling carriage, heated by a 5 ft tray of hot charcoal, passed over the seal with the object of slightly melting the surface in cold weather. When the carriage was travelling at anything more than a snail's pace the effect of this must have been precisely nil, while in the hot summer of 1846 the composition melted with most unfortunate results. Consequently a lime soap composition was first of all tried on the South Devon, but as this was found to form a hard skin on exposure to light and air it was replaced by a more viscous compound of soap and cod-oil. This, however, was sucked into the pipe by the rush of air when the valve was opened and had to be continually renewed. The effect of these sealing troubles was that the exhaustion of three miles of pipe by one engine took far longer than the three to five minutes which Brunel had calculated, while the horse-power expended in pumping was no less than three times the figure estimated. Such working was, of course, utterly uneconomical.

With the coming of winter yet another trouble manifested itself – the leather of the valve froze as hard as a board. The reason for this was that the action of the vacuum had drawn the natural oils out of the leather so that it had become pervious to water. Although the system struggled on for a few months more it must have been evident to Brunel that this was the beginning of an end which could not be far off. It came in June 1848 when the leather of the valve was found to be disintegrating throughout its entire length from Exeter to Newton Abbot. Apart from the effect of the vacuum, the chemical action of iron oxide and tannin had rotted the leather so that it began to tear away from its rivets at the hinge. For this there was no remedy except the complete renewal of the valve at an estimated cost of £25,000.

It was at this moment of most bitter failure to which all his unwearying efforts had led that Brunel displayed more fully than on any previous occasion those qualities of high courage and unfaltering decision which so distinguished him. It would have been fatally easy for him to have continued the atmospheric experiment in an attempt to save his face for, strange though it may seem, many people including the Chairman of the South Devon Company still believed passionately in the ultimate success of the system. Enthusiastic inventors and engineers were still busying themselves about it. No less than seventeen patents had been taken out for improvements on the continuous valve alone; Robert Mallet, the Irish engineer, suggested a number of developments including a method whereby the driver of the train could control the power exerted on the piston; Samuda designed a second 'weather valve' to protect the continuous valve; Brunel had himself proposed an hydraulic accumulator at each pumping station which would enable the engines to work more economically and the pipe to be exhausted very rapidly. But to him all these ingenious complications now only revealed the more clearly the practical defects of a system which in theory had seemed so attractive and to which he had pinned so much faith. Squarely he faced the fact that he had been responsible for the most costly failure in the history of engineering at that time. There could be no shirking the issue; the slate must be wiped clean. He therefore recommended to his directors that the atmospheric experiment should be pursued no further. True, he added the proviso that Samuda should be given the opportunity to renew the valve at his expense subject to a guarantee to maintain it in good order for an agreed period, but it must have been obvious to all that Samuda would never do this and that Brunel's recommendation was an irrevocable sentence of death. Already the two pumping stations at Dainton and Totnes were nearing completion, while the 22 in. pipes, which Brunel had intended to use on the Dainton inclines in conjunction with an expanding piston of his own design, had been laid from Newton Abbot as far as the Dart bridge. But now they were torn up along with the rest, leaving the field to the all-conquering steam locomotive.

That the sale of the atmospheric plant realized no less than £42,666 gives us a measure of the cost of this ill-fated experiment. The motive carriages were converted into brake vans. The pumping engines were sold, one of them to a lead mine near Ashburton, where it became known as the 'Brunel Engine' and worked for many years. The miles of atmospheric pipe were sold for scrap, most of them finding their way into the furnaces of Ebbw Vale. In 1912, however, a section of the 22 in. pipe was found at the Convent of the Sacred Hart, Goodrington House, Paignton, where it was doing duty as a surface drain. It was acquired by the Great Western Railway, who presented a length to the Science Museum. It is the only surviving relic if we except the Italianate engine houses with their tall campanile chimneys which survived as a forlorn monument to what Devonians called the 'Atmospheric Caper'.

Other atmospheric ventures were no more fortunate. The Croydon line soon followed the South Devon and the Kingstown & Dalkey, author of so much mischief, closed down in 1855. Last to go, owing to the optimistic efforts of Monsieur Mallet, was the five and a half-mile line from Nanterre near Paris to St Germain. With a 25-in. pipe and a maximum gradient of 1 in 28½ it struggled on until 1860.

One of the mysteries of the atmospheric experiment is this. An insuperable disadvantage of the system which at once springs to our minds is that the pipe completely precluded points and crossings. This was one of the reasons why the pipe was discontinued at each station. We might suppose that such an inherent drawback would have damned the atmospheric from the start for it was one which no amount of ingenuity could have overcome. Yet in all the contemporary writings on the subject it is hardly mentioned and even Robert Stephenson in his report does not stress it, although it necessitated the costly construction on the Croydon line of the world's first 'fly-over' crossing near Norwood. A partial explanation is that there was in any event very little locomotive movement within station limits at this early period. The function of the locomotive, as conceived in those days, was to draw a train from A to B and not to marshal it. Over the simple station layouts with their wagon turntables and traversing platforms the light rolling stock was manoeuvred by men and horses and in this way trains were made ready for locomotives to draw away. So, because the era of great marshalling yards and complex trackwork was still to come, this disadvantage of the atmospheric was not so apparent. It would still arise, however, at crossing stations on a single line and it is difficult to understand how the advocates of the system could have hoped to achieve that frequent single-line service which they so confidently predicted. Brunel referred to a 'simple mechanical contrivance to overcome the crossing difficulty' which he claimed to have evolved, but no trace of this is discoverable in his notes or sketch books. Perhaps he was referring merely to the crude expedient of the auxiliary piston.

The atmospheric railway which once raised such high hopes and wasted so much money is now looked upon by most people as a fantastic mechanical joke. We applaud the shrewdness of the Stephensons and marvel that an engineer so gifted as Isambard Brunel could ever have been so deluded. But it is notoriously easy to be wise after the event. Conservative engineers like the Stephensons have often decried inventions which perseverance has eventually brought to success. Before dismissing the atmospheric as an aberration we should forget for a moment the technical achievements of the past hundred years and try to see the invention through the eyes of our ancestors. If our experience of land travel had been limited to the stage coach or the first steam locomotives we, too, might have hailed the atmospheric railway as the transport system of the future. To be borne along by the powers of the air so smoothly, swiftly and silently, to accelerate so rapidly and with such a complete absence of any apparent effort must have been a breathtaking experience. It promised, moreover, entirely new standards of safety in high-speed travel because it imposed, by the very nature of its working, that 'absolute block' system which became recognized in after years as the only safe method of railway operation. Its performance was, indeed, so remarkable in its day that the solution of the technical problems which had to be overcome before it could become reliable and economical presented a challenge to the engineers of the day which few could resist. And of all the engineers in the country Brunel was by temperament and character the most likely to accept that challenge with both hands. The writer who included the atmospheric railway in a recent work on follies thus betrayed a lack of historical imagina-

1. Mary Brunel, the wife of I. K. Brunel. From the portrait by her brother, John Horsley, R.A.

2. I. K. Brunel at the time of the G.W.R. Survey. Engraved from a portrait by John Horsley, R.A.

3, 4 (*opposite*). Pageantry and Tragedy in the Thames Tunnel. In the oil painting by an unknown artist Sir Marc Brunel is being welcomed to the banquet by his son. The other picture is one of a series of water-colours by Goodall and depicts the recovery of a victim of the disastrous inundation two months later.

5, 6. The Telford and Brunel designs for Clifton Bridge. The former (top) is reproduced from the contemporary print; the latter is from the painting in watercolour by I. K. Brunel.

7 (*top left*). Brunel's office at 18 Duke Street, a minutely detailed water-colour by an unknown artist.

8 (*bottom left*). 'Making a Cutting, G.W.R.' (Sonning), from the water-colour by George Childs.

9, 10. Great Western Railway: Maidenhead Bridge and Box Tunnel, west portal: two of the famous series of lithographs by J. C. Bourne, 1846.

11, 12. The South Devon Atmospheric. These two pictures from a contemporary travellers' guide show the laying of the atmospheric pipe and one of the pumping stations.

13. Tregagle: a typical Brunel timber viaduct strides across a characteristic landscape.

14 (*top*). The *Royal Albert Bridge*, Saltash. The print betrays some artistic licence, but, more eloquently than any photograph, it conveys the romance and grandeur of Brunel's final railway masterpiece.

15. The swan song of the wide gauge. A photograph taken by the Rev. A. H. Malan in 1891 shows the *Rover*, one of Daniel Gooch's famous 8-ft singles of the 'Iron Duke' class in its final form.

16, 17 (*opposite*). Terminus New York: The P.S. *Great Western* clears the mouth of the Avon on her first epic voyage. Below, the s.s. *Great Britain* on her maiden voyage – the first passage of the North Atlantic by a screw steamer.

18. The s.s. *Great Britain* stranded at Dundrum Bay. This contemporary watercolour by an unknown artist shows the protective 'mattress' nearing completion, winter 1846-7.

19. Sketch of the proposed 'Great Ship' dated May 1854, from the Brunel Sketch Books. Sail planning appears to be the purpose of this sketch.

20 (*top right*). The *Great Eastern*: the bows of the hull before construction of the launching ways, from a special 'Leviathan Number' of the *Illustrated Times*.

21 (*bottom right*). The *Great Eastern*: the Grand Saloon. The large mirrors enclose the funnels and feedwater heaters. Picture from the *Illustrated London News*.

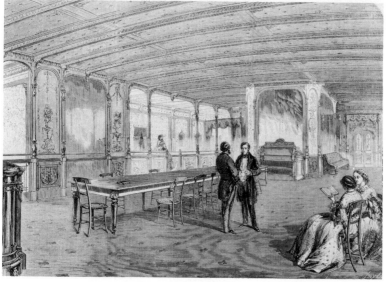

22 (*opposite*). I. K. Brunel, a photograph taken at Millwall shortly before the first attempt to launch the *Great Eastern*.

23. John Scott Russell.

24. I. K. Brunel: engraved from a photograph taken in the last year of his life.

25. A sight which her creator did not live to see: The P. & SS. *Great Eastern* sails from Southampton on her maiden voyage to New York, 1860. An *Illustrated London News* picture.

26. In Memoriam: The Clifton Suspension Bridge as completed by Brunel's fellow engineers after his death.

tion. Even if the atmospheric engine houses, Italianate on Brunel's South Devon, Gothic on Cubitt's London & Croydon, be compared with the nobleman's folly tower the analogy is false and instead of dismissing them with a patronizing smile we would do well to consider the motives of their architects a little more deeply. Brunel was a man of acute sensibility who understood very well what he was doing. He was proposing to introduce the first symbols of the new industrial age into a coastal landscape which had not altered materially since the Middle Ages. Therefore, as his sketch books reveal, he devoted great pains to the design of engine houses which he believed would not disfigure their setting. The result was far from being a folly. We may consider that he failed in his object and that if he had designed a purely functional box of a building and a chimney that looked like a chimney instead of a tower he would have served posterity better. That is not the point. The point is the intention and not our opinion of the result. It is that he cared whereas we do not. He may have been trying to achieve the impossible but we no longer make any pretence of conciliation in our total war upon the English landscape.

'I cannot anticipate the possibility of any inducement to continue the system beyond Newton.' '... Unless they [the patentees] can offer some guarantee for the efficiency of the valve, I fear that the Company should not be justified in taking that upon themselves, or incurring the expense attending the alteration of the engines.' Such cold, stilted sentences as these, addressed to the South Devon directors, represent Brunel's only surviving personal record of the atmospheric catastrophe. No allusion in private letter or journal entry appears to have survived. The failure was a blow to his reputation from which he soon recovered but we, knowing those early aspirations which he cherished throughout his life, may realize how much more shrewd and bitter a blow it must have been to his self-esteem. Until the railway was completed to Plymouth on 2 April 1849 he would accept for his services only a nominal retaining fee, but after such a disaster the wonder is that the Company continued to employ him at all. That they did so is a tribute not only to his magnetic personality but to his great abilities and, not least, to his courage in admitting failure. Had he lacked that courage and attempted to justify himself, the empty engine houses of South Devon could have marked the end of a promising career. As it was he was able to carry his broad gauge still farther into the west and so to leave behind him worthier memorials.

On the original main line of the Great Western, Brunel had carried the Sandford to Sonning by-road over the Sonning cutting by means of a timber bridge. He was so satisfied with this design that he used timber for a number of his early railway bridges, first in the bridge over the Avon at Bath, then on the Cheltenham & Great Western Union Railway at Stonehouse, St Mary's and Bourne and later on the South Wales line at Newport and Landore. He thus acquired very considerable experience of timber bridge building, and during these years he carried out many experiments to determine the strength of timber beams and the best methods of preserving them. The consequence was that no engineer in history has carried the art of timber bridge building to so high a pitch of perfection. While other railway engineers were using masonry or the treacherous cast iron, Brunel was building timber structures the like of which had never been seen before and will never be seen again. In the use of masonry or wrought iron he was also a master, but cast iron he avoided after one unfortunate experience with a skew bridge over the Uxbridge road at Hanwell where the cast iron girders broke as a result of the timber platform taking fire from a passing train. 'Cast Iron bridges are always giving trouble,' he wrote afterwards. '... I never use cast iron if I can help it; but, in some cases it is necessary, and to meet these I have had girders cast of a particular mixture of iron. .... The number I have is but few, because, as I have before said, I dislike them.'

The South Devon, Cornwall and West Cornwall Railways which comprised the broad-gauge route to Penzance, together with their various branches, set Brunel a difficult problem in civil engineering. For across their lines of route lie many deep and narrow valleys carved by the streams which, falling southwards from their high sources on Dartmoor Forest and Bodmin Moor, form the numerous inlets and narrow tidal creeks which indent the southern shores of Devon and Cornwall. The money to span such valleys by viaducts of masonry or wrought iron was not available, for in this far west capital expenditure was restricted because the companies concerned could not anticipate such heavy traffic returns as those serving the more populous counties. Brunel found the solution to this problem in the timber viaduct and so it is with Devon and Cornwall that these structures will always be associated.

So numerous were they – there were thirty-four on the Cornwall Railway between Plymouth and Truro alone – that they became a familiar feature of the western landscape and a very splendid one. There were five notable timber viaducts on the South Devon line between Totnes and Plymouth, of which the most celebrated was that carrying the line 114 ft above the waters of the little river Erme at Ivybridge. A charming lithograph of this viaduct with its tall twin piers linked by transverse arches was published by Angel of Exeter.

By the time the later viaducts on the South Devon's Tavistock branch and on the Cornish lines were built Brunel had perfected a standardized design of great beauty and simplicity. Viaducts of this pattern were constructed in two standard spans of 66 ft for the Cornwall and Tavistock lines and 50 ft for the more modest structures in West Cornwall. The object of this was to enable standard units of timber to be used, while the timber work was so designed that any unit could be replaced without interrupting traffic. Where the railway crossed tidal creeks, as between Devonport and St Germans, the superstructures were supported on timber trestles founded on piles, but inland they were carried on tall stone piers, each pierced transversely by Gothic-headed openings and featuring four buttresses which tapered from top to base. On the tops of these buttresses rested cast iron caps from which the timber beams sprang like the four outspread fingers of a hand to support the main longitudinal timbers. These main timbers, which carried the platform, consisted of two baulks set one above the other and joggled together so that their combined strength was almost equal to that of a continuous beam. Cross and diagonal timber ties and iron tie rods completed the structure. At the approaches there were no massive wing walls or abutments. King trusses at each end rested upon simple masonry platforms so that the bridges would not be affected by any settlement of the approach embankments. A simpler or more graceful design or one better calculated to become a part of the *genius loci* of the region it would be hard to conceive. In the primeval, storm-bitten landscape of western Cornwall the tapering piers of local stone looked as much at home as the gaunt chimney stacks of the tin mines. So nicely was the proportion of each beam and truss adjusted to the load which it would bear that the effect was one of a lightness and fragility so remarkable that it looked incapable of supporting the weight of even the lightest of locomotives. Indeed it was for this reason that, when the Cornwall Railway was opened, the local people displayed a marked reluctance to travel over it and time alone conquered their fear of the viaducts. They were certainly vertiginous. The greatest of them, at St Pinnock, near Liskeard, was 153 ft high, although Isambard III considered that Walkham Viaduct on the Tavistock line, 367 yds long and 132 ft high, was the finest example of his father's work in timber bridge building.

It has sometimes been suggested that such an extensive

use of timber was a mistaken policy and that this is proved
by the fact that not one of Brunel's timber viaducts survives
today. This criticism ignores the conditions prevailing at the
time they were built. Apart from the special need for economy in construction which has already been mentioned, it
was still possible in Brunel's day to buy very cheaply timber
of superb quality. All the viaducts in Devon and Cornwall
were built of yellow pine from Memel in the Baltic, the timber being treated by the old process of kyanizing as a precaution against fire. These spars of Baltic pine had an
average life of at least thirty years and some lasted as long
as sixty years. Such an expectation of life, coupled with the
ease of renewal which was a feature of the design, made the
timber bridge a decidedly economic proposition. The viaducts along the main route to Penzance were replaced by
steel and masonry structures when the line was doubled in
1908, but some of the many timber viaducts on branch lines
might still be with us today had it not been for the fact that
it became impossible to obtain timber of sufficiently high
quality even at prohibitive prices. By the end of the century
Baltic pine was no longer obtainable although a reasonably
good substitute was found in Quebec yellow pine. After 1914,
however, the only suitable timber available was Oregon pine,
which had an average life of only eight years. This, of
course, was hopelessly uneconomic and so the timber viaducts had to go.

Brunel can be said to have originated a new regional craft
and one highly suited to the men of western England with
their long maritime tradition. Highly skilled bridge gangs
examined each viaduct four times a year and when repairs or
replacements were necessary they would lower themselves
from the decks in bowline loops, swinging dizzily, perhaps
one hundred feet or more above the ground. Until the turn
of the century a large number of these gangs, each consisting of fourteen men, a chargeman and two look-outs,
were kept constantly at work. But by 1931 only three timber

viaducts, Collegewood, Ringwell and Carnon, remained. When they were replaced soon after, the last surviving gang was disbanded and a special skill which had become traditional was no longer required.

In view of their number and their long life it is remarkable that the history of Brunel's timber viaducts records no instance of accident or failure. The only case of damage by fire occurred in South Wales, where the main span of the Usk viaduct at Newport was destroyed before it had been completed and was replaced by wrought iron. On one occasion two goods trains met in head-on collision on the St Germans viaduct on the Cornwall line. Not only did the timber parapet prevent the trains from falling, but the structure was quite unstrained and continued to carry main line traffic to Penzance until 1908.

By far the most formidable obstacle which barred the extension of the broad gauge west of Plymouth was Cornwall's historic eastern boundary moat – the Tamar. As an alternative to a steam train ferry Brunel originally designed for this crossing what would have been the greatest timber bridge in the world with six spans of 100 ft and one of no less than 250 ft. For reasons already given, such a bridge could never have survived until today so we have to thank the Admiralty, whose headroom and minimum-span stipulations forced Brunel to abandon the idea of using timber and to design in wrought iron the last and the greatest of all his railway works – the Royal Albert Bridge at Saltash.

Nothing could be more false than the notion that Brunel was a theorist who tried out his ideas at the expense of others. On the contrary he was pre-eminently a master of that school which believes that a pinch of practice is worth a pound of theory. The supreme example of a theorist amongst his contemporaries was Doctor Dionysius Lardner and we know the value Brunel attached to his pronouncements. Like his timber viaducts, his wrought iron bridges were evolved not by untested theory and calculation but by the results of patient and costly experiments in which large girders of different forms were constructed and tested to destruction. All his major wrought iron bridges represent the successful outcome of these experiments and that each was not only larger than, but also an improvement upon, its predecessor reveals how Brunel never stood still, never rested content with past achievements but always learnt and applied the lessons of experience. So it could be said that his bow and string spans on the reconstructed Usk viaduct and on the Thames bridge at Windsor were the 'prentice work' and the bridge over the Wye at Chepstow the prototype for the final masterpiece at Saltash.

Nearly all Brunel's designs reveal his love of symmetry but his Chepstow bridge was an exception dictated by the unusual lie of the land. On the east a limestone cliff 120 ft high rises sheer from the waters of the Wye, whereas on the west side lies an area of near level washland of clay and gravel only a few feet above high-water mark. Here, too, as at Saltash, the interests of navigation had to be considered for there was then a considerable traffic on the Wye, trows and other small coastal craft trading to Brockweir, Redbrook and Monmouth. Brunel therefore planned to bridge the river with a single high-level span 300 ft long and 100 ft above high water and to cross the washland on the western side by means of three approach spans of 100 ft each. The cliff provided a natural abutment for one end of the main span, but the other end, and also the girders of the three approach spans, had to be carried on lofty piers. For these Brunel decided to use cast iron cylinders 8 ft and 6 ft in diameter respectively. These were sunk through the soft ground until they found a solid foundation and were subsequently filled with concrete. He had used similar cast iron piers previously at Windsor and in both cases he adopted the same method for sinking them which his father had first employed in sinking the Rotherhithe shaft of the Thames tunnel. The bottom of the section to be sunk was given a

cutting edge and heavy weights were applied to the top while
the workmen excavated the ground inside. At Chepstow Brunel had an experimental cylinder cast with an external screw
thread of 7 in. pitch with the idea that it could be driven
home like a screw pile. This worked well in the gravel and
clay, but beds of running sand were encountered in which
the thread would not grip and the idea was abandoned.
The influx of water through these sand beds gave considerable trouble and Brunel had to introduce the 'pneumatic
method' as it was then called, in order to sink the large
columns for the main span. The upper ends of the cylinders
were sealed and compressed air introduced to keep the water
out, workmen and materials entering through air-lock
doors.

Brunel came to the conclusion as a result of his girder
experiments that semicircular, pear-shaped or circular sectional forms were best fitted to withstand the compression
stresses to which the upper flange of any bridge girder is
subject. Consequently a 'Brunel girder' may be recognized as
readily as if it bore his name upon it. The two 300-ft trusses
of the Chepstow bridge which carry each of the two lines of
rail are, in effect, huge girders of this typical form. The
main girders, 7½ ft deep, below the rail platform represent
the lower flange. Fifty feet above them, supported on
masonry piers arched for the passage of the trains, the upper
flange consists of a circular wrought iron tube 9 ft in
diameter. From the reinforced ends of this tube run the
chains of suspension bridge pattern which support the girders below. They do so through rollers and saddles so that
the girders shall not be affected by any movement of the
truss under load.[2]

The importance which Brunel attached to practical tests
may be judged from the fact that before permanent erection
was begun, one of these great trusses weighing 460 tons was

2. Since this paragraph was written Chepstow bridge has been rebuilt (1962).

assembled on temporary piers on the river bank and tested
with a distributed load of 770 tons.

Brunel planned to float the tubes into place below the
piers and then hoist them with chain tackle attached to winches or 'crabs' which he had specially designed for the purpose. The delicacy and hazard of such an operation in a river
where the tides run with great violence and rise to a height
of as much as 40 ft may be imagined. Success depended upon
the whole operation being carried out speedily and with perfect precision. Brunel was commander-in-chief, his first lieu-
tenants being his chief assistant Brereton and Captain
Claxton, the latter, as we should expect, being in charge of
the waterborne manoeuvres.

The first tube was slewed at right angles to the river and
pushed forward on trolleys until its end overhung the water.
Beneath it was a pontoon formed of six iron barges. This
pontoon took the weight of the tube as the tide rose and was
then guided across the river by a crab which hauled on two
chain cables anchored to the cliff opposite. As soon as it was
in position the lifting tackles were attached. So smoothly was
the whole operation carried out that by nightfall the tube
had been lifted to rail level and by the end of the next day it
was in its place upon the piers. The bridge was opened for
single-line traffic on 14 July 1852 while the second tube and
truss was being dealt with in the same way.

We may appreciate from all this how much Brunel benefited from his experience at Chepstow when he confronted
the far greater task of bridging the Tamar. Here, the river
to be crossed is 1,100 ft wide and 70 ft deep at high water,
while the Admiralty required a headway of 100 ft. Forced
to abandon his original idea of a timber structure, he
prepared two tentative designs, one of four spans and
another with a single immense span of 1,000 ft. Both were
discarded, the first on account of the difficulty of sinking
so many piers in deep water and the second on the score
of expense. Finally he determined upon two main spans

of 465 ft each which would involve only one deep water pier.

The formation of this central pier was Brunel's greatest problem, for soundings showed that it would be necessary to go down through sand and mud to a rock foundation no less than 80 feet below high-water level. He decided to try the idea of using a cast iron cylinder as a coffer-dam within which a masonry pier could be built. In 1848 a trial cylinder 6 ft in diameter and 85 ft long was made, towed out to the site between two hulks equipped with special tackle, and lowered into the Tamar. When its bottom end had sealed itself, the water was pumped out and no less than 175 borings were made through the mud into the rock below. From the information so obtained Brunel was able to make a model of the invisible rock showing its exact geological nature and its profile. Finally, in January 1849, the mud was excavated down to the rock and a small piece of masonry constructed to demonstrate the practicability of the scheme. At this juncture the Cornwall Railway Company suspended operations for three years owing to lack of capital.

When it was eventually decided to proceed, the need for economy had become paramount and Brunel reported as follows to the Board of Trade:

This bridge had been always assumed to be constructed for a double line of railway as well as the rest of the line. In constructing the whole of the line at present with a single line of rails, except at certain places, the prospect of doubling it hereafter is not wholly abandoned, but with respect to the bridge it is otherwise.

It is now universally admitted that when a sufficient object is to be attained, arrangements may easily be made by which a short piece of single line can be worked without any appreciable inconvenience. ... This will make a reduction of at least £100,000.

So the Saltash bridge was built for a single line only and remains so to this day.

The great cylinder, 35 ft in diameter at the base, which

Brunel designed for constructing the underwater portion of the central pier was a most ingenious piece of apparatus which embodied experience which he had gained in his youth in the sealing of the Thames tunnel breaches. It was actually a tube within a tube, the former incorporating a diving bell within which the masons worked. This was kept clear of water by pumps and ventilated by means of a pipe which extended from the top of the dome of the bell to the full height of the outer cylinder. But the influx of water into this inner working compartment could never be great because the annular space, 4 ft wide, between the inner and outer cylinders was sealed and pressurized, the same 'pneumatic apparatus' being used for this purpose as had been previously employed at Chepstow. By this device only the workmen clearing the mud in the annulus as the cylinder descended had to work under air pressure. The masons in the central chamber were spared this inconvenience; also the necessity of passing all the material for the pier through air lock doors was avoided. The lower edges of this great double cylinder were shaped to fit as closely as possible the profile of the submerged rock as it had been plotted three years before.

The cylinder was built on the river bank, floated off on the tide, guided to the site by pontoons and lowered into position in June 1854. By February 1855 it had reached the rock, some difficulty and delay having been caused by the presence of dense beds of oyster shells in the mud through which it had to sink. The rock was intensely hard greenstone trap which proved extremely difficult to work. Moreover, an unsuspected fissure in the rock caused water to be forced through under pressure into the central compartment which had to be strengthened in situ and pressurized. However, by the end of 1856 the great pier had been completed to the temporary cap which would receive the ends of the main trusses. The cylinder, which now encircled the pier, had done its work. Brunel had designed it in halves with an eye to its eventual removal, so it was now speedily unbolted and towed ashore.

While the pier was building, the first of the main trusses was taking shape on the Devon shore. Its principle was exactly the same as that of the main span of the Chepstow bridge, and apart from its vastly greater size it differed from it only in detail. Whereas the tubes of the Chepstow bridge are for practical purposes straight, being only very slightly arched for the sake of appearance,[3] the Saltash tubes are arched to an extent equal to the fall of the suspension chains. Again, instead of the circular section used at Chepstow the Saltash tubes are oval, being 12 ft 3 in. high and 16 ft 9 in. broad. Brunel's purpose in depressing the tubes into ovals was that their breadth was thus made equal to that of the bridge platform below so that the suspended chains would fall vertically. This was an improvement on the Chepstow bridge, where the chains are inclined. Other design details were modified to suit the different circumstances. At Chepstow, owing to the restricted site and to the fact that the Wye navigation could not be obstructed for more than a single tide, Brunel had designed a truss whose different portions could be quickly raised separately and assembled aloft. The Tamar, on the other hand, allowed more room for manoeuvre, while as the bridge was in two spans the question of obstruction did not arise. He therefore designed the Saltash trusses to be prefabricated complete on the shore and floated bodily into position.

There was one feature of these trusses which must have saddened their designer. This was that the tension chains which were used on this, his last bridge, rightly belonged to his first. They had been bought by the Cornwall Railway from the Clifton Bridge Company when that most unfortunate concern again stopped work for lack of funds in 1853. Obviously they cannot have been of sufficient length

3. In his sketchbook, Brunel gives the cant of the Chepstow tubes as 1 ft 6 in. only.

to equip both the Saltash trusses. But their makers, the Copperhouse Foundry of Hayle, were near at hand to provide additional links and to adapt them to their new purpose. How long it must have seemed to Brunel since, as a young man filled with enthusiasm by his first important commission, he had travelled down to Hayle to see these very links tested by an hydraulic press.

The weight of each truss complete was over 1,000 tons so that their floating was no mean undertaking. Brunel had obtained valuable experience not only at Chepstow but also at the Menai, where he had assisted Robert Stephenson to float the tubes of his great Britannia bridge. But this did not fill him with any false confidence; on the contrary he planned the whole operation with meticulous care down to the smallest detail. Nothing was to be left to chance or to the hurried improvisation of the moment. In his notebooks we find the pencilled notes which he jotted down as he brooded over the problem; thus:

Saltash

Weight to be floated, say 1,000 T.
Height of c. of g., say 45 ft
Pontoon at each end 500 Tons.
4 Pontoons 100 × 20 × 10

Oblique bows would also probably make them more manageable; each £1,500.

Floating

Signals. Numbers to distinguish tow lines. The order is given and flags for the order seem to be the best. Numbers about 30″ high on placard, standards or poles and flags of about the same size with a stiffening rod thus: [a small sketch].

Signals by flag:
Heave in – red.
Hold on – white.
Pay out – blue.
Waved gently means gently.
Waved violently means quickly.

The pontoons were built at Plymouth and launched sideways into the river. They were virtually floating tanks designed so that water could be rapidly let into or pumped out of them in order to vary their draft. The truss had been built parallel to the river bank and two docks just wide enough to admit the pontoons were cut beneath each end. The plan was that when the pontoons had floated the truss they would be guided into position by cables, each bearing a number for signalling purposes, attached to crabs on the shore and on ships which would be anchored in the Tamar. Every move was carefully rehearsed and printed instructions issued to all those taking part.

The day fixed for the floating, 1 September 1857, was brilliantly fine and the whole neighbourhood was en fête. Church bells pealed, flags hung from every house in Saltash, a general holiday was declared and from all the country round the people flocked to see the wonder performed until every field and vantage point on both banks of the Tamar was crowded to capacity. Out in the river the five naval vessels under the command of Captain Claxton lay ready at their moorings. Beyond their field of operations the water was packed with crowded, flag-bedecked craft. In the morning the expectant throng watched the pontoons being manoeuvred into position, two in each dock, and the cables attached. As the tide rose the water was pumped out of the pontoons and at a quarter past one there sounded a murmur like the sudden sighing of a wind as the great truss lifted slightly and the thousands of awestruck spectators whispered 'she floats'.

At this moment, like the conductor of an orchestra, Brunel moved to his place upon a platform mounted high in the centre of the truss. Directly above him were his signallers, standing ready with their numbers and flags. He had insisted that the whole operation must be carried out in complete silence and his wishes had been widely publicized. Consequently, no sooner had he taken up his position than there fell a dramatic stillness like that which follows the tap of a conductor's baton, and every eye in the vast crowd was strained towards the distant figure of the engineer. Numbers whose purport was unintelligible to the crowd were displayed; flags flickered and then the huge truss swung slowly and majestically out into the Tamar. 'Not a voice was heard,' wrote an eye-witness –

... As by some mysterious agency, the tube and rail, borne on the pontoons, travelled to their resting place, and with such quietude as marked the building of Solomon's temple. With the impressive silence which is the highest evidence of power, it *slid*, as it were, into its position without an accident, without any extraordinary mechanical effort, without a 'misfit', to the eighth of an inch.

Just as the time of high water came at three o'clock, the ends of the tube were secured in their positions on the piers from which they would be raised by hydraulic presses as the masonry was built up beneath them. As soon as the truss was safely in place the tension was broken. A band of the Royal Marines struck up 'See the conquering hero comes' and Brunel stepped down from the platform to the accompaniment of a storm of cheering. It was a moment of triumph which must have sweetened the bitter memory of the atmospheric disaster. But not one of the thousands of west-countrymen who cheered themselves hoarse that day realized that their tribute was also a valediction, that their hail was also a farewell.

It was Brunel's chief assistant, Brereton, who superintended the floating of the second Saltash span in July 1858 and who saw the work through to its successful completion in the following spring. When Prince Albert, as Lord Warden of the Stanneries, travelled down from Paddington to open the Royal Albert Bridge in May 1859 amid fresh scenes of wild enthusiasm, the last link in the broad-gauge route to the west was completed. Wrote the ballad monger:

From Saltash to St Germans, Liskeard and St Austell,
The County of Cornwall was all in a bustle,
Prince Albert is coming the people did say
To open the Bridge and the Cornish Railway.
From Redruth, and Cambourne, St Just in the west
The people did flock all dressed in their best.
From all parts of England you'll now have a chance
To travel by steam right down to Penzance.

But the engineer was not there. No flags flew, no bands played, no crowds cheered when he took his first and last look at the completed bridge. He lay on a specially prepared platform truck, while one of Gooch's locomotives drew him very slowly beneath the pier arches and over the great girders. For his railway career was ended. Broken by the last and the most ambitious of all his schemes – his great ship – Brunel was dying.

BOOK III

[11]

Terminus New York

In November 1854, when he was forty-eight, Brunel noted that he had been responsible for building 1,046 miles of railway. 'When the Birmingham & Dudley is open,' he wrote, 'there will be: Worked by the Grt Western 500 miles, and under my direction 600 miles.' It was then just over eighteen years since the first Great Western contract had been let. Even when due credit is given to his staff of assistant engineers, the completion of an average of 58 miles of line every year over so long a period is a very remarkable performance, particularly so when we remember the close personal supervision upon which he always insisted. Moreover, when we think of this construction programme going steadily forward notwithstanding protracted Parliamentary contests, the struggle for the adoption of the broad gauge, the gauge war and the atmospheric disaster, it does not seem humanly possible for one man to have accomplished so much. Yet the truth is even less credible. While Brunel was in the midst of this stormy railway career which would alone have immortalized his name, he was also staking an equally valid claim to fame in a totally different field of engineering – trans-Atlantic steam navigation.

'I have been making half a dozen boats lately, till I've worn my hands to pieces,' wrote the schoolboy from Hove, while we have already seen how in youth Brunel had dreamed of launching a fleet of powered warships against the Turk as he sat by the fireside in his Rotherhithe lodgings. Such nautical ambitions were only to be expected. Not only was his father a sailor who had first made his engineering reputation through his block-making machinery; Marc Brunel had also made a positive contribution to the history

of steam navigation. As early as 1814 he had made the first
of a number of experimental voyages between London and
Margate in a small steamer whose paddles and double-acting engines incorporated ideas of his own. In 1822 he had
patented various improvements in marine engines which included the double-acting engine of 'Gothic' form as subsequently built by Boulton & Watt, an improved form of
governor, a surface condenser to obviate the use of sea water
in boilers and the constant blowing down which this involved, and lastly a form of hopper-fed mechanical stoker.
Yet it seems strange that Marc apparently had no faith in
the possibilities of steam power for trans-oceanic navigation.
Asked if he would act as consulting engineer to a scheme for
operating steamships between England and the West Indies
he replied bluntly: 'As my opinion is that steam cannot do
for distant navigation, I cannot take part in any scheme.'

In fairness to the memory of Sir Marc Brunel, however, it
must be said that he was only expressing the view which was
universally held at that time. The insuperable difficulty, it
was believed, was that no steamship could possibly carry
enough fuel for an ocean crossing. When, in 1819, the American ship *Savannah*, after coaling at Kinsale, reached
Liverpool from Savannah to become the first steamship in
the world to cross any ocean, this belief was in no way
shaken. For it is not strictly correct to call the *Savannah* a
steamship. More truthfully she was a sailing ship fitted with
a small single cylinder auxiliary steam engine of only 90
i.h.p. The paddle wheels which this engine drove were made
collapsible so that they could be brought up on deck when
not required and thus not interfere with her sailing qualities.
On a voyage which lasted 27 days 11 hours, the engine was
only run for a total of about 85 hours so that it could not
fairly be claimed that the Atlantic had been crossed by steam.[1]

1. Ocean crossings were made by other steam vessels between 1818
and 1832 but in every case they relied upon sail and used steam power
in adverse conditions only.

Although we find no references to the fuel problem in his early writings, it is obvious that Brunel did not accept his father's conclusion, and that he gave the matter a great deal of thought. It had always been assumed that the construction of larger ships was no solution to the difficulty. If the size of hull was doubled, it was argued, then twice the power would be needed to drive it and therefore double the weight of coal would have to be carried to generate the power. Brunel felt sure that there was a flaw in this argument somewhere, and when he discovered it he formulated a simple rule which has formed part of the alphabet of steam shipbuilding ever since. This is that whereas the carrying capacity of a hull increases as the cube of its dimensions, its resistance, or in other words the power required to drive it through the water, only increases as the square of those dimensions. Once stated, it seems blindingly obvious that the tonnage or capacity of a ship is a question of volume whereas its resistance is a matter of surface area, but in the 1830s it was not appreciated or, if it was, the lesson had not been applied to ocean steam navigation. Only Brunel realized that owing to this simple fact it was possible to design a steamship which could carry enough coal for any given length of voyage. A trans-Atlantic steamship was not an impossibility, therefore, but only a problem of correct proportion. So convinced of this did he become that in October 1835 he flung down his most celebrated challenge.

It was at an early meeting of the directors of the Great Western Railway at Radley's Hotel in Bridge Street, Blackfriars. Someone expressed serious misgivings at the enormous length of the proposed main line from London to Bristol and their engineer retorted: 'Why not make it longer, and have a steamboat go from Bristol to New York and call it the *Great Western*?' Uneasy laughter broke the silence that followed this absurd suggestion. Of all those present only Thomas Guppy took Brunel's question seriously, and buttonholed him as soon as the meeting was over. The two discussed the project until a late hour that night and we may imagine how easily Brunel won over his enthusiastic Bristol friend. He was always a hard man to resist once a new scheme had fired his imagination. Not only was there the potent influence of his own personality; in an age of wonders when yesterday's dream was today's achievement, to speculative men like Guppy the schemes themselves were irresistibly attractive. What had at first seemed sheer folly, a grandiose but eccentric dream, would soon become sublimely simple and logical whilst losing nothing of its grandeur as Brunel coolly explained the carefully reasoned engineering propositions from which his scheme had been evolved. In this case it was the elementary little fact of the difference between the cube and the square, working like a magic spell upon his imagination, which was to produce three of the greatest ships that the world had ever seen. Guppy was convinced, and the immediate outcome of that nocturnal discussion was the formation in Bristol of the Great Western Steamship Company with Peter Maze, another Great Western Railway pioneer, as chairman and Brunel's good friend and nautical adviser Captain Christopher Claxton as managing director.

It was resolved that the Company's first ship should be built to Brunel's designs by William Patterson of Bristol under the supervision of a Building Committee consisting of Guppy, Claxton and Brunel. On 28 July 1836 Patterson set up the sternpost of the ship and whenever Brunel was in Bristol on railway business the little committee would meet and discuss the details of their trans-Atlantic challenger in a haze of cigar smoke until far into the night. Although she was to be built of oak by traditional methods, she would be larger than any steamship so far built and Brunel insisted that her hull be given great strength, particularly in the longitudinal plane, so that she could weather the worst Atlantic storms without stress. Her framing was constructed on an improved principle which had first been evolved by Sir

Robert Sepping, Surveyor to the Navy, and her ribs were as massive as those of a ship of the line. She was closely trussed with iron and wood diagonals and shelf pieces, while a series of iron bolts 1½ in. in diameter ran the whole length of her bottom frames. All her timbers were kyanized and her hull copper-sheathed.

Marine engines of unprecedented size would be required to drive such a hull. Several firms were invited to tender, but Brunel advised the Company to accept the specification submitted by Messrs Maudslay, Sons & Field because, as he rightly pointed out, this firm were the most celebrated and successful marine engine builders of the day. Maudslay's engines were of the conventional side lever type having two cylinders, each of which could drive one paddle wheel or both; hence the term 'a pair of engines'. At the working steam pressure of 5 lb per sq. in. a total of 750 i.h.p. was produced, approaching double that of any ship afloat. Steam was supplied by four flue boilers together weighing 100 tons empty.

With the engines contracted for and the hull well under way in Patterson's Dock, there appeared in Bristol that strange character who, like the chorus in some burlesque of Greek tragedy, might have existed for no other purpose than to provide the improbable comic relief in Brunel's life story. In August 1836 Doctor Dionysius Lardner arrived to address the British Association on the subject of trans-Atlantic Steam Navigation. As usual, the learned man produced figures which proved beyond fear of contradiction that the thing could not be done. 'Take a vessel of 1,600 tons, provided with 400 horse-power engines,' he urged his audience. 'You must take 2⅓ tons for each horse-power, so the vessel must have 1,348 tons of coal. To that add 400 tons, and the vessel must carry a burden of 1,748 tons. I think it would be a waste of time, under all the circumstances, to say much more to convince you of the inexpediency of attempting a direct voyage to New York. . . .' No sooner was the lecture over than

Brunel, who was in the audience, jumped up and engaged the doctor. Unfortunately no record of this battle of wits survives but it is said that for once even his powers failed to counteract the effect upon the audience of Lardner's dogmatic pronouncements and the figures – always those magic figures which no one but Brunel dared question – with which he backed them up. Once again it would be left to the engineer to prove the theorist wrong by practical demonstration.

This episode in no way shook the confidence of the Steamship Company, and while the controversy continued to rage, work on the *Great Western* went steadily forward. On 19 July 1837 she was floated out of Patterson's dock at Wapping in Bristol harbour and a company of 300 were entertained to a ceremonial luncheon in her main saloon. Accompanied by the tug *Lion* and the steamers *Herald* and *Benledi*, she left Bristol under sail on 18 August for London River to take on her engines and to have her saloon decorated. While she lay at Blackwall crowds flocked down to the river to admire what the press called 'her magnificent proportions and stupendous machinery'. Her hull, so huge to the eyes of those beholders, must have looked all the more impressive for being painted a sombre black, only relieved at her graceful bow, where there glittered the gilded figure of Neptune escorted by two dolphins. Of the other source of wonder, her grand saloon, no pictorial record appears to have survived and we have only this tantalizing contemporary description:

The saloon of this vessel is 75 feet in length, longer than any other steam ship or vessel in the World, and 21 feet wide, except where there are recesses on each side when it is 34 feet & 9 feet high clear of the beam, which is increased by the lantern light. It will be needless to enter into a minute description of all the ornamental paintings, decorations, etc., which are so lavishly expended on this saloon, but it will suffice to say that all the taste and skill of the first rate upholsterers, decorative and landscape painters, etc., etc., have been employed on it.

But while the majority came to admire, there must have been others who regarded the *Great Western* with jealous eyes. She represented Bristol's attempt to reassert her old maritime supremacy. Because the two sides of Brunel's career have always been considered in isolation, the connexion between his Atlantic challenger and the determined efforts of the 'Liverpool Party' to discredit him and seize control of the Great Western Railway is not properly appreciated. Certain it was that Bristol's rivals dared not passively await the outcome of Brunel's experiment. If it succeeded, such a delay might prove fatal to any prospective competitor. So the formation of the Bristol Company had been closely followed by that of two rivals, the British & American Steam Navigation Company of London and the Transatlantic Steamship Company of Liverpool. Curling & Young of Limehouse laid down the *British Queen*, a slightly larger ship than the *Great Western*, for the London Company, while the Liverpudlians purchased before completion the *Liverpool*, a somewhat smaller vessel of 1,150 tons register which Messrs Humble & Milcrest of Liverpool had laid down to the order of Sir John Tobin. These, then, were to have been the representatives of the rival ports in the race to New York, but soon after the *Great Western* entered the Thames it became obvious that she would be ready to make her maiden voyage long before her rivals. Determined not to be beaten, however, both the other companies resorted to charter, the *Sirius* of the St George's Steam Packet Company's[2] fleet deputizing for the *British Queen*, and the *Royal William* of the City of Dublin Steam Packet Company for the *Liverpool*. Although both these charters were new ships, launched in 1837, they were intended only for Anglo-Irish service and required much modification, particularly the provision of increased bunker space, before they could venture into the Atlantic. The *Sirius* was being prepared for her bid in the Thames and desper-

2. Now the City of Cork Steam Packet Company.

ate efforts were made to complete the *Great Western* first
and get her back to Bristol so that she could start her maiden
voyage before her rival. Imagine then the feelings of Brunel
and all the loyal Bristolians aboard his ship when, on Wednesday 28 March 1838, the little *Sirius* slipped away down
river carrying a crew of 35 and 40 passengers bound, they
hoped, for New York. Although little more than half the
size of the *Great Western*, she was a smart vessel with her
two tall masts and her dazzling white figurehead of a hound
holding the Dog Star between its paws.

The *Sirius* would not have it all her own way, however.
After running her trials on the 24th and 28th of the month,
the *Great Western* left her moorings at eight minutes past
six on the last morning of March and steamed away in pursuit with Guppy, Claxton, Brunel and other members of the
Company on board. Their ship still stood a good chance, for
they knew that the Master of the *Sirius*, Lt Roberts, R.N.,
intended coaling at Cork. Then, after only two hours' steaming, that outrageous fortune which so often seemed to dog
Brunel's enterprises dealt a blow which might have proved
mortal.

The party were up on deck that bright morning admiring
their splendid ship, listening to the strong heartbeat of her
engines and the surge of her great cycloidal paddle wheels. It
was 8.15 a.m. and the ship was off Leigh when satisfaction
turned to consternation as flames and dense clouds of acrid
smoke suddenly belched from the for'ard boiler room. Let
George Pearne, the chief engineer, a man of great gallantry
and presence of mind, give his own account of an occurrence
which was so nearly disastrous. In his log for the day he
wrote:

The fore stoke-hole and engine room soon became enveloped
in dense smoke, and the upper part in flames. Thinking it possible the ship might be saved, and that it was important to save
the boilers, I crawled down, after a strong inhalation of fresh air,
and succeeded in putting on a feed plunger and opening all the

boiler feed cocks, suffering the engines to work to pump them up, as the steam was generating fast from the flames round the upper part of the boilers.

The cause of the outbreak was immediately apparent. The boiler lagging of felt and red lead had been carried up too close to the furnace flues beneath the base of the funnel and had ignited spontaneously, setting fire to the deck beams and the under side of the deck planking. Captain Claxton rushed to the boiler room to help Pearne while the ship's captain, Lt James Hosken, R.N., fearing the worst, headed his ship towards Canvey Island where she took the ground on the Chapman Sand.

Claxton was standing beneath the boiler-room hatchway in a choking fog of smoke playing a hose on the blazing decks when he was suddenly bowled over by some heavy object falling upon him from above. Picking himself up and peering through the smoke, Claxton discovered what had struck him down. It was the body of a man which now lay very still on the boiler-room plating, the head submerged up to the ears in the water from the pumps and fire-hoses which had collected there. Instantly Claxton dragged the man's head clear of the water and yelled for a rope to be lowered from the deck, whereupon the body was hauled away. It was not until the fire was mastered and he climbed on deck that Claxton discovered that the victim was Brunel and that he had unwittingly saved the life of his friend.

Brunel had started to descend the ladder into the boiler room when a rung, which had been partially burnt away, collapsed under his weight and he fell 18 ft to what would have been almost certain death had not Claxton chanced to be there to break his fall and then to save him from drowning. Although he was in great pain and unable to move, Brunel insisted that whatever might happen to him the fire must be put out and the ship go forward as quickly as possible. He was laid upon a sail until the fire had been extinguished, when he was gently lowered into one of the boats and taken to Canvey Island, where he had to remain for several weeks. This, Brunel's second hairbreadth escape from accidental death, occurred while the first section of broad-gauge permanent way was being laid between Paddington and Maidenhead. It was to his enforced absence at this crucial time that a great deal of the initial trouble with this permanent way was due.

Having done her best to destroy both herself and her creator, the *Great Western* lifted on the evening tide and the water thundered again in her paddle boxes as she swung away from the Chapman Light on an arc of white water to resume her westward voyage. But the accident had wasted twelve vital hours and until the ship was broken up the evidence of it remained in the charred deck planking over the boilers. She made a fast passage round the Longships, averaging nearly 13 m.p.h., but the rumour that she had been destroyed by fire in the Thames travelled even faster, so that it was with delighted incredulity that the people of Bristol saw their *Great Western* anchor at Kingroad on the afternoon of Monday 2 April.

Within two days of arrival, Captain Claxton received a two-page list of matters to be attended to before the ship sailed which had been dictated by the indefatigable engineer from his bed on Canvey Island. Seventh April was the date fixed for the sailing and there was more high pressure activity as Brunel's instructions were carried out while the ship was coaling and victualling. Bad weather delayed departure by 24 hours so that it was not until 10 a.m. on Sunday 8 April that the *Great Western* left for New York. The hopes and the goodwill of a whole city went with her, but commercially her departure could not have been less auspicious. She carried only seven passengers in her spacious saloon. No less than fifty of those who had booked passages at 35 guineas a time from Claxton's little office at 19 Trinity Street had cancelled them when they heard the exaggerated news of the fire. As for the trans-Atlantic race, if any one of the crowd who watched the *Great Western* draw slowly away through the Avon gorge that morning had known the exact position of the *Sirius*, it would have seemed to them a foregone conclusion.

Lt Roberts took the little *Sirius* out of Cork harbour, a whole day's steaming nearer New York than Bristol, on 4 April. It was a venture of the highest courage to hazard a vessel so small – only 703 tons gross register – in the open Atlantic at the period of the spring equinox and in defiance of the gloomy croakings of Doctor Lardner and his fraternity. When the ship encountered strong head winds off the Blaskets, both passengers and crew began to take an extremely poor view of their survival prospects and implored the captain to put the ship about and abandon such a rash attempt. But in Lt Roberts the *Sirius* possessed a master of iron determination and discipline worthy of such an historic occasion. The ship would go on and go she did, day after day until, as the log neared the 3,000 miles mark, coal began to run short precisely as Doctor Lardner had forecast. There is a long-standing legend that the *Sirius* completed her voyage only by burning her cabin panelling and furniture and even (a typically fatuous journalistic touch this) a child's doll. In fact, by good fortune, careful management and the sacrifice of four barrels of resin from the cargo such desperate expedients were avoided, but it was a very close call. She arrived off New York on 22 April. When she docked next morning after 19 days at sea she had only 15 tons of coal left in her bunkers. Such a risky feat had done absolutely nothing to prove the practicability of ocean steam navigation, but as an epic of courageous seamanship the *Sirius* and her crew most richly deserved their victory and the tumultuous tribute accorded them by the citizens of New York.

Not many hours elapsed before New York realized how extremely narrow the victory of the *Sirius* had been. In the early hours of that morning, St George's Day, unheralded and unsung, the *Great Western* dropped anchor off Sandy Hook, 15 days and 5 hours from the port of Bristol. She had encountered the same adverse winds as the *Sirius*, but there had been nothing heroic about the crossing, Lt Hosken and his crew displaying no more than ordinary good seamanship as the *Great Western* forged steadily westwards, gaining on the *Sirius* at the rate of over two knots an hour. The man who deserved the victory which the ship so narrowly failed to win was Brunel, for he had proved his point. The *Great Western* still had nearly 200 tons of coal on board when she docked.

The achievement of the *Sirius* might have been dismissed as a bold stunt, but the appearance of the *Great Western* on the afternoon of the same day convinced America that this was no freak exploit but the inauguration of a new era of rapid and reliable ocean transport between the old world and the new. In her great size, her superior speed and in the new standards of comfort which she set, the *Great Western* was fittingly named, for she was indeed the ocean counterpart of Brunel's broad gauge. Among the crowds who watched her steam to her berth in that ultimate western terminus which had been her creator's goal from the beginning was the pioneer journalist, James Gordon Bennett, Senior, and this was his description of the scene as he recorded it in his *Morning Herald*[3]:

The approach of the *Great Western* to the harbour, and in front of the Battery, was most magnificent. It was about four o'clock yesterday afternoon. The sky was clear – the crowds immense. The Battery was filled with the human multitude, one half of whom were females, their faces covered with smiles, and their delicate persons with the gayest attire. Below, on the broad blue water, appeared this huge thing of life, with four masts and emitting volumes of smoke. She looked black and blackguard ... rakish, cool, reckless, fierce, and forbidding in sombre colours to an extreme. As she neared the *Sirius*, she slackened her move-

3. Soon to become the *New York Herald*.

ments, and took a sweep round, forming a sort of half circle. At this moment, the whole Battery sent forth a tumultuous shout of delight, at the revelation of her magnificent proportions. After making another turn towards Staten Island, she made another sweep, and shot towards East River with extraordinary speed. The vast multitude rent the air with their shouts again, waving handkerchiefs, hats, hurrahing!

Meanwhile, a passenger, leaning over the rail of the *Great Western*, was no less deeply moved by the historic occasion. 'Myriads were collected,' he wrote afterwards,

boats had gathered round us in countless confusion, flags were flying, guns were firing, and cheering rose from the shore, the boats and all around loudly and gloriously as though it would never have done. It was an exciting moment, a moment of triumph.

But the occasion was also one of tragedy though none of the cheering crowds knew it. While the passenger watched from the rail, down below George Pearne was paying his own tribute to the ship in a letter to Maudslays. 'The engines,' he wrote, 'I am proud to say, have performed even beyond my expectations, which was at all times sanguine.' They were, he went on, 'a piece of magnificent perfection'. But his tribute was never finished. As the ship ran alongside Pike Street Wharf, Pearne, who had saved her from destruction only three weeks before, was now, upon the instant of success, fatally scalded in the act of blowing down her boilers.

We may think it a pity that Brunel was not present to witness the triumphant vindication of his theory of proportion, but even had his accident not prevented it, he had had no intention of making the voyage. When he had launched one engineering rocket upon the world he was always much too busy preparing the next to waste any time in gazing after it or admiring its effect upon beholders. The act of creation was all, the finished work valued only for the lessons it could teach, the improvements it could suggest.

The *Great Western*'s effortless crossing so far restored

public confidence that she carried 68 passengers when she left New York on her return passage on 7 May. Notwithstanding the breakage of a connecting rod brass which put one engine out of action for 48 hours, the ship reached Bristol in 15 days. So impressed were the passengers that they signed a testimonial to Captain Hoskens. As we should expect, the return of the *Great Western* to her home port was marked by similar scenes of bell-ringing, flag-flying and general rejoicing. 'The joy and pleasure announced by all classes,' declared the *Bristol Mirror*, '. . . has been unequalled in the city for many years, and they almost stand upon a level with the tidings from the Nile, Trafalgar Bay, and the plains of Waterloo.'

Meanwhile, what of the third competitor in the trans-Atlantic contest? The *Royal William* arrived in New York on 24 July after a passage of 18 days 23 hours from Liverpool. By doing so she achieved two distinctions; she became the first steamer to cross from Liverpool and the smallest steam vessel ever to make the passage between the old world and the new. The *Royal William* made two more voyages to New York and the *Sirius* one more before the *Liverpool* and the *British Queen* entered the service and enabled these gallant little ships to return to their cross-channel duties.

Neither the *Liverpool* nor the *British Queen* could match the performance of the *Great Western*. The former left Liverpool on 20 October but was forced to put into Cork by stress of weather. The passage from Cork to New York took her 16 days 17 hours. She was sold to the P. & O. Line in July 1840 when the Liverpool Company was wound up. The *British Queen* was more successful. She made nine Atlantic crossings before she was sold to the Belgian Government in 1841 and broken up in the following year. Her first crossing, from Portsmouth in July 1839, was made at a mean speed of 8.4 knots as compared with the *Great Western*'s 8.8 knots on her maiden voyage against the spring equinoctials.

On her return voyage on 1 August there was intense excitement when it became known that the *Great Western* intended sailing on the same day. The *Great Western* narrowly defeated her rival by docking at Bristol on the evening of 14 August while the *British Queen* arrived at Portsmouth next morning. It was subsequently proved that the *Great Western* had the advantage of her rival in all conditions but particularly in rough weather.

In 1840 there was a tragic sequel to the stirring transAtlantic race of '38. After his epic voyages with the little *Sirius*, the British and American Steam Navigation Company gave Lt Roberts the command of their *President*, a sister ship to the *British Queen* from the same builders. On his second return voyage to Liverpool from New York with his new command this very gallant master went down with his ship when the *President* foundered in a mid-Atlantic gale with the loss of all the 135 souls on board.

This was only one of the all too common tragedies which marred the early days of steam power on the deep seas and which revealed that the marine engineer was by no manner of means the master of the North Atlantic. Yet, year in, year out, fair weather or foul, the *Great Western* continued, like some ferry boat, to shuttle between Bristol and New York with almost contemptuous regularity and at speeds which even the lightly laden mail steamers on the shorter route to Halifax could not equal.[4] She made no less than 67 crossings in eight years and thus most truly earned that honour of the Atlantic Blue Riband which she was the first to wear. Her outstanding success was due to three factors: her proportions, her ample power and, above all, the great longitudinal strength which her bridge-builder designer had built into her hull.

The *Great Western* had not completed her second voyage in 1838 before Brunel had started to plan a larger and better ship. Another wooden paddler of not less than 2,000 tons bur-

4. Her best crossings were: Westbound 13 days; Eastbound 12 days 6 hours.

den was at first suggested, but by October he was investigating the merits of iron. It was at this juncture that the little iron steeple-engined paddle steamer *Rainbow* entered the port of Bristol to load for Antwerp. Brunel at once urged Claxton and William Patterson to make the voyage to Antwerp and observe the behaviour of the *Rainbow* in a seaway. Their findings were combined with his own observation and calculations in a report so favourable that the directors of the Great Western Steamship Company resolved forthwith to build in iron. The same building committee, consisting of Claxton, Guppy and Brunel, was authorized to draw up plans and superintend construction at Patterson's Yard.

Four tentative designs were prepared and discarded. The fifth, showing a vessel of the unprecedented size for those days of 3,270 tons gross, was finally approved and on 19 July 1839 the keel plates of the steamship *Great Britain* were laid down by the Company themselves in the capacity of Patterson's employers. This was Brunel's idea, for he had argued that in constructing an iron hull of such size the preconceived notions of a timber shipbuilder might otherwise prove more of a hindrance than a help. But that the Company also decided to construct their own engines was due entirely to a decision of the directors taken against the advice of their engineer.

Tenders for the paddle engines had been received from Messrs Seaward & Capel, Maudslay, Sons & Field and from a young man named Francis Humphrys, the patentee of a trunk engine[5] for which Messrs Halls of Dartford held the manufacturing rights. Although their figure was the high-

5. One of the problems confronting the designers of early marine engines was how to reduce the excessive height (or length) imposed by the orthodox layout of piston rod, crosshead and connecting rod. The trunk engine was one of a number of expedients adopted. Instead of a piston rod, the piston itself had a hollow extension or 'trunk' which worked through the gland on the cylinder cover. The connecting rod was connected directly to this trunk, thus eliminating both piston rod and crosshead.

est, Brunel again urged the Company to accept Maudslay's tender. He considered that Humphrys's estimate was far too low to be at all feasible and advised that young man to think again. The directors disagreed with him, however, and accepted Humphrys's tender, but having done so it soon became apparent that the latter must have prepared his figures without fully consulting Messrs Hall. This firm now pointed out that the construction of a pair of engines of such prodigious size would involve the purchase of special tools and equipment which they might never require again and which, therefore, would have to be charged to the one pair of engines. They urged the Great Western Steamship Company to manufacture Mr Humphrys's engines themselves. It is quite evident from this that the firm were not at all anxious to participate in the *Great Britain* adventure, but the Steamship Company accepted their advice, set about installing the necessary plant and appointed Humphrys engineer in charge.

Poor Francis Humphrys! He was, perhaps, the most pathetic victim of Brunel's ruthless pursuit of perfection. His great engines were never completed and he himself lacked that spring-steel temperament which had carried Brunel through his early disappointments. Yet Humphrys did succeed in leaving a permanent mark upon the engineering world. His engine design required an intermediate paddle-shaft of such dimensions that he could find no firm who would undertake to forge it. All heavy forging work was at this time carried out by the tilt or helve hammer which had scarcely changed at all in design since the seventeenth century even though it might now be steam instead of water powered. The hammer head was carried on one end of a long ash beam or 'helve', the other metal-shod end connecting with a cogged wheel. As this wheel revolved, each cog in its passage raised and then released the hammer. But Humphrys's paddle-shaft was so large that it would 'gag' the largest tilt hammer in existence, that is to say when the billet stood on the anvil the helve would no longer have an effective lift. On 24 November 1839 the distracted Humphrys addressed a letter, begging for advice and help, to James Nasmyth, once one of Maudslay's pupils and now partner in the famous firm of Nasmyth, Wilson & Company, 'I find,' wrote Humphrys, 'that there is not a forge hammer in England or Scotland powerful enough to forge the intermediate paddle-shaft of the *Great Britain*. What am I to do? Do you think I might dare to use cast-iron?' Nasmyth's answer to this appeal was historic. It was to design the steam hammer which would for many years remain one of the engineer's most formidable pieces of heavy armament. 'In little more than half an hour after receiving Mr Humphrys's letter narrating his unlooked-for difficulty,' Nasmyth writes in his autobiography, 'I had the whole contrivance, in all its executant details, before me in a page of my Scheme Book.'

While Francis Humphrys was thus occupied with his engine problems, Brunel had still not decided upon the final form which the hull of the new ship should take. On his orders Captain Claxton's son Berkeley, who had become one of his assistants, made no less than six voyages on the *Great Western* during which he did little else but record the degree and frequency of roll and pitch under different conditions, the performance of her engines on varying steam cut-offs and the effect of such expansive working on fuel consumption. And this had to be carefully set down in accordance with Brunel's minute instructions. Then, in May 1840, there entered the port of Bristol the *Archimedes*, a very remarkable vessel indeed, and this event initiated a further series of recordings and experiments.

The *Archimedes* was a three-masted topsail schooner engined by the Rennie Brothers and launched at Millwall in November 1838. What was so remarkable about her was that she was driven, not by paddle wheels, but by the new form of screw propeller recently invented by Sir Francis Pettit Smith. She was, in fact, the first considerable sea-going vessel in the world to be screw propelled. Guppy took a trip round the coast to Liverpool in her and was so favourably impressed that the directors decided to suspend work on the paddle engines and to charter the *Archimedes* for six months so that Brunel could carry out further experiments. His long report to the directors on the results of these experiments reveals the masterly thoroughness of his investigation and the correctness, in the light of history, of his conclusions. While he had many faults to find with the *Archimedes*, her performance was such that these very faults emphasized the excellence of the principle and he expressed himself strongly in favour of adopting screw propulsion on the *Great Britain*. On the strength of this report Francis Humphrys was ordered to abandon in their partially completed state the mighty paddle engines with which he had hoped to make his name and to design screw engines with the least possible delay. 'Mr Humphrys,' wrote Nasmyth, 'was the man of the most sensitive and sanguine constitution of mind. The labour and anxiety which he had already undergone, and perhaps the disappointment of his hopes, proved too much for him and a brain fever carried him off after a few days' illness.'

Having installed so much costly plant, it went without saying that the screw engines must be built by the Company in Bristol and with the death of their engine designer the responsibility for designing and building them fell upon the little three-man Building Committee. They were now faced with the task not only of building a ship of unprecedented size in an almost untried material but also of introducing a novel form of propulsion which involved the design of special engines. No one but Brunel would have dared to combine so many novelties in one vessel. The progress of the ship was slow, but this was not due to technical problems alone. The Company's financial position was not all that had been hoped and the costly abandonment of the paddle engines cannot have improved the position. The *Great West-*

ern was the Company's only source of revenue and it was no fault of hers that she was not proving a great commercial success.

There was a complete failing upon the part of the Bristol Dock Company to grasp what Brunel had realized from the outset: that the finest ships in the world could not regain for the city her lost pre-eminence unless the port itself kept pace with the times. The improvements which he had carried out in 1833 had only toyed with the problem. Even so, the recommendations and provisions he had made then for keeping the harbour clear of mud had been almost totally neglected with the result that the Cumberland Basin and the Float were shoaled up just as badly as ever. He pointed this out in a report to the Dock Company which he made in 1839 but his criticisms were ignored and his advice was not acted upon. Moreover, his two great ships were driven away to Liverpool by exorbitant harbour dues. His last bid to save the day for Bristol was his plan for a floating pier at Portishead where the largest steamers could lie alongside as at Liverpool, but this, like Joshua Franklyn's far-sighted Sea Mills project, came to nothing. It is true that Brunel built a new entrance lock and made other improvements shortly after the launch of the *Great Britain*, but again it was a case of too little and too late.

If any optimistic Bristolian supposed that the winding up of the Transatlantic Steamship Company marked the end of Liverpool's bid for the American trade he was woefully mistaken. Apart from the man-made asset of her magnificent docks, Liverpool's geographical situation was far more advantageous and it was inevitable that where one had failed another would succeed. In February 1840 there was launched on the Clyde the paddle steamer *Britannia* and she was very soon followed by three Clyde-built sister ships, *Acadia*, *Caledonia* and *Columbia*. All four had been ordered by the newly formed North American Royal Mail Steam Packet Company, which the world would soon know by the simpler title of the Cunard Steamship Company. These ships worked the shorter passage from Liverpool to Halifax and Boston whereas the *Great Western*, still, despite her age, a larger and more comfortable vessel, continued to ply direct to New York. But Edward Cunard had secured for his company a commercial advantage which was to prove absolutely decisive; this was the American mail contract. For it soon became apparent that without the Post Office subsidy which was the fruit of a mail contract, the operation of inter-continental passenger services with the small steamers of the day was not an economic proposition.

Meanwhile, despite this threat from Cunard, in her Bristol dock the *Great Britain* slowly took shape, a shape large enough to swallow two of Cunard's steamships which the Liverpool press had called 'immense'. There can be no doubt that Brunel's big-ship policy was commercially as well as technically sound provided the passengers could be found to fill them, which, at this period, was doubtful.

In his hull design for the *Great Britain* Brunel again showed a much greater concern for longitudinal strength than any shipbuilder had so far displayed. Some of the earliest builders of iron hulls had tended to follow timber shipbuilding practice by putting the strength into the knees and into transverse frames. Brunel, however, realized that in designing a hull of exceptional size in a new material a fresh start had to be made from first principles. Ten iron girders ran along the bottom of the ship for her entire length and to the upper flanges of these an iron deck was secured. This deck was not watertight, but it is easy to see how Brunel evolved from this that cellular form of construction which he later used to great effect and which would become a commonplace of modern shipbuilding technique. In addition to being divided into six compartments by five watertight transverse bulkheads, two longitudinal bulkheads divided the ship up to main deck level. The engines and boilers occupied the large central portion thus formed, while the side

compartments, triangulated for additional strength, formed
the coal bunkers. Above, strong diagonal iron deck beams
gave the whole hull the strength of a box girder while additional longitudinal strength at this level was provided in
two ways: by iron stringer plates in the sides of the ship and
by two massive scantlings of Baltic pine fitted at the angle
between the ship's ribs and the main deck shelfpieces. Certainly it was a hull which must have made many an old
Bristol shipwright rub his eyes or shake his head in be-
wilderment.

There was little more orthodoxy to be found in the engine
room. Brunel's decision to adopt the screw propeller posed a
technical problem the precise opposite of that which the
marine engineer confronts today. Nowadays, turbines have
to be geared down to the propellers; then, the crankshaft
speed of the marine steam engine was so low that it had to be
geared up to the screw shaft because direct drive would have
required a propeller of impossibly large diameter and coarse
pitch. The engine of the *Archimedes* drove the shaft
through straight gears, an arrangement which Brunel condemned as inefficient, unreliable and hideously noisy. The
engines of the *Great Britain* consisted of four inclined cylinders, 7 ft 4 in. diameter by 6 ft stroke. These were mounted
low down in the hull and their connecting rods drove an overhead crankshaft whose bearings were carried in two massive
transverse frames. Water was pumped through waterways in
the crank to cool the bearings. Brunel had at first thought of
driving the screw shaft by multiple belts or 'straps', as he
called them, but after further thought he decided to take the
drive from a large drum on the crankshaft by means of four
sets of toothed chains, the ratio being one to three. After
much careful consideration and experiment the six-bladed
screw which he decided upon was 15 ft 6 in. diameter and 25 ft
pitch, the normal shaft speed being 54 r.p.m. Supplied with
steam at 15 lb per square inch, the engines of the *Great
Britain* indicated 1500 h.p. at 18 r.p.m., and what an impres-

sive sight and sound this must have been! The double-ended boiler was 34 ft long and divided into six compartments each with a furnace fore and aft, making twelve in all. There was a feed-water heater of the type which David Napier had patented in 1842. This consisted of an annular casing round the base of the funnel through which the feed-water was pumped, thus turning to useful purpose the excessive amount of waste heat which had been the cause of the fire on the *Great Western*. The top of the feed heater was open to the atmosphere through a standpipe beside the funnel, steam pressure being so low that water could be fed from the heater through the clacks into the boilers simply by gravity. There can be no doubt that Brunel decided to adopt this device as a result of the accident on the *Great Western* which had so nearly cost him his life.

The *Great Britain* was named and launched on 19 July 1843 by the Prince Consort, who travelled down by a special train from Paddington driven by Gooch with Brunel accompanying him on the footplate. After a company of 600 had sat down to a banquet on board, the ceremony took place. It consisted merely of opening the dry dock sluices at the moment the Prince crashed the bottle against her bows. Bristol, needless to say, held high holiday. Church bells rang, flags flew everywhere and in the packed main streets a series of triumphal arches linked gable to timbered gable of the old city. The greatest moment came when the new ship was towed slowly out of the dock into the floating harbour, when the tumult of gun-firing, martial music, bell-ringing and cheering became positively deafening. Huge and very splendid that long black hull must have looked with the golden Arms of England shining upon her bow. But handsome is as handsome does, and who could then say how she would perform? 'If all goes well,' wrote Brunel to Guppy, a few days later, 'we shall all gain credit, but *quod scriptum est manet*, if the result disappoint anybody, my written report will be remembered by everybody, and I shall have to bear the storm – and all that spite and revenge can do at the Admiralty will be done!'[6]

The *Great Britain* measured 289 ft between perpendiculars and 51 ft extreme breadth so she was both too long and too wide for the Bristol Dock lock chambers. The question of length was of no consequence because prompt handling could work the ship out into the river while the flood of a spring tide made a level. As to the breadth difficulty, the Dock Company agreed to the temporary removal of masonry from the sidewalls of the locks to allow her to pass. In this fashion the ship was moved down from the Float into the half-tidal Cumberland Basin. It was arranged that she should pass out on to the river at high water of the morning tide on 10 December 1844. Claxton, who was aboard the tug, looked back over the stern at the high bows of the *Great Britain* as she shouldered her way into the lower lock and realized that there was going to be precious little room to spare. A moment later the ship had stuck fast. Happily Claxton and his men, acting with great promptitude, just managed to free her, haul her back into the basin and close the lock gates before the tide ebbed. Had they not succeeded the entire contents of the basin would have drained away on the ebb to leave the ship wedged in the lock chamber and probably badly damaged. Claxton and Brunel worked all the short winter day directing the removal of the offending masonry in the hope that the ship could be got out on the night tide, which was the last of the springs. They just succeeded and the tricky operation of getting the ship through the lock in pitch darkness was successfully performed. Excusing himself for his failure to keep an appointment in Merthyr next day, Brunel wrote: 'We have had an unexpected difficulty with the *Great Britain* this morning. She stuck in the lock; we *did* get her back. I have been hard at

6. Brunel is referring to his report on the screw propeller. His reference to the Admiralty is explained by the next chapter.

work all day altering the masonry of the lock.[7] Tonight, our last tide, we have succeeded in getting her outside, and I confess I cannot leave her till I see her afloat again, and all clear of her difficulties. I have, as you will admit, much at stake here, and I am too anxious about it to leave her.'

On 23 January 1845 the *Great Britain* sailed for the Thames to complete her protracted fitting out. She encountered very heavy weather round the Longships and coming up Channel so that her average speed of $12\frac{1}{3}$ knots seemed to bode well. While she lay at Blackwall she created an immense amount of interest. It was like the Thames Tunnel days all over again as London society, led by the Queen and her Consort, flocked down east to admire her spacious saloons, and her 64 staterooms as they perambulated the thousand and more yards of Brussels carpeting. It was obviously unthinkable that the Queen could penetrate below decks, so the appearance and action of the engines was demonstrated to her by Brunel with the aid of a model which had been specially made for the occasion.

At long last the *Great Britain* left Liverpool for New York on her maiden voyage on 26 August. She went out in 14 days 21 hours and returned in $15\frac{1}{2}$ days. On her second voyage she broke her propeller on the return passage which she completed under sail (she carried six masts) in 18 days. The experience of these two voyages had shown her to be deficient in steam-raising power, and over the winter, while she was being fitted with a new propeller, the opportunity was taken to modify her boilers. How effective these modifications

7. The present disused south entrance lock, known as the 'Brunel Lock', is often pointed to as the scene of the *Great Britain's* misadventure. Indeed it is said that evidence of the hurried work of 10 December 1844 can still be seen. Unfortunately this is not correct. The larger of the two locks at this time was the northern one and this was naturally the scene of the incident. Brunel subsequently rebuilt and enlarged the smaller southern lock which still bears his name, while all trace of the old northern lock disappeared when the present north lock was built.

were may be judged from the fact that on her first voyage in 1846 she averaged 11¾ knots for the first five days until she had the misfortune to be delayed by the breakage of an air-pump strap. In spite of other minor engine-room troubles, the *Great Britain* made the return passage of this trip and both the out and home passages of her July voyage in around thirteen days. For so unconventional a design, mechanical teething troubles were only to be expected and this increasing regularity promised well. Brunel and his colleagues of the Great Western Steamship Company must have just acquired confidence in the successful future of their fine new ship when disaster struck as swiftly and unexpectedly as the fire that first morning on the *Great Western*.

On 22 September 1846 the *Great Britain* left Liverpool for New York with 180 passengers, the largest complement any trans-Atlantic steamer had ever carried up to that time. A few hours later when night had fallen the ship struck. Passengers rushed from their cabins in terror as the *Great Britain* suddenly shuddered, lurched and then could be heard pounding over rocks with a hideous grinding noise. 'I cannot tell you of the anguish of that night!' an emotional lady passenger told the *Illustrated London News* reporter afterwards.

The sea broke over the ship, the waves struck her like thunder claps, the gravel grated below. There was the throwing overboard of coal, the cries of children, the groans of women, the signal guns, even the tears of men, and, amidst all, the Voice of Prayer, and this for long dark hours. Oh! what a fearful night!

Quite obviously this good lady was determined to wring the last drop of anguish out of her situation. 'The newspapers by no means represent the extent of the danger,' she insisted tartly. In fact there was none, although in any other ship but the *Great Britain* the danger would have been so great that no tale-tellers might have survived. For it is doubtful whether any wooden ship could have withstood such stresses without breaking up. But, thanks to that immense strength which Brunel had built into her hull, the *Great Britain* lay intact and as strong as a bridge girder while the panicking passengers were ordered back to their cabins to await daylight.

To us the most incredible thing about this bygone disaster is that no one on board, from Captain Hosken downwards, had the faintest idea where they were. At the moment the ship struck, Hosken had supposed that she was rounding the Isle of Man; perhaps he thought she had run on the island. He was subsequently exonerated from blame because serious errors were discovered in a new chart for the area with which he had been issued. Also, as his later correspondence reveals, Brunel suspected that the iron hull may have affected the ship's compass, a point he had not apparently thought of before the accident.

When dawn broke over the stranded ship, her crew saw to the south-west across a wide bay a range of high mountains sweeping down majestically almost sheer to the sea. They were the mountains of Mourne. The *Great Britain* was well and truly aground in Dundrum Bay in the County Down. At low tide a procession of Irish carts carried all the disconsolate passengers and their baggage to safety over the sands. Immediate efforts to get the ship off by lightening her failed and there she remained until Captain Claxton arrived from England. Although the bay appeared to consist only of sand, rocks were concealed beneath it and on these the ship had holed herself in two places. She lay with her stern and port quarter fully exposed to the prevailing weather from south and west. Nevertheless, Claxton decided to make an attempt to get her off on the high spring tide of Monday 28 September. On Sunday, however, a violent southerly gale sprang up, raising such seas that at high water they were breaking right over the ship. He was forced to abandon any hope of refloating; instead he set sail on her and drove her farther up the beach where the waves assailed her less savagely. Yet it was obvious to Claxton that no ship, however strong, could survive a whole winter on that exposed shore.

Next to arrive at Dundrum were William Patterson and an expert in ship salvage work named Alexander Bremner. These two attempted to construct breakwaters to protect the ship, but as fast as they did so the seas carried them away. It was at this juncture that the directors of the Great Western Steamship Company appear to have given up all hope of protecting or salvaging the ship and to have resigned themselves to writing her off as a total loss.

It was in the autumn of this year that Brunel was struggling with the installation of the atmospheric system in South Devon as well as with an overwhelming amount of Parliamentary business. It was quite impossible for him to get over to Ireland for a time, bitter though his feelings must have been. When at length one bleak December day he stood on the sands of Dundrum and saw his splendid ship lying abandoned to wind and weather, a cold fury seized him, that sudden intolerant rage against the incompetence and stupidity of his fellows which can sometimes overmaster the man of pride and intellect. Having examined the ship and her situation he decided how she must be protected until the weather served and ordered the crestfallen Captain Hosken to proceed with the work immediately in accordance with his instructions and on his personal responsibility. Not a moment was to be lost. To Claxton he wrote:

... I was grieved to see this fine ship lying unprotected, deserted and abandoned by all those who ought to know her value, and ought to have protected her, instead of being humbugged by schemers and underwriters. ... The result ... is that the finest ship in the world, in excellent condition, such that £4,000 or £5,000 would repair all the damage done, has been left, and is lying, like a useless saucepan kicking about on the most exposed shore that you can imagine, with no more effort or skill applied to protect the property than the said saucepan would have received on the beach at Brighton. Does the ship belong to the

Company? For protection, if not for removal, is the Company free to act without the underwriters? If we are in this position, and if we have ordinary luck from storms for the next three weeks, I have little or no anxiety about the ship; but if the Company is not free to act as they like in protecting her, and in preventing our property being thrown away by trusting to schemers, then please write off immediately to Hosken to stop his proceeding with my plans . . .

As to the state of the ship, she is as straight and as sound as she ever was, as a whole. . . . I told you that Hosken's drawing was a proof, to my eye, that the ship was not broken: the first glimpse of her satisfied me that all the part above her 5 or 6 feet water line is as true as ever. It is beautiful to look at, and really how she can be talked of in the way she has been, even by you, I cannot understand. It is positively cruel; it would be like taking away the character of a young woman without any grounds whatever.

The ship is perfect, except that at one part the bottom is much bruised and knocked in holes in several places. But even within three feet of the damaged part there is no strain or injury whatever. . . . There is some slight damage to [the stern], not otherwise important than as pointing out the necessity for some precautions if she is to be saved. I say 'if', for really when I saw a vessel still in perfect condition left to the tender mercies of an awfully exposed shore for weeks, while a parcel of quacks are amusing you with schemes for getting her off, she in the meantime being left to go to pieces, I could hardly help feeling as if her own parents and guardians meant her to die there. . . . What are we doing? What are we wasting precious time about? The steed is being quietly stolen while we are discussing the relative merits of a Bramah or a Chubb's lock to be put on at some future time! It is really shocking . . .

Having thus relieved his feelings, Brunel ended his letter by describing what should be done to protect the ship until the following spring, when the weather would enable salvage operations to begin.

I should stack a mass of large strong fagots lashed together, skewered together with iron rods, weighted down with iron, sandbags, etc., wrapping the whole round with chains, just like a huge poultice under her stern and halfway up her length on the sea side. ... I have ordered the fagots to be begun delivering. I went myself with Hosken to Lord Roden's agent about it, and I hope they are already beginning to deliver them.

This result of Brunel's whirlwind visit to Dundrum galvanized everyone into activity from the directors in Bristol to the sailors on the spot and the local Irish. The directors sent Claxton out again post haste to superintend the work Brunel had proposed, and soon strings of Irish carts began to creak to and fro across the strand whenever the tides allowed carrying bundles of faggots and beech spars cut not only from Lord Roden's wood but from the neighbouring demesnes of Lord Downshire and Tyrella. At first, matters did not go too well. It was difficult to establish a foundation for the huge poultice. Despite all efforts to weight and lash the bundles down, the seas scoured beneath them and swept them away, sometimes as many as 200 at a time. Even Claxton, a determined and resourceful man, wise in the ways of the sea, began to lose heart. He told his tale of woe in a letter to Brunel, but if he expected sympathy he was disappointed. The reply brought no shred of comfort. It read:

You have failed, I think, in sinking and keeping down the fagots from that which causes nine-tenths of all failures in this world, from not doing quite enough. ... I would only impress upon you one principle of action which I have always found very successful, which is to stick obstinately to one plan (until I believe it wrong), and to devote all my scheming to that one plan and, on the same principle, to stick to one method and push that to the utmost limits before I allow myself to wander into others; in fact, to use a simile, to stick to the one point of attack, however defended, and if the force first brought up is not sufficient, to bring ten times as much; but never to try back upon another in the hope of finding it easier. So with the fagots – if a six-bundle fagot won't reach out of water, try a twenty-bundle one; if hundredweights won't keep it down, try tons.

Shamed and goaded into further effort by this admonition Claxton tried again and this time succeeded in getting a firm footing upon which the great protective screen was built up. He was astonished by the effectiveness of this mattress. No sooner had its foundation been secured than the force of the seas was far more effectually broken than by the breakwater of heavy timber baulks which Patterson and Bremner had attempted to build and which had been repeatedly swept away. Moreover the foundations became so consolidated in the sand that it was only with the greatest difficulty that they were cut away when the time came to move the ship next May. It was, wrote Claxton afterwards, 'a mass ... more difficult to move than granite rock'.

In the late spring the *Great Britain*'s head was laboriously raised over 8 ft by lightening her and by ramming wedges under her at high water. In this way it was possible to get at her bottom and to repair temporarily the leaks to an extent sufficient to be mastered by pumping. The scheme then was to await the highest of spring tides before making resolute efforts to warp her seawards. She had been lightened to such an extent that it was hoped that the tide would lift her sufficiently to clear the reef which barred her passage to the sea. But there followed weary weeks of soul-destroying delay and disappointment for Claxton and his men as the tides repeatedly failed to reach their anticipated height. Then, at last, on 27 August, the tide flowed full and strong. 'Huzza! Huzza!' wrote Claxton to Brunel,

you know what that means. .... The tide rose to 15 ft 8 inches. She rose therefore easily over the rock, but was clear of it by only just five inches, which shows how near a squeak we had – it was a most anxious affair but it is over. I marked 170 yards in the sand and on our warp, and at that extent I stopped her. ... I have no doubt that tomorrow we shall see her free.

Even then, Claxton's anxieties were not at an end. He had hoped when he wrote to Brunel that he would get the ship to

Liverpool next day. Her Majesty's ships *Birkenhead* and *Victory* were standing by in the bay ready to take her in tow as soon as she floated off. Finding that she was still making a great deal of water notwithstanding the temporary repairs, Claxton had engaged a force of 120 Irish labourers to man the pumps. She had, of course, no steam auxiliaries because everything possible had been dismantled and removed from her engine-room. Unacquainted with the ways of the Gael upon his native soil, imagine poor Claxton's dismay on the morning of the 28th when only 36 of the pumping gang turned up and these few, instead of getting to work as the tide rose, held a long and animated discussion amongst themselves as to how much money they should or should not accept. As a result, by the time she was taken in tow, the *Great Britain* had six feet of water in her engine-room. Although, with the help of volunteers from the two naval vessels, the pumps began to gain at the rate of four inches an hour, the idea of making straight for Liverpool had to be abandoned. Instead it was proposed to run the ship aground in the shelter of Strangford Lough so that she could be cleared of water. They had just sighted Killard Point when this plan, too, was foiled as a dense sea fog suddenly rolled down and completely obscured the coast. To have attempted the narrow entrance channel to Strangford in such circumstances would have been folly, so the little convoy with their water-logged charge wallowing heavily in their wake struggled on round the peninsula to Belfast Lough. Here the *Great Britain* was grounded and cleared of water while Claxton recruited another and more reliable pumping gang in Belfast docks. Thanks to their continuous efforts the ship was towed safely to Liverpool next day. But it was touch and go, for no sooner had she been manoeuvred into place over the gridiron where she would be repaired than she sank immediately the pumps were stopped.

So the *Great Britain* was saved to sail again, but at a cost which was more than the already straitened resources of the

Great Western Steamship Company could stand. It was wound up and the two ships were sold. The gallant *Great Western* was acquired by the Royal Mail Steam Packet Company. For ten years thereafter she voyaged between Southampton and the West Indies with the same exemplary regularity which she had shown on the North Atlantic. She was broken up at Castles' Yard, Millbank, in 1856/7, Brunel himself visiting her as she lay there awaiting her end.

Although the grounding of the *Great Britain* at Dundrum was so commercially disastrous, no more convincing practical demonstration of the correctness of Brunel's theories of iron shipbuilding could possibly have been devised. By this long ordeal the *Great Britain* showed that she was as nearly indestructible as such a ship could be, but if there were any who still doubted this they must have been convinced by her subsequent career. When she had been repaired she did not, so to speak, leave the Great Western family, for she was bought by the firm of Gibbs, Bright & Company of Bristol and Liverpool, who fitted her out for the Australian trade. She was fitted with new oscillating engines of 500 nominal horsepower by John Penn of Greenwich and a new screw. This reduced her power by nearly fifty per cent, yet apparently she was still able to steam at ten knots without sails. She was re-rigged, her original six masts, one square and the rest schooner rigged, being replaced by four, of which two were square rigged. Evidently this was still not satisfactory, for she was soon altered to a three-masted full rigged ship in which guise she became famous and was described as 'a very majestic ship'.

In 1852 the *Great Britain* steamed out of Liverpool on her first voyage to Melbourne with 630 passengers and a million pounds' worth of specie on board. It was the first of 32 voyages in 23 years during which time she became one of the most popular and most celebrated ships in the Australian trade. It was a long record of continuous service which was only interrupted by the Crimean War and the Indian Mutiny, when she served as a troopship. In 1875 the ship was laid up and advertised for sale. In 1882 she passed into the ownership of the firm of A. Biggs, Sons & Company and to the astonishment of all sailormen, who thought they had seen the last of her, she reappeared in the Mersey in November of that year as a sailing ship with her iron hull sheathed in wood. In this new guise she made two voyages to San Francisco. Then, on 6 February 1886, the gallent old ship put out from Penarth for Panama on what was to be her last voyage. When rounding the Horn, her cargo of coal was found to be on fire and her crew forced her captain to put back to the Falkland Islands, which she reached on 25 May. Here she was surveyed and condemned. The Falkland Islands Company purchased her as a hulk for the storage of wool and coal, which duty she performed until 1933. In that year the then Governor of the Falkland Islands, Sir Henniker Heaton, launched an appeal for the preservation of the famous ship, but the estimated cost of restoration was found to be too high and the project was dropped. In 1937 the *Great Britain* was towed away to a quiet grave in Sparrow Cove, Falkland Islands, where the indestructible hull of Brunel's splendid ship could be seen until 1970, when a privately funded rescue mission brought it back to Bristol to be restored.

Like the seven-foot gauge, the Great Western Steamship Company was a magnificent failure. The luckless accident to the *Great Britain* merely hastened an end which only the landing of a mail contract could conceivably have averted. Even this might not have saved them for the Company put themselves in a hopelessly weak position as soon as they decided to build the engines of the *Great Britain* themselves at Bristol and it is worth remembering that this decision was taken against Brunel's advice. The building of the hull of the ship by the Company involved no great capital commitment in fixed plant because William Patterson's dock was used for the purpose, but the engines were a different story altogether. On their completion the Company was left in the absurd position of owning an extensive plant for the construction and maintenance of its fleet situated at a port which that fleet had already outgrown. Such a plant was a doubtful investment in the best of circumstances; and to instal it at Bristol seems on the face of it sheer lunacy. One can only suppose that, just as Brunel had believed that other railway companies must adopt his broad gauge, so the directors of the steamship company supposed that the success of the *Great Western* would persuade the city of Bristol to rebuild its docks upon the grand scale.

The unfortunate Company had a life of only ten years and owned only two ships. But what ships they were! Both the largest ships afloat in their day, one was the first true transAtlantic steamship, while the other was at once the first screw ship and the first iron ship to cross any ocean. And just as the feats performed upon the broad gauge stimulated railway development, so the performance of the *Great Western* and the *Great Britain* on the high seas very probably advanced the progress of ocean steam navigation by a quarter of a century.

Brunel against
Bureaucracy

EVEN in an age of individualism, Brunel's public life was remarkable for his roundly expressed hatred of government officials, and of any law, rule or regulation which interfered with individual responsibility or initiative. The Patent Laws were one of his anathemas, for it was his belief that, by enabling astute firms or individuals to take out patents of principle, they stifled invention instead of encouraging it. He himself obstinately refused to protect any of his ideas, with the consequence that his design for a rifle with a polygonal barrel, which was made for him by Westley Richards in 1852, was subsequently covered by Joseph Whitworth's patent. The latter had seen Brunel's rifle when he visited Richards' shop in 1854. Four years later Richards pointed out to him that his rifle infringed the Whitworth patent. 'I have never seen Whitworth's patent,' Brunel retorted. 'What is it exactly he does patent? It cannot be merely the polygon.' But it was, and it could be said that from that small beginning there grew a great armament business.

When in 1841 a Select Committee was set up to consider the desirability of appointing Government Inspectors of railways, Brunel was almost alone in flatly opposing any form of government supervision whatsoever. Railway engineers, he said, 'understood very well how to look after the public safety, and putting a person over them must shackle them. They had not only more ability to find out what was necessary than any inspecting officer could have, but they had a greater desire to do it.'

In 1848, when he was asked for his views by the Royal

Commission on the Application of Iron to Railway Structures, he was equally forthright. He said:

> If the Commission is to inquire into the conditions '*to be observed*', it is to be presumed that they will . . . lay down, or at least suggest, 'rules' and 'conditions' to be observed in the construction of bridges, or, in other words, embarrass and shackle the progress of improvement tomorrow by recording and registering as law the prejudices or errors of today. . . . No man, however bold or however high he may stand in his profession, can resist the benumbing effect of rules laid down by authority. . . . Devoted as I am to my profession, I see with fear and regret this tendency to legislate and to rule . . .

Thus the Commissioners, who had been expecting a technical discourse on bridge-building, received instead a dusty answer.

Brunel's especial scorn, however, was reserved for government departments and their officials, the Admiralty in particular. The innate caution of the civil service mentality, its inability to take unequivocal decisions or accept personal responsibility, represented the very opposite of all that Brunel stood for. Moreover, he was always mindful of the way his father had been treated by Government and, apart from the abortive Woolwich dock project in his early years which had petered out in the usual futile way, he managed to steer clear of Government schemes until 1841.

In that year Sir Edward Parry, Controller of Steam Machinery for the Navy Board, suggested to the Lords of the Admiralty that when an engineer of the eminence of I. K. Brunel contemplated using the screw propeller in his new ship it was time their Lordships investigated the matter. A request for a copy of the report on the *Archimedes* experiments was followed by an interview at which it was agreed that a screw vessel should be built for the purpose of further experiment under Brunel's superintendence. It was a condition of his consent that he should have full responsibility without the intervention of any government official and that he should communicate direct with the Lords or with Sir Edward Parry only.

Within a fortnight of this interview, Brunel had drawn up specifications for the engines of this vessel and obtained a tender from Maudslay, Sons & Field which the Admiralty accepted. Meanwhile he and Claxton carried out some experiments at Southampton with the paddle steamer *Polyphemus* to obtain figures which could be compared with the otherwise similar screw vessel. So far so good. When the engines, which were of the Siamese[1] type, and of 200 n.h.p., were practically completed, Brunel inquired about the hull which the Navy Board were supposed to be building. It could not be found. Oh yes, the ship had been ordered, but evidently it had not been laid down. At this point it became clear that Brunel had enemies at the Admiralty who were prepared to put every obstruction they could in his path. Exactly what their motive was is not clear, but in any case he was the last man to prove popular with civil servants.

The mystery of the missing ship caused quite a flutter in Whitehall and Brunel was summoned to attend Sir George Cockburn, the First Sea Lord. He found Sir George in that condition of apoplexy which, so far as its high-ranking officers are concerned, seems to be part of the tradition of the Senior Service. 'Do you mean to suppose that we shall cut up Her Majesty's ships after this fashion, sir?' he roared, pointing a quivering finger at an extraordinary model on his desk which showed an old three-decker with a large slice cut out of its stern in a fashion quite ridiculous. Examining this curious object with amused interest, Brunel saw that it bore a neat label inscribed: 'Mr Brunel's Mode of Applying the Screw to Her Majesty's Ships.' Quite unruffled, he smilingly disclaimed any responsibility and while Sir George

1. In the Siamese engine the piston rods of the two vertical cylinders were connected to a common T-shaped crosshead and a single connecting rod which drove the overhead paddle shaft.

fumed and messengers scurried off in search of subordinates who might explain matters, he sat unconcerned, scratching off the offending label with his pocket knife. It appears that the Surveyor of the Navy was, as he continued to be, the villain of the plot, but that gentleman had prudently absented himself on this occasion and was nowhere to be found.

The next move in this protracted farce was an intimation that a ship named the *Acheron* would be prepared to receive the engines and screw. To this Brunel replied that the hull of the *Acheron* with its full after-body was utterly unsuitable for the purpose. An inscrutable silence fell over the Admiralty which even Brunel's inquiries failed to penetrate, so after four months he sent in his resignation. This awoke the sleepers in Whitehall, for he received an immediate summons to another meeting which he afterwards described in a letter as follows:

Not a word was said about my complaint of the past, but they said they wished me to continue the experiments, and that my screw was to be tried first. I said that was not at all what would suit me; that I would, if they wished it, conduct an experiment as originally proposed; that I had no screw, that I was no competitor, but an arbitrator in whom the Admiralty had perfect confidence; that I was this or nothing.

Evidently Britannia, though wakened, was still a trifle fuddled. Just because she had mistaken him for another of those tiresome inventors there was no need for this Mr Brunel to be so brusque. 'Then commenced,' Brunel went on, 'a tedious fencing . . .

However, it ended in all parties being written to, and told that they were to follow my directions, and that I was to proceed to give such instructions as should enable a full experiment to be made of all screws generally. I then requested that this time I might have my instructions in writing . . .

A few days later he was notified that the sloop *Rattler* of

888 tons burden would be adapted for screw propulsion under his direction. This ship was little better than the *Acheron* for the purpose and that Brunel's enemies were still active in Whitehall was revealed very clearly by Captain Claxton when, in his old age, he recalled the affair in a letter to Henry Brunel:

Every effort that could be made by H.M. Surveyor was made to prevent the success of the screw [the old man wrote], and the vessel was not such as I.K.B. approved. He had often (through me, C.C.) tried to find out where the ship was building, but never could get any information and never was consulted. All at once there was a flare-up at the Admiralty and *Ratter* [sic] was named.

At her trials no fair play was given. She came out uncoppered – not even dubbed down, but roughened as much as a vessel could be with lumpy pitch. She answered well nevertheless and is a pattern to this day of the successful results of his calculations.

The *Rattler* was actually built at Sheerness. As in the case of the *Great Britain* her engines were geared up to the propeller shaft, but here the drive was so arranged that the ratio could be changed in the course of experiments with propellers of different design and pitch. One of the numerous tests carried out consisted of a 'tug-of-war' between the *Rattler* and the otherwise similar paddle sloop *Alecto*. With both vessels steaming full ahead, the *Rattler* pulled her rival away at the rate of 2.8 knots. Altogether, despite every machination and delaying tactic of Whitehall, the *Rattler* tests proved the superiority of the screw and led to its rapid adoption in the British Navy. After this experience it is not surprising that it was only the exigencies of war which induced Brunel to have any further dealings with government.

This book has contained practically no reference to any extraneous national or international event since the Bristol riots of 1831. There is good reason for this. In the years of peace which followed the battle of Waterloo no other historic event occurred which had any bearing upon Brunel's life and work. On the contrary, throughout these years it was Brunel and his fellow engineers, not the soldiers or the statesmen, who were shaping history. They found England still an agricultural country and they left her an industrial one. It was a black and ugly business, but Brunel was far too fully occupied in driving his iron roads and building his great ships to become involved in the social upheavals which followed in their wake. Only in 1848 does a ripple of the panic roused by the conjunction of the second French Revolution with the Chartist movement ruffle for the moment the ordered and opulent calm of 18 Duke Street, when Brunel enrolled as a special constable for the Westminster district. It had seemed then that the English throne might fall as others had done, but nothing happened, the peace of Westminster was not disturbed, and with the death of O'Connor a movement which had included Marx and Engels among its members petered out. But in 1853 the peace of Waterloo came to an end at last. Europe went to war again and England decided that if the balance of power was to be preserved the ambitions of Imperial Russia must be curbed. The despatch of a British expeditionary force to the Crimea and the campaign which followed had an effect even upon Brunel's busy life.

In the conduct of the Crimean campaign the new and the old Englands attempted to combine operations with catastrophic and shocking results. Only steam power at sea and the use of the electric telegraph, by vastly extending the range of military striking power, made possible the expeditious transport of an entire army to the remote Bosphorus and thence from Varna to Sevastopol. But the steam transports, the *Great Britain* and her sisters, alone belonged to the present. The army which paraded through London to board them with its gleaming cavalry charges, its brilliant uniforms and plumed shakoes may have presented a superb spectacle but it was one which had become as out of date as a scene in some village pageant. For it was an army whose organization, strategy and supply services had not changed since Waterloo and which was to prove itself totally unfitted to wage long-range warfare.

The protracted carnage before Sevastopol and the futile bombardment of the famous Five Forts of Kronstadt by combined British and French naval forces filled Brunel with impotent rage. It appeared to him that the flower of the British army was being wantonly sacrificed to that very arrogance, incompetence and obstructionism of government officials which he had hated and scorned all his life. If only these government departments would use the engineering resources of the day, these things would never happen, was his argument. 'I should like a contract for taking Cronstatt', he wrote bitterly to Claxton, 'I will find a Company to do it.' In answer to the obvious question, Brunel designed his extraordinary monitor or 'floating siege gun' and even went so far as to construct a model of it.

When afloat, the model of this remarkable craft resembled a submarine with only its conning tower above the surface. In the gunboat this conning tower consisted of a heavily armoured hemispherical turret housing a single 12-inch gun firing three rounds a minute, the ammunition being raised from the submerged hull and breech-loaded mechanically. So that there should be no vulnerable screw or rudder to be damaged or fouled, the hull was jet propelled by steam. There were, in fact, three jets, one main jet right aft for forward propulsion and two smaller lateral jets for manoeuvring. The controls of these jets would be under the hand of the gun-layer in the turret so that by their aid he could bring the gun to bear on its target. The jet propulsion machinery was intended only for short-range use, and the method by which Brunel proposed bringing the gunboats to the theatre of war was prophetic of the scenes which would take place on the shores of his ancestral Normandy nearly a century later.

'I propose', he wrote, 'a ship of the class of small screw colliers made to open at the bows and its contents floated out ready for action.'

Brunel had considerable correspondence on the subject of the gunboats with General Sir John Fox Burgoyne, who became deeply interested and suggested a number of modifications and improvements in the light of his experience. Sir John was at this time without doubt the greatest living military engineer. He had been chief of engineers under Sir John Moore at Corunna, had been engaged on the construction of the famous defence lines of Torres Vedras and became commanding engineer on Wellington's staff. He filled a similar post under Lord Raglan, the commander of the British force in the Crimea, and was responsible for organizing the disembarkation at Old Fort and the first dispositions for the Siege of Sevastopol. Although he was by this time a veteran of seventy-two, Sir John's mind was evidently more open to new ideas than those of many of his colleagues. He mentioned the scheme to Lord Palmerston and advised Brunel to take up the matter with the Admiralty. This Brunel, remembering his previous experience, stubbornly refused to do with the consequence that Claxton, who believed passionately in the merits of his friend's invention, surreptitiously purloined plans and model from the cupboard in Duke Street to which they had been relegated and bore them off to the Admiralty himself.

What Brunel had to say to Claxton when he found out what he had done we do not know, but he shortly afterwards wrote again to General Sir John Burgoyne in such a strain that that eminent soldier could no longer entertain any doubts as to his correspondent's opinion of the Admiralty. 'You asssume', wrote Brunel,

that something has been done or is doing in the matter which I spoke to you about last month – did you not know that it had been brought within the withering influence of the Admiralty and that (of course) therefore, the curtain had dropped upon it and nothing had resulted? It would exercise the intellects of our acutest philosophers to investigate and discover what is the powerful agent which acts upon all matters brought within the range of the mere atmosphere of that department. They have an extraordinary supply of cold water and capacious and heavy extinguishers, but I was prepared for and proof against such coarse offensive measures. But they have an unlimited supply of some *negative* principle which seems to absorb and eliminate everything that approaches them. ... It is a curious and puzzling phenomenon, but in my experience it has always attended every contact with the Admiralty.

That, of course, was the end of the gunboat and with the fall of Sevastopol any immediate need for such a weapon disappeared. There is, however, a postscript to this story. Shortly after Brunel's death, his old friend, loyal to the last, stumped along to the Admiralty to retrieve the model of which he had thought so highly. Claxton's inquiry was met at first with the usual blank incomprehension and later, as the old Captain stubbornly persisted, with an ant-like scurrying to and fro of irritated bureaucrats. The model could not be found. But at last a light of comprehension dawned upon one official's face. 'Oh, *I* know,' he exclaimed. 'It is a duck-shooting thing, is it not, painted white?'

It was during the winter of 1854–5 that the shortcomings of British military organization in the Crimea became most shockingly apparent. On the night of 14 November a violent storm sank no less than 33 ships laden with precious winter stores. Even worse in its effect was the loss in this one night of fourteen days' supply of fodder for the draft horses on which the transport of all supplies from Balaclava to the troops investing Sevastopol was utterly dependent. This disaster, however, serious though it undoubtedly was, cannot absolve the authorities from responsibility for the appalling privation and hardships which the troops suffered on the exposed plateau of the Upland which confronted the Russian fortress. An efficient and elastic organization would have adjusted itself to such a disaster as to any other hazard of war whereas this archaic, rigid and therefore brittle system broke.

So appalling did the situation become that, at its worst, there were only 11,000 British troops left before Sevastopol and no less than 12,000 in hospital. And what hospitals! This was the greatest scandal of all. Britain's Turkish allies had offered her the huge quadrangular barrack building at Scutari as a base hospital and, perhaps because of its impressive appearance, the offer was accepted without any consideration for its grave natural disadvantages as a hospital and its complete lack of amenities. Nor was any effort made to remedy its deficiencies. The sick and wounded from the Crimea must necessarily be brought to Scutari by sea, but there were no facilities for landing them. They had to be brought ashore in caiques or rowing boats and thence carried up precipitous slopes on stretchers. In the hospital itself there was no proper water supply or drainage system. An inadequate number of open privies without any means of flushing or cleansing stood beside the main water storage tank, the supply to which was eventually found to be flowing through the rotting carcass of a horse. Small wonder that a wounded man stood a better chance of recovery on the Upland than he did by being brought to Scutari. The whole vast barracks became a charnel house packed with dying men too numerous to receive any effective aid. And for every one who died of his wounds, three died from the effects of the fearful miasma of ordure, bodily corruption, filth and disease which pervaded every crowded corner of this nightmare building.

Fortunately, as is well known, there was at Scutari one person whose inexhaustible energy and withering contempt for red-tape and officialdom was equalled only by Brunel. Her name was Florence Nightingale. But for that energy which, like his, seemed able to dispense with the need for sleep, the situation at Scutari would have been even more catastrophic than it was; but for that contempt, which she expressed so pungently in her reports, the true state of affairs at Scutari might never have become known in England. As it was, by January 1855 the condition of the troops in the Crimea had become a national scandal. The country was profoundly shocked and when, on the 26th of the month, John Roebuck, the radical member for Sheffield, moved a vote of censure on the Government and proposed the appointment of a committee of inquiry it was carried in an uproar. The Government fell, Lord Palmerston succeeded Lord Aberdeen as Prime Minister, and at the end of February a Sanitary Commission was sent out to Scutari by Lord Panmure, who combined the hitherto separate offices of Secretary of War and Secretary at War.

Yet this storm and the government upheaval which it provoked left undisturbed and secure in his permanent office the one government official above all others whom Florence Nightingale hated. He was Permanent Under Secretary at the War Office and to her he appeared to be the arch obstructionist concerned only to oppose by every unscrupulous means the introduction of those reforms of the army medical services to which she was to devote her life. It was of him that Florence wrote:

He was a dictator, an autocrat irresponsible to Parliament, quite unassailable from any quarter, immovable in the midst of a so-called constitutional government and under a Secretary of State who *is* responsible to Parliament.

Now here we come upon one of those fascinating and apparently insoluble contradictions with which history will sometimes present us. For Florence Nightingale's arch villain was none other than Benjamin Hawes – now Sir Benjamin Hawes, K.C.B. – Brunel's brother-in-law, his nearest and dearest friend and the only government official for whose energy and initiative he always cherished a profound admiration. The idealistic man full of reforming zeal whose election to the first reformed Parliament Brunel had sup-

ported; the capable man of affairs whose drive, ability and
power to command Brunel so much respected and had more
than once called to his aid – how are we to reconcile this
character who emerges so clearly from Brunel's diaries and
letters with Florence Nightingale's reactionary monster?

It becomes even more difficult to believe her description
when we find this autocrat writing to his friend Brunel on 16
February 1855 to ask if he would be willing to design an improved hospital for the Crimea which could be quickly built
in England and then shipped out for assembly on some predetermined site. We can only suspect that the answer to this
contradiction may possibly lie in Florence's own character.
Like Brunel's, it was a character quite impossible to ignore;
she was either loved or hated; people either became her devoted disciples or else, in an age which still believed that a
woman's place was in the home, they intensely disliked her
as a busybody interfering in men's affairs. It is clear that Sir
Benjamin Hawes belonged to the second group and his re-
sentment may have found expression in an active and possibly unjust way. His invitation to Brunel may have been
prompted by a desire to show this tiresome woman what a
man could achieve without her interference. Certainly the
name of Nightingale is never mentioned in the correspondence between the two friends, while for her part, Florence
appears to have ignored completely the very remarkable
fruits of their collaboration.

Brunel's reply to Hawes, written on the day he received the
inquiry, shows that he must already have given considerable
thought to the matter. The revelation of the scandalous conditions at Scutari caused so much controversy that Hawes's
formal invitation in all probability merely confirmed
officially a project already discussed and agreed in private
conversation between them. Be this as it may, Brunel replied: 'This is a matter in which I think I ought to be able to
be useful and therefore I need hardly say that my time and
my best exertions without any limitations are entirely at the

Service of Government,' and having thus pledged himself he set to work with a speed remarkable even for him. Only six days later he was writing to Hawes outlining the conditions desirable and requesting that contoured sketch maps of suggested sites should be submitted to him for consideration. An initial contract for buildings for a hospital of 1,000 beds was already placed on his own initiative. An outraged squeak of protest from the War Office Contracts Department at this unorthodox and precipitate behaviour produced only the following retort:

Such a course may possibly be unusual in the execution of Government work, but it involves only an amount of responsibility which men in my profession are accustomed to take. . . . It is only by the prompt and independent actions of a single individual entrusted with such powers that expedition can be secured and vexatious and mischievous delays avoided. . . . These buildings, *if wanted at all*, must be wanted before they can possibly arrive.

To Hawes on 5 March he explained the idea behind his design for the hospital. Because it has governed the layout of similar temporary buildings ever since, it reads now like a statement of the obvious, but it was not so then. It was

that the aggregate of the buildings should consist of such parts as might be conveniently united into one whole under great variations of conditions of the form and nature of the site. That the several parts must be capable of being formed into a whole united by covered passages, and that it should be capable of extension by the addition of parts to any size.

He then went on to explain the layout in detail. Each standard unit would consist of two wards each for twenty-four patients and it would be completely self-contained with its own nurses' rooms, water closets, outhouses and other details 'so that by no accident can any building arrive at its destination to be erected without having these essentials complete'. Each patient was allowed 1,000 cubic feet of air space and one large ventilator fan was provided for each unit. This fan, he was careful to point out, was designed to force air *into* the wards and not to extract it as that might draw smells from the closets into the wards. There were fixed wash basins and invalid baths of his own design, while each unit would be sent out with its own wooden trunk drainage system. Surgery, dispensary and officers' rooms would consist of the same standard units; only the kitchen, laundry and bakehouse would be metal instead of wooden buildings because of the fire risk. In the wooden ward buildings, all heating and lighting would be performed by candles which Brunel considered the safest means then available.

It was at first proposed to erect the new hospital at Scutari, but this plan was soon dropped in favour of Renkioi. Doctor Parkes was appointed Chief Medical Superintendent of the new hospital, one of Brunel's assistants, John Brunton,[2] was despatched in advance to prepare the site and superintend erection, and in March Hawes was warned that 1,800 tons of shipping space would have to be provided in about a fortnight.

Brunel's letters to Parkes and Brunton at this time provide a classic example of that scrupulous attention to detail which was the secret of his success as an organizer. To Brunton on 2 April he wrote:

All plans will be sent in duplicate. ... By steamer *Hawk* or *Gertrude* I shall send a derrick and most of the tools, and as each vessel sails you shall hear by post what is in her. You are most fortunate in having exactly the man in Dr Parkes that I should have selected – an enthusiastic, clever, agreeable man, devoted to the object, understanding the plans and works and quite disposed to attach as much importance to the perfection of the building and all those parts I deem most important as to mere doctoring.

The son of the contractor goes with the head foreman, ten

2. John was the son of Brunel's old rival William Brunton who died in 1851.

carpenters, the foreman of the W.C. makers and two men who worked on the iron houses and can lay pipes. I am sending a small forge and two carpenter's benches, but you will need assistant carpenters and labourers, fifty to sixty in all. ... I shall have sent you excellent assistants – try and succeed. Do not *let anything induce you to alter the general system and arrangement that I have laid down.*

On 13 April we find him writing again to Brunton and here, as in his letter to Parkes written on the same day, he reveals his passionate concern for cleanliness, which shows how clearly he understood the reason for the catastrophe of Scutari. It also shows that he was alive to another problem: that all his precautions could be defeated by the patients themselves, the great majority of whom would never have seen a water closet in their lives.

'Materials and men for the whole will leave next week,' he writes. 'I will send you bills of lading for the five vessels: the schooner *Susan* and barque *Portwallis*, the sailers *Vassiter* and *Tedjorat* and the *Gertrude* and *Hawk* steamers. By the first named steamer, a fast one, the men will go with Mr Eassie's son.

I would only add to my instructions attention to closet floors by paving or other means so that water cannot lodge in it but it can be kept perfectly clean. If I have a monomania it is a belief in the efficiency of sweet air for invalids and the only point of my hospital I feel anxious about is this . . .'

Five days later he sends the bills of lading for the complete hospital and adds the note:

I trust these men will pull all together, but good management will always ensure this – and you must try while you make each man more immediately responsible for his own work to help each other – and to do this it is a good thing occasionally to put your hand to a tool yourself and blow the bellows or any other inferior work, not as a display but on some occasion when it is wanted and thus set an example. I have always found it answer.

Meanwhile he had written to Parkes:

All the vessels with the entire hospital will I believe have left England before the end of next week, that is before 21st. Finding that none of the Ordnance Stores were likely to be ready, and indeed that no positive time could be ascertained for their being ready, I obtained authority yesterday to purchase one third of the required quantity of bedding and some other similar stores and they are now going aboard with the buildings. I have added twenty shower baths, one for each ward and six vapour baths. You will be amazed to find also certain boxes of paper for the water closets – I find that at a cost of a few shillings per day an ample supply could be furnished and the mechanical success of the W.C.s will be much influenced by this. I hope you will succeed in getting it used and not abused. In order to assist in this important object I send out some printed notices or handbills to be stuck up, if you see no objection, in the closet room opposite each closet exhorting the men to use the apparatus properly and telling them how to do so. If you do not approve of such appeals the paper can be used for other purposes and perhaps impart some information in its exit from this upper world.

The buildings will be very quick after you; I almost fear you cannot have satisfied yourself about the site by the time they arrive. If you depend on Government Officers and if they at all resemble those at home, with one or two exceptions, your patience will be well tried.

That he could not resist this parting shot at his King Charles's head, the government official, is surely forgivable. He had designed, built and delivered on shipboard an entire portable hospital, unique of its kind, only to find that officialdom had been unable in that time to assemble the perfectly ordinary small stores with which to stock it. His wrath was such that it is doubtful whether a certain Major O'Brien, who was responsible for these stores, was ever the same man again. In the event, however, the missing stores were available by the time the hospital was ready to receive patients. Erection began on 21 May and took longer than Brunel had expected because Brunton was unable to find any local labour which he could trust, with the consequence that the hospital was built entirely by the little gang of eighteen men who had been sent out from England. Nevertheless, by 12 July the hospital was ready to admit 300 patients and by 4 December it was equipped with its full quota of 1,000 beds. Thanks to Brunel's foresight the work of erection went perfectly smoothly and the Renkioi Hospital was a complete success. In the short time it was operating before peace was declared close on fifteen hundred sick and wounded men passed though its wards, of whom only fifty died, a very different record from that of the fearful charnel house at, Scutari. Materials for the extension of Renkioi had already been contracted for when peace came and Brunel, in his usual methodical way, sent Brunton disposal instructions.

I don't want the thing to be flung into a ditch when done with [he wrote], but should prefer a useful end; that each part should be made the *most* of and methodically and profitably disposed of. Everybody here expresses themselves highly satisfied with Everybody there and what we have done. I should wish to show that it was no *spirit* but just a sober exercise of common sense . . .

If the Renkioi Hospital was only a monument to common sense it was certainly a shining one and the tragedy is that Brunel was not called upon to exercise that faculty much sooner. At least the calamity of Scutari might then have been very much mitigated; as it was, Renkioi, with its spacious wards, its modern sanitation and its air conditioning, although Brunel conjured it up like a magician in a mere two months, came too late.

*

From Mississippi and from Nile –
From Baltic, Ganges, Bosphorus,
In England's ark assembled thus
Are friend and guest.
Look down the mighty sunlit aisle,

> And see the sumptuous banquet set,
> The brotherhood of nations met
> Around the feast!

The Crimean war was the first blow to the fond hopes of those who believed that the improved communications of the machine age must inevitably promote the universal brotherhood of man. Of this optimistic belief, so eloquently expressed in the above lines by William Makepeace Thackeray, the Great Exhibition of 1851 had appeared to be the supreme affirmation. It was the first truly international exhibition, the first display of that commercial rivalry between nations which, so the theory ran, would soon take the place of the physical combats of earlier and less enlightened days. So far as Britain was concerned the exhibition was a kind of mid-century pause for breath and for mutual admiration. The workshop of the world was closed for stock-taking; for a few brief months its racking din seemed to grow fainter and there was an atmosphere of holiday in the air.

Needless to say, Brunel was involved in various ways: as a member of the Machinery Section Committee and of the Building Committee and also as chairman and reporter of the judges for Class 7 on Civil Engineering, architecture and building contrivances. His most eloquent representative in the exhibition itself was Gooch's latest and largest broadgauge flyer *Lord of the Isles*, which, raised upon a plinth, successfully dominated the machinery hall to the chagrin of narrow-gauge rivals.

The Exhibition Commission assigned the Building Committee the difficult task of producing a suitable home for the exhibition. Designs were invited, but of the two hundred and forty-five submissions not one was considered suitable and the Committee set to work to produce a design of its own. This, the two hundred and forty-sixth, was a good deal worse than many of its predecessors. Exhibiting all the faults inseparable from 'design by committee', when published it was very rightly howled down by an indignant public. It featured a triple range of squat and insignificant-looking brick buildings surmounted by a truly immense iron dome 200 ft in diameter and 150 ft high. This dome, it is scarcely necessary to remark, was Brunel's contribution to the design and it merely made the rest look ridiculous. Perhaps this was the effect intended, for he expressed strong objections to housing a temporary exhibition in substantial and expensive brick buildings. As an alternative he advocated what he called the 'railway shed style', but what exactly he meant by this will never be known, for all argument on the building question was decided once for all by the arrival of Joseph Paxton's inspired design for that fairy-tale palace of iron and glass which so concisely expressed in architectural terms the spirit of its age. Of it Brunel reported:

As regards Mr Paxton's claim, amid the competition of the whole of Europe, he proposed that mode and form of construction of building which appeared on first sight ... the best adapted in every respect for the purpose for which it was intended.

He was indeed a most fervent admirer of Paxton's building. He took a deep interest in its subsequent removal to Sydenham and designed the two water towers, each carrying 3,000 tons of water at a height of 200 ft, which flanked it there.

It is obvious, too, that Paxton's Crystal Palace deeply influenced Brunel in his design for the new passenger station at Paddington. Although a station on the present site had been intended from the first, the temporary terminus which utilized the arches of the Bishop's Road bridge lingered on for many years and it was not until the end of 1850 that the Great Western directors sanctioned the new building. It was a decision very welcome to Brunel for Paddington was, so to speak, the front entrance to his broad-gauge empire and that temporary station at Bishop's Road must have seemed a poor thing beside Hardwicke's great narrow-gauge citadel at Euston Square. Now was his chance to redress the balance and on 13 January 1851 we find him writing to Digby Wyatt about the project with all his old infectious enthusiasm:

I am going to design, in a great hurry, and I believe to build, a Station after my own fancy; that is, with engineering roofs, etc. etc. It is at Paddington, in a cutting, and admitting of no exterior, all interior and all roofed in. . . . Now, such a thing will be entirely *metal* as to all the general forms, arrangements and design; it almost of necessity becomes an Engineering Work, but, to be honest, even if it were not, it is a branch of architecture of which I am fond, and, *of course*, believe myself fully competent for; but for *detail* of ornamentation I neither have time nor knowledge, and with all my confidence in my own ability I have never any objection to advice and assistance even in the department which I keep to myself, namely the general design.

Now, in this building which, *entre nous*, will be one of the largest of its class, I want to carry out, strictly and fully, all those correct notions of the use of metal which I believe you and I share (except that I should carry them still farther than you) and I think it will be a nice opportunity.

Are you willing to enter upon the work *professionally* in the subordinate capacity (I put it in the least attractive form at first) of my *Assistant* for the ornamental details? Having put the question in the least elegant form, I would add that I should wish it very much, that I trust your knowledge of me would lead you to expect anything but a disagreeable mode of consulting you, and of using and acknowledging your assistance; and I would remind you that it may prove as good an opportunity as you are likely to have (unless it leads to others, and I hope better) of applying that principle you have lately advocated.

If you are disposed to accept my offer, can you be with me this evening at 9½ p.m.? It is the only time this week I can appoint, and the matter presses *very much*, the building must be half finished by the summer. Do not let your work for the Exhibition prevent you. You are an industrious man, and night work will suit me best.

I want to show the public also that *colours* ought to be used.

I shall expect you at 9½ this evening.

Secretary of the Executive Committee of the Great Exhibition, Digby Wyatt was at this time as busy a man as Brunel, but who could resist such an invitation as this? Wyatt came, and the result was the Paddington which we see today except for the addition of the northernmost bay. Whatever we may think of Wyatt's detail, an incongruous echo of the Exhibition's Moorish pavilion and executed under the influence of Mr Owen (Alhambra) Jones, it is of small account beside the splendid architectural adventure of Brunel's great roof. With its lofty aisles and transepts the new station made a worthy gateway to the broad gauge. It was completed in 1854 and it speaks well for Brunel's foresight that it proved adequate for the needs of the railway's expanding traffic without any alteration for over fifty years.

His work for the Great Exhibition brought Brunel into contact with another remarkable character with whom he was soon to become closely associated. This was John Scott Russell, joint secretary of the Royal Exhibition Commission with Stafford Northcote, later Lord Iddesleigh, and co-founder of the Exhibition with Henry Cole. Two years younger than Brunel, Scott Russell was a Glaswegian, a student of Edinburgh and Glasgow Universities and a very gifted man. He had carried out considerable research, using tanks and models, into the nature of waves with a view to improving the design of ships' hulls. The results of this research he embodied in what he called the 'waveline' hull form, and between 1835 and 1839, while he was manager of a shipyard at Greenock,[3] he built four ships according to this system, the *Wave*, *Scott Russell*, *Flambeau* and *Fire King*. At the same period he constructed six steam road coaches which operated between Glasgow and Paisley. In 1844 he removed to London, where he took over from Ditchburn & Mare the shipyard at Millwall on the Isle of Dogs which had been founded by William Fairbairn. He was a Fellow of the Royal Society, Vice-President of the Institution of Civil

3. Subsequently Caird & Company.

Engineers, Secretary of the Society of Arts and Co-founder and Vice-President of the Institute of Naval Architects.

In short, Scott Russell was a brilliant marine engineer and it may have been this very brilliance which blinded Brunel to his defects when he chose him as his partner in his last great enterprise. Until now, Brunel's gift of character judgement had never once failed him; Thomas Guppy, Christopher Claxton, Charles Saunders, Daniel Gooch – all these had not merely proved reliable business associates but most loyal friends. But in John Scott Russell that seemingly infallible judgement betrayed him at last.

[13]

That Great Leviathan

THE orgy of commercial opportunism and wild speculation which followed upon the success of the first railways sickened and disgusted Brunel. Railway construction became for him no longer the great adventure it had seemed in 1835 but a drudgery. There were far too many calls upon his services, while commercial considerations dictated a more utilitarian aproach to railway schemes. No longer could he exercise his architectural powers as he had done on the line to Bristol; that brief golden age of railway building was over. Speculators thinking only of return on capital would no longer allow their engineer to indulge his sense of fitness and grandeur.

Brunel foresaw the railway mania and the recession which must surely follow it. In September 1844 he wrote to his friend Charles Babbage:

Things are in an unhealthy state of fever here, which must end in a reaction; there are railway projects fully equal to £100,000,000 of capital for next year, and all the world is mad. Some will no doubt have cause to be so before the winter is over.

To another old friend of his youth, Adolph d'Eichthal, he wrote a few months later:

Here the whole world is railway mad. I am really sick of hearing proposals made. I wish it were at an end. I prefer engineering very much to projecting, of which I keep as clear as I can. . . . The dreadful scramble in which I am obliged to get through my business is by no means a good sample of the way in which work ought to be done. . . . I wish I could suggest a plan that would greatly diminish the number of projects; it would suit my interests and those of my clients perfectly if all railways were stopped for several years to come.

In 1848 there began that reaction to the mania which quickened and became a collapse with the downfall of George Hudson, 'The Railway King', in the following year. This recession forced Brunel to dismiss a number of his young assistant engineers. Remembering his own early disappointments when he was struggling to establish a reputation it cut him to the quick to have to do this. In August he wrote:

I have generally anxieties and vexations of my own, and at present they are certainly not *below* the average but they are completely absorbed and overpowered by the pain I have to undergo for others. For some weeks past, my spirits are completely floored by a sense of the amount of disappointment, annoyance, and – in too many cases – *deep distress* inflicted by me, though I am but an instrument, in dismissing young men who have been looking forward to a prosperous career in their profession, unsuspicious of the coming storm which I, and others mixing in the world, have foreseen. It is positively shocking to see how many of these young engineers have looked upon their positions as a certainty, have been marrying and making themselves happy, and now suddenly find themselves in debt and penniless. You can hardly imagine what I have to undergo in receiving letters of entreaty which I have no power to attend to. Everywhere we are reducing ...

This situation in the railway world, the disastrous atmospheric experiment and the stranding of the *Great Britain* all combined to disillusion Brunel. He began to contemplate retirement and in that retirement the fulfilment of another youthful ambition – to build a house for himself.

During the summer of 1847, when he was working on the South Devon Railway, Brunel took a furnished house at Torquay and that autumn, after much deliberation, he purchased some land at Watcombe overlooking Babbacombe Bay. It consisted of some sloping fields at a point near Barton, three miles north of Torquay on the Teignmouth turnpike where he had often, in earlier days, stopped his carriage to admire the view. Each summer thereafter the Brunels visited Watcombe whenever his business allowed and a special series of notebooks reveals his preoccupation with his country estate. The family became well known in Barton, and Brunel built at his own expense the small church at Church Road, which is now converted to secular purposes.[1]

It was Brunel's plan to refashion the prospect and lay out the gardens before building the house and to this end he sought the advice and assistance of the best known landscape gardener of the day – William Nesfield the elder, who had laid out Regent's Park, St James's Park and remodelled Kew Gardens. Young trees were moved to new positions by means of specially designed tools and numerous ornamental trees and shrubs were planted in strict accordance with plans carefully thought out with an eye to vistas and contrived prospects. Brunel at first sketched for himself a Gothic castle with a great turret tower brooding over the sea, but, as he probably realized, this was better suited to some colder and more savage coast than that of South Devon. It was soon relegated in favour of a design for an Italianate villa with a belvedere and a colonnaded terrace. From this terrace a double flight of balustraded steps led down to an Italian garden whose central feature was a great stone fountain basin. Although this design comes from Brunel's pencil, he was obviously influenced here by Nesfield, for it bears a strong resemblance to the latter's work at Great Witley in Worcestershire for Lord Ward. But this house of which Brunel dreamed was never built.[2] No more than the foundations were laid; no fountains ever plashed in their great stone

1. A drawing in one of the sketch books showing an altar and a reredos featuring the scene of the Last Supper may conceivably have been intended for this church.

2. A house, now known as 'Brunel Manor', was built on the site after his death. In the style of a French chateau, it has no connexion with the family or with any design of Brunel's.

basin; and the woman and the child who appear in one of his drawings remain pathetic ghosts, pacing a terrace which never existed. After 1855 the notebooks contain fewer and fewer references to Watcombe, for by this time Brunel's last enterprise had become all consuming and the dream of a peaceful retirement to the West Country was driven ever further into the background of his thoughts.

Although they betray none of his private thoughts and emotions, Brunel's sketch and notebooks do reveal very clearly the mind of the engineer at work. They show that the same disillusionment with railways which led the private man to contemplate retirement drove the engineer to dwell more and more upon problems of marine engineering. What could be more natural? With England a tom-tiddler's ground of rival railway speculators, the oceans of the world offered far freer scope for a man of his restless temperament. The sketch books faithfully reflect this change of emphasis. Amidst drawings for the new Paddington and station layouts for the broad-gauge route into Birmingham, the eye is suddenly arrested by a sketch of an extraordinary steamship as long as the page is wide and bristling with funnels and masts. It is dated 25 March 1852. The drawing is headed 'East India Steamship' and beneath it is scribbled casually the note: 'Say 600 ft × 65 ft × 30 ft', dimensions which any contemporary shipwright would have regarded with absolute incredulity. Thereafter the pages of the sketch books are haunted by the apparitions of gigantic ships. They appear with different masts and rigs; sometimes with both screw and paddles; sometimes with two sets of paddles. What technical problem they represent is disclosed by the mention of Australia and by such jottings as this:

Question: Whether ship should carry her coals out and home or only out. Both ways, 15,000 Tons of coal; one way, 8,000 Tons.

Design A. 600 × 67 × 28   20,500T.
      „   B. 660 × 83 × 30   30,000T.

700 × 67 beam are very fair. As far as it affects speed through
smooth water, I think these proportions will bear a considerable
increase of beam – the question is rather one of comparative advantage of extreme steadiness obtained by great length and narrow
proportional beam.

Following the same principle that had governed the design
of the *Great Western* and the *Great Britain*, Brunel was now
working out the best proportions for a ship which could
carry enough coal to steam to the Antipodes and back.

The discontents and miseries of the 'Hungry Forties', the
discovery of gold in Australia in 1851 and the revelation of
her agricultural wealth at the Great Exhibition of that year
all combined to produce a sudden wave of emigration to
what seemed a new promised land and the result was a boom
in the Australian shipping trade. As we have already seen,
Gibbs, Bright & Co. put the *Great Britain* on to the Australia
run as soon as she had been repaired. Although her steaming range exceeded that of any ship afloat when she was
built, she was only intended for the North Atlantic and
Gibbs, Bright & Co. had to send bunker coal out for her in
specially chartered ships from Penarth to the Cape. In 1851
the Australian Mail Company consulted Brunel on the type
of ship which would be most suitable for their Australian
mail service. He advised ships of from 5,000 to 6,000 tons
burden which need coal only at the Cape. Although this
seemed a most startling and impracticable proposition to
some of the directors, it was finally agreed, and under Brunel's direction two ships, the *Victoria* and the *Adelaide*, were
laid down in the following year at John Scott Russell's yard
at Millwall.[3]

This was the background of events which had led Brunel
to consider the idea of building a steamer large enough to

3. Brunel had nothing to do with the design of these ships, which
was left entirely to Russell; and it is significant that they were framed
transversely and lacked the great longitudinal strength which distinguished all three Brunel ships.

carry her own coals round the world, for that is what a voyage to Australia and back involved. He argued that this would not only obviate the expense of establishing and supplying overseas bunkering stations, but that a ship of such a size would be superior in speed and cheaper to operate, taking into account fuel costs, crew's wages and maintenance, than two ships of half the size.

Brunel evidently decided that the conception of a ship six times the size of anything afloat was unlikely to appeal to the conservative directors of the Australian Mail Company, but during the late spring of 1852 he discussed the idea with Claxton and Scott Russell. Claxton was the only friend of the old Great Western Steamship Company days to follow him into this new venture, for Thomas Guppy had permanently removed to Italy in 1849, where he had established a large mechanical engineering business in Naples and bought a villa at Portici where he lived until his death in 1882. Brunel must have felt Guppy's loss very keenly, particularly when he embarked on this new adventure. On Scott Russell's suggestion the idea of the 'great ship', as it was soon called, was next mentioned to some of the directors of the Eastern Steam Navigation Company. This concern had been formed in January 1851 to open new passenger mail steamer routes between England, India, China and Australia, but in March 1852 the British Government had decided to grant the mail contract exclusively to the Peninsular & Oriental Steam Navigation Company, with the result that the unfortunate Eastern Steam directors found themselves, to use a modern Americanism, 'out on a limb'. They were ready to listen to anyone with a scheme which promised to revive their fortunes and at the same time strike a blow at their lucky rivals. Events now moved swiftly. An investigating committee which the directors at once appointed reported themselves unanimously in favour of the scheme. It was adopted at a general meeting of the Company in July, when Brunel was appointed Engineer and it was decided to invite tenders for the hull of the ship and her engines from John Scott Russell, James Watt & Co. and others.

The way ahead, however, was to be by no manner of means so smooth. There were disagreements and resignations on the Board and capital was inadequate, with the result that by the time the Company was in a position to place the contracts for the ship early in 1854 it resembled the original concern of 1851 in name only. It had become to a very large extent a flotation of Brunel's, the majority of the directors being his nominees. The trouble had begun when Brunel presented his detailed plans to the Board. The chairman and several of the directors objected on the grounds that the scheme was too antagonistic to the General Screw Company in which, presumably, they had an interest. This was a rival of the Ship Propeller Company formed to exploit F. P. Smith's propeller which Brunel proposed to use. The outcome of this row was the resignation of the General Screw Company faction; at which point Brunel himself writing in 1854, takes up the story:

I set to work and got Mr Hope to be Chairman and Messrs Talbot, Miles, Baker and Potter directors of the new Board who adopted the plans. This Board was strengthened by Bayley of Manchester, Betts and McCalmont, S. Russell, Robinson and Davidson per Blake. In December '52 Goschen, the only doubtful director remaining, resigned and we got Geach appointed in his place. Geach and I then set to work to get the Capital and by December 1853 after great exertion 40,000 shares were taken and £120,000 paid up. . . . Betts, Peto, Brassey, Dargan and Wythers taking 1,000 apiece. More than once we have nearly failed and broken down. . . . After two years' exertions we are thus set going, contracts entered into and work commenced 25 February 1854.

In addition to persuading others to take up shares, Brunel had a very considerable stake in the Company himself. Never before, indeed, had he so deeply involved both himself and others in any enterprise of his undertaking. That he felt this very keenly was revealed a few months later in his reply to a resolution passed by the directors requesting him to nominate a resident engineer. He said:

... The heavy responsibility of having induced more than half of the present Directors of the Company to join, and the equally heavy responsibility towards the holders of nearly half the capital, must ensure on my part an amount of anxious and constant attention to the whole business of the Company which is rarely given by a professional man to any one subject, and, as it seems to me, ought to command a proportionate degree of confidence, or rather command entire confidence, in me. ... The fact is that I never embarked in any one thing to which I have so entirely devoted myself, and to which I have devoted so much time, thought and labour, on the success of which I have staked so much reputation, and to which I have so largely committed myself and those who were disposed to place faith in me; nor was I ever engaged in a work which from its nature required for its conduct and success that it should be entrusted so entirely to my individual management and control. ... I cannot act under any supervision, or form part of any system which recognizes any other adviser than myself, or any other source of information than mine, on any question connected with the construction or mode of carrying out practically this great project on which I have staked my character; nor could I continue to act if it could be assumed for a moment that the work required to be looked after by a Director, or anybody but myself or those employed directly by me and for me personally for that purpose. If any doubt ever arises on these points I must cease to be responsible and cease to act.

On receipt of this missive the directors at once rescinded their resolution and no more was heard about resident engineers.

In a letter to John Yates, the Secretary of the Eastern Steam Navigation Company, dated 16 June 1852, Brunel had tentatively estimated the cost of his great ship at £500,000. The actual figure tendered, however, was £377,200, this being made up as follows: hull, £275,200; screw engines and boilers, £60,000; paddle engines and boilers, £42,000. To these figures there was to be added an unspecified sum for steam auxiliary engines at the rate of £40 per horsepower. Before Scott Russell submitted his tender, Brunel had written in a letter to him:

The wisest and safest plan in striking out a new path is to go straight in the direction we believe to be right, disregarding the small impedimenta which may appear to be in our way – to design everything in the first instance for the best possible results ... and without yielding in the least to any prejudices now existing ... or any fear of the consequences.

Faced with the prospect of building a ship of fantastic magnitude for an engineer who expressed these perfectionist sentiments, Russell would, had he been a wiser man, have added a substantial sum for 'contingencies' to his contract price. As it was, his figure for the hull was a gross underestimate. He even offered to reduce it to £258,000 if, as was at first intended, a contract was placed with him for a sister ship. Consider the terms of the final contract which Russell signed on 22 December 1853. This contract

Provided for the construction, trial, launch and delivery of an iron ship of the general dimensions of 680 ft between perpendiculars, 83 ft beam and 58 ft deep according to the drawings annexed signed by the engineer, I. K. Brunel.

All vertical joints to be butt joints and to be double riveted wherever required by the engineer. Bulkheads to be at 60 ft intervals. No cast iron to be used anywhere except for slide valves and cocks without special permission of the engineer. The water tightness of every part to be tested before launching, the several compartments to be filled one at a time up to the level of the lower deck.

The ship to be built in a dock.

After the launch such trials and trial trips at sea will be made with the engines and probably under sail as in the opinion of the engineer may be necessary to ascertain the efficiency of the work. Any defects then discovered in workmanship or quality of materials to be forthwith remedied by, or at the expense of, the contractor to the satisfaction of the engineer.

All calculations, drawings, models and templates which the contractor may prepare shall from time to time be submitted to the Engineer for his revision and alteration or approval. The Engineer to have entire control over the proceedings and the workmanship.

In the light of subsequent events, these terms are of considerable interest. We may wonder why Brunel did not query Russell's figure, but, once again, it is easy to be wise after the event. Russell was, after all, reputedly a practical shipbuilder and it is clear that at this time Brunel entertained the highest opinion of his abilities. Russell, for his part, may have been tempted by the fame which would be his as builder of the ship into quoting an uneconomic price in order to be sure of the contract.

The quotations for the ship's engines must also have been shots in the dark, for no one had attempted to build engines of such a size before and no one could know the technical difficulties which would be encountered in the production of castings and forgings of such dimensions. Brunel's specifications called for not less than an output of 4,000 i.h.p. at 45 r.p.m. from the screw engines and 2,600 i.h.p. at 17 r.p.m. from the paddle engines, both at 15 lbs steam pressure.[4]

At this point it will be as well to explain the considerations which led Brunel to adopt both screw and paddle propulsion. First, it would not have been possible at that time to apply to a single paddleshaft or a single screw sufficient shaft horsepower to propel a hull of such size at the required speed. Secondly, after his experience with the unsatisfactory chain drive on the *Great Britain*, Brunel was anxious to drive the propeller shaft direct from the engine crankshaft. In his preliminary calculations, he had assumed

4. The actual maximum working steam pressure was 25 lbs giving a total of nearly 10,000 i.h.p.

that this would mean a screw 25 ft in diameter, the diameter of the screw actually fitted being 24 ft. Now the Eastern Steam Navigation Company had stipulated that the ship must be capable of entering the Hooghly on her home run from Australia and this limited her draught. Brunel calculated that if the maximum draught was restricted to 30 ft, then she would carry enough coal for a return passage from the Hooghly on 23 ft draught or, if she coaled at Trincomalee, she could work out of the Hooghly with a good cargo of goods and passengers on only 20 to 21 ft draught. This satisfied the Company but it meant that on 22 ft draught a 25 ft diameter propeller would be 5 ft out of water, allowing 2 ft for the keel and clearance. Thus Brunel had to reckon with a serious loss of power from the screw at lighter draughts and it was the purpose of the paddle unit to compensate for this. It was after long consultation with Joshua Field of Maudslay, Sons & Field that the screw to paddle power ratio of approximately two-thirds to one-third was decided upon. The paddle wheels, 60 ft in diameter, were designed to have a 6 ft hold of the water at 20 ft draught.

In the design of the hull of his great ship Brunel developed to its logical conclusion the double-hulled or cellular system of construction which he had first introduced in embryonic form in the *Great Britain*. What had been merely an iron deck in that ship now became an inner water-tight skin which extended, not only across the bottom of the ship but also up the sides to the level of the lower deck, 35 ft above the keelplate or 5 ft above deep loadline. The space between these two hulls was 2.8 ft, and the longitudinal girders and transverse webs which united them divided this space into a series of cells. The main deck was constructed in the same fashion so that the hull had the strength of a huge box girder. There was no transverse framing, but iron watertight bulkheads divided the ship into 10 compartments, each 60 ft long. As on the *Great Britain*, there were also two longitudinal bulkheads, 36 ft apart, which extended throughout the 350 ft length of the two engine and boiler rooms and up to the loadline, this same section above the main deck being occupied by the passenger accommodation. At the lowest level, two tunnels, one for access and the other housing main steam pipes, connected the two engine rooms, the bulkheads being otherwise continuous up to loadline.

This, in very brief and non-technical terms, was the hull design in which Brunel embodied all his previous experience not only as a shipbuilder but also as a builder of bridges. The problems, indeed, are not dissimilar except that in the ship they are more complicated because the stresses vary very rapidly as the support given by the water changes. Considered as a girder, a ship's hull in a seaway is not only subjected to severe racking strains as she rolls, but the upper member will be alternately in compression and in tension as the ship tends to sag or to hog. All these factors Brunel had taken into consideration in a hull design which, in the words of Sir Westcott Abell, 'stands out as a milestone in the progress of building ships of iron and later steel'.

In describing how he checked Scott Russell's detailed hull drawings, Brunel explained his own guiding principles very clearly. He said:

I found for instance an unnecessary introduction of a filling piece or strip such as is frequently used in shipbuilding to avoid bending angle irons – made a slight alteration in the disposition of the plates that rendered this unnecessary and found that we thus saved 40 *tons weight* of iron, or say £1,200 of money in first cost!! and 40 tons of cargo freight – at least £3,000 a year!!

The principle of construction of the ship is in fact entirely new if merely from the rule which I have laid down and shall rigidly persevere in that no materials shall be employed on any part except at the place and in the direction and in the proportion to which it is required and can be usefully applied for the strength of the ship and none merely for the purpose of facilitating the framing and first construction. In the present construction of iron ships the plates are not proportioned to the strength required at different parts and nearly 20 per cent of the total weight is expended in angle irons and strainers which may be useful or convenient in the mere *putting together* of the whole as a great box, but is almost useless or very much misapplied in affecting the strength of the structure as a ship. All this misconstruction I forbid and the consequence is that every part had to be considered and designed as if an iron ship had never been built before; indeed I believe we should get on much quicker if we had no previous habits and prejudices on the subject.

Notwithstanding this meticulous equation of proportion with function, a very remarkable degree of standardization was achieved in the construction of the ship. Throughout the entire hull only two sizes of angle iron and two thicknesses of plate, 1 in. and 3/4 in., were used. The plates measured 10 ft by 2 ft 9 in., this being the largest size which could be economically rolled at this time. The ironwork was supplied by the firm of Samuel Beale & Company, Parkgate Ironworks, Rotherham, of whom Charles Geach, the director who assisted Brunel to raise the necessary capital, was principal partner.[5] Geach was one of the most enthusiastic advocates of the scheme and his death a year after construction began was a serious blow to Brunel.

This somewhat long and fully documented account of the inception of the ship has been necessary, in the light of after events, in order to establish beyond any question the part played by Brunel in bringing to birth his great brain-child. For in the protracted and ultimately tragic drama which was about to be played, one of the important elements which previous accounts have failed to recognize was the influence of that most corrosive of human weaknesses – jealousy. The first portent of this appeared in the shape of a newspaper article as early as November 1854. The effect of the recent abolition of the stamp duty on newspapers had been greatly to increase the circulation and consequently the power of the popular press. Hence the building of the great ship took

5. A very few of the plates, those requiring an exceptional curvature, were of Lowmoor iron.

place under a glare of publicity which, though it might seem a feeble glimmer to us, was quite without precedent then. As a standing wonder the ship was a godsend to journalists and news editors, for it could always be relied upon to fill a column. Paragraphs could be devoted to those asinine comparisons in which journalists have ever delighted to indulge: the ship would reach from corner to corner of such-and-such a London Square; she was larger than Noah's Ark, or her masts would be higher than such-and-such a steeple, and so forth *ad nauseam*. But the article which caught Brunel's eye on this occasion contained no make-weight paragraphs of this kind. On the contrary it was so well informed that it could only have been based upon an interview with somebody connected with the Company; somebody who was, moreover, well versed in the construction plans which had not, at this time, proceeded very far. It possessed one particularly odd feature, however. Although of considerable length, the name of Brunel appeared only once. This was the context: 'Mr Brunel, the Engineer of the Eastern Steam Navigation Company, approved of the project, and Mr Scott Russell undertook to carry out the design.'

Brunel at once wrote to Secretary John Yates. He said:

I have taken the trouble to read through the long article in 'The —' and am much annoyed by it.[6] I have always made it a rule, which I have found by some years' experience a safe and profitable one, to have nothing to do with newspaper articles; but then, if on the one hand the works I have been connected with have rarely been puffed (never by me), they have also been rarely affected by misstatements; as such notices, when not inserted by interested parties, are always slight and unauthentic, and drop without producing any effect. This article . . . however, bears rather evidently a stamp of authority [and] . . . may acquire the character of being an authorized statement.

6. As quoted by Isambard Brunel, who withheld the name of the paper concerned.

... Although from system I have never interfered with newspaper statements, it has not been from any affected or real indifference to public opinion, perhaps it was more from pride than modesty, and therefore I am by no means indifferent to a statement which would lead the public, and perhaps by degrees our own friends, to forget the origin of our present scheme, and to believe that I, happening at that time to be the consulting Engineer of this Company, which I was not, and having had no peculiar connexion with previous successful improvements in steam navigation, allowed them to adopt some plan suggested by others, who I suspect, if even were such the case, would never appear to share with me the responsibility if any failure should result. Of this certainly I have no fear but at the same time I am desirous of something more than mere immunity from blame.

I not only read this article once, but I was so struck by the marked care shown in depreciating those efforts which I had successfully made in advancing steam navigation ... that I read the paper a second time. ... The objectionable points that I refer to evidently did not strike you, and that is a strong proof how easily incorrect impressions insinuate themselves unawares; but I feel strongly that a judicious friend would not have failed to do justice to the spirited merchants of Bristol, who, in spite of the strongest condemnation of the plan by the highest authorities, and the ridicule of others, persevered in building and starting the first transatlantic steamer. ... A writer wishing success to our enterprise would not have omitted to mention that I had a claim to public confidence on this occasion, for the reason that I was at least the principal adviser in those previously successful attempts.

And lastly, I cannot allow it to be stated, apparently on authority, while I have the whole heavy responsibility of its success resting on my shoulders, that I am a mere passive approver of the project of another, which in fact originated solely with me, and has been worked out by me at great cost of labour and thought devoted to it now for not less than three years.

Whether Brunel or John Yates ever succeeded in establishing positively who had given the interview upon which this offending article was based we do not know. Yet it becomes pertinent at this point to quote in full a curious missive which Brunel received three months later from John Scott Russell following a visit of the Prince Consort to his yard at Millwall.

My Dear Mr Brunel,

I was very sorry to find that you could not meet the Prince this morning. It was, however, of less consequence than I thought or than you thought as the visit was mainly designed for the Battery [*sic?*] in which he takes a deep interest and he bestowed comparatively little time on the Great Ship.

I took the opportunity of explaining what I supposed he and everybody else knew, that you are the Father of the great ship and not I and to say many other things which I should not have done if you had been present. Allow me to take this opportunity once for all while on this point of saying that you may always trust your reputation as far as the Big Ship is concerned to my care in your absence for many reasons: 1st I have as much reputation as I desire or deserve. 2nd I think it much wiser to be just than unjust. 3rd I would much rather preserve your friendship (which I think I possess) than filch your fame (if I could) and forfeit your friendship which I should. *Verbum Sat.* I will never trouble you on this point again but remain

Always Faithfully Yours,
J. Scott Russell.

Upon the protestations of this letter there is no need to comment; the reader can draw his own conclusions as well, no doubt, as did Brunel.

Such was the unhappy atmosphere, unsuspected by the public, in which there was begun upon the Isle of Dogs the building of the mightiest ship the world had ever seen. 'Could I have foreseen the work I have to go through', wrote Brunel later, 'I would never have entered upon it, but I never flinch, and do it we will.'

Before construction could commence a building site had to be found and prepared. The river frontage of Scott Russell's yard was quite inadequate, indeed the whole area of the premises would have been insufficient to contain the leviathan. Fortunately, however, there was another shipyard adjoining which had been standing vacant for twelve months. This was the yard which had been equipped and opened in 1837–8 by David Napier, the engineer who proved the practicability of steam navigation on the narrow seas and so paved the way for Brunel's *Great Western*. Napier was not only a great engineer but an extremely canny Scot who realized long before anyone else that the Thames was no place for iron shipbuilding. 'London never will be a place for building steamers,' he wrote when he retired in 1852, 'on account of everything connected with their production being higher there than in the North.' Acting on this belief Napier's sons, after a visit to Australia, set themselves up in business on their native Clyde and the yard lay deserted until 1853, when it was leased by the Eastern Steam Navigation Company. Now there was space enough. The ship could be built almost entirely in the Napier Yard, leaving Scott Russell's premises clear to carry out the manufacturing work, and a railway was laid between the two for the better transport of materials.

The next question to be decided was how the ship was to be built and launched. The terms of Russell's contract had stated that it was to be built in a dock, but, almost before its ink was dry, that idea was abandoned. Russell had quoted from £8,000 to £10,000 to make a dock – another estimate which was obviously wildly optimistic – and subsequently Brunel reported that it would not only be difficult to find a suitable site, but that the nature of the soil would make construction hazardous. To the idea of building the ship on an end-on slip, Brunel raised the objection that if such a slip was constructed on the gradient of 1 in 12 to 15 which he thought necessary, then the forefoot of the ship would be forty feet in the air. This would be a great inconvenience when building the ship, especially as he planned to install the engines and boilers before launching. There remained the third alternative which was to build the ship parallel with the river bank and launch her sideways down an inclined way. This was Brunel's choice, and to effect it he proposed designing and constructing a mechanical 'patent slip' fitted with wheels or rollers which could afterwards be moved and re-erected at the port to which the ship intended to trade so that she could be readily hauled out for examination and repair.

The building site at the Napier Yard was accordingly laid out for a sideways launch, and the whole area was piled with 12 to 15 in. square oak piles from 20 to 38 ft long driven 5 ft apart. By the time Russell reported this work complete on 29 May 1854 two 600-ton cargoes of timber had been consumed. The heads of the piles were left standing 4 ft above ground level and it was upon them that the bottom of the ship would rest. Packing could be added to the heads of these piles as necessary, but the midship section of the hull was flat-bottomed and there was no keel. Brunel had decided on this design of hull, partly to simplify the design of the launching cradles and partly so that the ship would be capable of settling evenly and supporting herself on a gridiron for repairs, there being no dry dock in the world big enough for her.

While the site was being piled a huge moulding floor was completed and roofed where the ship's lines could be laid out full scale. In Russell's own yard new plate-bending and punching machines were installed and coffer-dams were sunk in the foundry floor to enable the enormous paddle engine cylinders to be cast. A new shop had to be built in which to erect the paddle engines because, when assembled, they would stand 40 ft high. These preparations completed, Russell reported to Brunel on 9 July that: 'The keel[7] has been laid and the whole of the plans of the various additional works you have proposed are fully occupying our attention.' Thenceforward throughout the following nine months, work

7. Russell is referring here to the flat keel plate which formed the backbone of the ship. As previously stated, there was no keel in the true sense of the word.

on the hull went forward slowly. In comparison with modern methods the small plates used increased the number of riveted joints, while every one of the millions of rivets had to be closed by hand, the rivet boys and holders-up labouring for months on end amid a deafening clangour in the confined space between the two hulls. Moreover, the whole hull was built from the ground with only tackles to lift the plates into position. Scott Russell had proposed constructing a crane with a 6o-ton lift and a 400 ft traverse for £2,951, but Brunel described this quotation as 'very frightening' and the idea of the crane was dropped.

Meanwhile work on the engines had begun. In August Brunel went up to Birmingham to witness the casting of the first screw engine cylinder in Watt's famous Soho Foundry. There were to be four of these cylinders, horizontally opposed, each 7 ft in diameter by 4 ft stroke. Then in October Scott Russell wrote: 'I expect to cast the last great cylinder Friday afternoon 4½ o'clock. Will you come and see it done? Two have been bored and faced and the third is boring.' The bore and stroke of these four paddle engine cylinders was 6 ft 2 in. by no less than 14 ft, and each mould swallowed 34 tons of molten iron. If the crankshaft for the proposed paddle engines of the *Great Britain* had daunted the forgers of the day, it was a bent pin to a crowbar compared with the shaft required for this engine, and notwithstanding James Nasmyth's steam hammer, Russell experienced the same difficulty that poor Francis Humphrys had done in finding any contractor to undertake the work. Finally, by coincidence, it was Messrs Fulton & Neilson who agreed to make the attempt at the Lancefield Forge, Glasgow, David Napier's old stamping ground before he moved to Millwall. Special furnaces had to be built and a battery of Condie moving cylinder steam hammers installed before the job could be attempted. Twice they failed, the cranks containing flaws in the webs, but the third time the Lancefield smiths succeeded in producing in perfection the largest forging the world had yet seen. It weighed over 40 tons and was charged for at the rate of over £100 a ton, but even at this price it must have represented a loss to Lancefield.

To suppose that in the matter of these engines Brunel merely specified the required performance and left the rest to the builders would be quite wrong. Notes, sketches and Journal entries all give the lie to this. Thus at the end of February he records a long discussion with Dixon, Scott Russell's assistant, on the design of the paddle engine valve gear, which ended in the adoption of his suggested modifications. These engines, apart from their great size, were of orthodox design whereas the screw engines were much more original and could almost be said to have been designed by Brunel, so many discussions did he have with Blake of James Watt & Company and so many were the modifications that he suggested. In this respect the following diary entries are typical:

3 March 1854. Blake called with drawings of screw engines. Insufficiency of bearing surface. Crank pins 18″ × 18″ not enough; 24″ × 24″ barely sufficient. Problem necessity of working crank between the two piston rods limits dimensions.

7 March 1854. Have been thinking a great deal over Blake's arrangements and have come to conclusion that double piston rods carried through to the opposite engine *cannot* unless with a very low steam pressure leave room for a properly proportioned crank. Wrote long letter to B. on necessity of resorting to two distinct connecting rods[8] – shall I not succeed at last by degrees in getting two oblique cylinders? I shall try.

9 March 1854. Blake here again. Found, as I had thought, a mistake of ten to one in his calculations of bearing surface.

Brunel decided to equip these screw engines with the

8. As built there were three connecting rods to each pair of opposed cylinders, all working on a common crank-pin. The starboard cylinder had a single rod with its big-end working centrally on the crank while the port cylinder crosshead carried two connecting rods whose big-ends worked one on each side of the former.

governor which his father had invented and he wrote to the Patent Office requesting them to trace the original drawings.

I remember the experimental governor being made [he wrote], and my working at it, and I made the drawings for a patent which I believe my father took out – it must have been about the year '23 or '24.[9] I should like to see the old drawing on which I worked when young.

Other features which Brunel proposed were the steam jacketing of all cylinders, the jacket steam being supplied from an auxiliary boiler, and the provision, as on the *Great Britain*, of annular feed water heaters round the funnels.[10] This last arrangement served the dual purpose of conserving waste heat and insulating the funnels where they passed through the saloons.

It was on New Year's Day 1855 that coming events first began to cast their ominous shadow. Scott Russell informed Brunel that he was in serious financial difficulties and that Martin's Bank would not allow him further credit. To overcome this difficulty Brunel proposed that the sum still due to Russell under his contract should be paid in ten instalments of £8,000 as the work progressed, each such payment to be sanctioned by him by certificate so soon as he was satisfied that the equivalent amount of work had been done. To this arrangement both Russell and his banker agreed.

The next problem to exercise Brunel's mind was that of the launch, for the directors had rejected the idea of his special roller-launching gear on the score of expense. He was committed now to a sideways launch so the question was how, without his roller gear, he was going to move an inert mass weighing 12,000 tons a distance of 200 ft from the building site to low water level in the river. While he was pondering over this, he was much heartened to receive in

9. It was actually dated 1822.
10. The steam jacketing of the cylinders was not carried out.

March a letter from a certain G. W. Bull, of Buffalo, in the United States. It read as follows:

I have just been reading your report to the Directors of the E.S.N. on the state of your mammoth steamship, and in relation to your mode of launching her, I wish to say that we have on Lake Erie six steamers of from 324 to 380 ft and of 1,900 to 2,400 tons measurement which have all been launched sideways – in fact since the first steamer of 210 ft keel was launched in that way (in 1837) no vessel here has been built for longitudinal launching of any size. It is by all odds the safest and preserves the vessel without strain in the slightest degree. Two large steamers (the *Plymouth Rock* and the *Western World*) 330 ft keel were launched here last summer with their boilers and nearly the whole of their engines in place and with their main deck and hurricane deck cabins all complete. ... Since 1837, hundreds of vessels have been launched here in this way and in no instance has the slightest accident happened or injury to the vessel. Wishing you most complete success in getting your mammoth into the water I have thrown this together in the hope it may aid or at least strengthen you in the complete feasibility of the plan you have adopted.

To this Brunel replied asking numerous questions as to the gradients of the launching ways and the methods employed. Bull's informative reply which he received at the end of April concluded by saying: 'In my opinion you need not have the slightest fear of (what I ardently wish you) *complete success*. I wish I could be there to see.'

Now Brunel discussed this problem long and earnestly with Scott Russell and showed him the American correspondence with the only result that an irreconcilable difference of opinion arose between the two engineers. From the first Brunel had insisted that the launch must be controlled whatever the precise means employed. He maintained that it would be madness to let such an immense bulk slide freely into a restricted waterway where it would certainly stick with disastrous results. No, it must be lowered carefully down to low water mark and then floated on the tide. To this Scott Russell would not agree. He remained firmly in favour of a free launch, pointing out that all the American launches which Bull described in his letter were free. Such an attitude is understandable when we recall that Russell had committed himself to launch the ship under his contract and that, as Brunel was the first to admit, a controlled launch would be very much more costly.

This disagreement, which occurred during April 1855, marked the beginning of the end. From this date forward Brunel would be confronted by a remorselessly rising tide of troubles against which he would contend as valiantly as Cuchulain in his fight with the sea. A profound change in the hitherto cordial relationship between Brunel and Russell becomes evident. It was marked by offhandedness, by a deliberate lack of cooperation on Russell's side and by a growing impatience on the part of Brunel. It is clear that Russell's jealousy, hitherto studiously concealed, now erupted in active and bitter enmity. Yet the policy of non-cooperation which now led Russell to neglect the great ship contract was not dictated simply by personal animosity. The tone of Brunel's correspondence during the next few fateful months suggests that he found Scott Russell's conduct inexplicably foolish. Had we been in his place we might have been similarly misled, for the fact that it was the first move in a clever plan to evade a commitment which was becoming embarrassing, was by no means obvious; had it been so then the plan must have failed.

The first intimation that all was not well comes in a letter from Brunel to Russell dated 24 April in which he writes: 'I begin to be quite alarmed at the state of your contract – four months are gone and I cannot say that even the designs are completed or even sufficiently settled to justify a single bit of work being proceeded with – we shall get into trouble.' On 8 May he writes again about some photographs of the ship which Russell had been asked to supply: 'The Directors ask me for them constantly and I have only the silly answer to give that it is not my fault. You and I are ready enough to abuse Government Offices – it seems we are either impotent or neglectful ourselves.'

So far the tone, though exasperated, is still cordial while the phrasing of the letters indicates the efforts that Brunel was making to preserve a friendly relationship and sense of joint responsibility. But now the tone alters, while so far as Russell is concerned the old effusive style of address, which he last displayed in March in the curious letter which has already been quoted, is replaced by the formal 'Dear Sir' and 'Your Obedient Servant'. On the day after his letter about the photographs, Brunel wrote: 'Your reply this morning to my long list of complaints is an admirable specimen of an Under Secretary's reply in the House to a Member's motion – it does not satisfy one single honest craving for information and for assurance of remedy.' He goes on to complain about some indicator cards from the paddle engines which Russell had sent him. '. . . I do not want better indicators than usual. . . . Those made on this occasion and to which I objected were absurd – like the attempts at writing of a two-year-old baby – and I take credit to myself that I did not resent the insult of showing them to me as indicator cards.'

In May there was an outbreak of fire in Russell's yard. Brunel wrote to Yates advising him to insure the ship and sent a friendly note to Russell suggesting the appointment of watchmen in future: 'I have great faith in professional policemen, *Punch* notwithstanding, but possibly you may think the men might not like it.' In July, in the course of a letter to Blake, he wrote on the subject of the controlled launch: '. . . Mr R. finally concurs in my determination that, *coûte que coûte*, that which was safest and surest *must* alone be considered.' But it soon became evident that 'Mr R.'s' concurrence was so much lip service; also that things were moving rapidly from bad to worse.

On 12 August, Brunel sent Scott Russell a private letter

from Haverfordwest, whither he had gone to superintend the provision of coaling jetties for the great ship at Milford Haven. In this he pointed out the gravity of the situation. Only £65,000 out of the total contract sum remained to be paid and yet the ship was far behind with 4,500 tons of iron-work still to erect.

I have tried gentle means first [he wrote], I must now streng-then the dose a little. If you do not see with me the necessity of shaking off suddenly the drowsiness of sleep that is upon us and feel it so strongly that like the sleepy man just overcome with cold you feel that unless done instantly you are lost – In fact unless, as I say, on *Monday next* we are busy as ants at ten different places now untouched I give it up – but you will do it.

By this time Brunel had finally decided upon the method of launching the ship, but before he could work out the de-tails he required certain information, the centre of gravity, the centre of flotation and so on, which he requested Russell to furnish. Once again he could obtain no satisfaction. On 23 August he wrote:

I feel it most essential to have these particulars at once and am rather surprised to find you less anxious to possess precise know-ledge on such points than I am – Experience may make you quite easy, you are wrong, I think – at all events I am not going to trust to chance and must satisfy myself at once on these points which will influence the arrangements to be made for launching.

Scott Russell had just forecast that he would be in a posi-tion to launch the ship next March but he still made no move to supply the necessary information. Instead he in-formed Brunel that he estimated that the sum of £37,673 18s. additional to the contract figure would be necessary to complete the ship. This began a long haggle over alterations which Brunel had made in the designs and which Russell claimed were all 'extras' not covered by contract. He followed this by submitting an obviously incorrect tonnage figure for the completed portion of the ship. At this Brunel lost patience. 'How the devil can you say that you satisfied yourself on the weight of the ship,' he retorted, 'when the figures your Clerk gave you are 1,000T less than I make it or than you made it a few months ago – for *shame* – if you are satisfied. I am sorry to give you trouble but I think you will thank me for it – I wish you *were* my obedient servant, I should begin by a little flogging . . .'

On 12 October Russell wrote to say that his banker, Martin, was demanding an immediate payment of £12,000 out of a total of £15,000 said to be due, and requested Brunel to let him have a certificate authorizing the payment of this sum. Reluctantly, in view of the state of the works, Brunel authorized a payment of £10,000 but a few days later he gave Russell a second and unmistakable warning:

I must beg of you to look back to the several letters I have written to you for weeks past. . . . The time is arrived at which the work of the former [launching ways] should have been commenced. At all events, I cannot, of course, include the cost of these works in my estimate of work done until it is done, and I must caution you that the future periodical payments must be regulated by this as well as other considerations. You know my views – let the plans be prepared immediately and something done before you apply for your next payment.

This seems to have had little effect for a week later Russell wrote glibly:

Mr Martin has been here again about the reduction of the last certificate from £15,000, as formerly agreed with him, to £10,000 and I fear I shall get into trouble unless we can see our way to a *definite* arrangement for the future. I am keeping up an enormous establishment of people working night and day. I must either have payments with certainty or reduce my number of hands.

Now this was, on the face of it, true, but by no means all Russell's 'enormous establishment' was now devoted to the great ship. Russell had begun laying down other ships and was diverting more and more labour to them. Moreover, to add insult to injury these new craft had been laid down in the Napier Yard, which was not his property, and in such a way that they positively prevented the completion of the stern of the great ship. This, so far as Brunel was concerned, was the last straw. Added to this, Russell had still not let him have the information he required for his launching calculations despite endless requests. So far, believing in his sincerity, he had 'covered' Russell by giving no inkling of the true state of affairs in his reports to the directors on the progress of the ship, but on 14 November he dropped the first hint of impending trouble by warning them that Scott Russell might become 'embarrassed in his proceedings'.

On 2 December Brunel sent Russell what amounted to an ultimatum on the launching question:

I must beg you to let me have with the least possible delay the correct position of the centre of flotation at the 15 ft draft line and let this calculation be made and tested with the greatest care that no possibility of a mistake can arise. I cannot stand any longer the anxiety I have felt ever since we commenced the ship as to her launching and having now calculated myself her weight and centre of gravity at time of launching I must have her centre of flotation and this your people can do better than mine.

But that he no longer had any confidence in Russell is revealed by the fact that he wrote simultaneously to William Jacomb, his personal assistant at Millwall, as follows:

I have requested Mr Russell to have made a careful recalculation of the position of the centre of flotation at the 15 ft draft, but I should like to have this and some other calculations as to flotation made by ourselves also to check this. How are we off for the means of getting these calculations made? If you want further help you must have it. ... To satisfy me they must be made by two different sets of people at two different places, say Millwall and Duke Street. You will see at once the importance of all this when you think that *we have not an inch* to spare in the launching when she must be on a perfectly even keel ...

In this way the vital information was at long last obtained and by 19 December the detailed plans for the launching cradles and ways had been prepared.

Russell's reaction to these plans was to complain to the directors that they represented a scheme never intended in the first place and, so far as he was concerned, financially out of the question. This brought matters to a head at last by provoking from Brunel a report to the directors in which he threw all his cards on the table. 'Mr Russell', he declared,

states that the mode of launching *now* proposed will be much more costly than he had contemplated and brings this forward as one of the reasons why his means are insufficient. I cannot say what Mr R's estimate may now be or what he now believes to have been his former expectations, but I can say quite plainly and decidedly that I know nothing of any alteration of plan whatever but that which was always intended . . .

I know what is now passing in Mr R's mind and therefore shall not affect ignorance of it. . . . He has once or twice thrown out the idea of letting the ship go, leaving it to descend itself and take its chance as in an ordinary launch. I believe the operation simply impracticable, that is to say that a body of that length drawing 15 ft of water could not launch itself broadside by impetus into such a depth of water – it would stick half-way, even if it went straight and was uninjured, and would probably cost £50,000 to get out of the mess and I have told him so. But even if such a thing were possible, the proceeding would in my opinion be simple madness unless the vessel were well insured and you wished to get rid of it . . .

Brunel then went on to disclose to them for the first time the true state of affairs at Millwall; the deficiency of 1,200 tons of iron; the six small ships which Russell was building on the site leased by the Company, one of them preventing the erection of the stern of the great ship which might, for this reason, be delayed two months. He concluded:

Mr Russell, I regret to say, no longer appears to attend either to any friendly representations and entreaties or to my more formal demands and my duty to the Company compels me to state that I see no means of my obtaining proper attention to the terms of the contract otherwise than by refusing to recommend the advance of any more money unless and until those terms are complied with ...

Brunel realized what his decision to withhold further payment might mean, and it was for this reason that he had forborne for so long. But now there was no alternative.

A week after his fateful report to the directors, Brunel wrote to secretary John Yates suggesting that the Company should itself carry on the work of the ship, that all materials should be moved on to the company's property and such property extended if possible. 'I have', he goes on,

been thinking over the little I *know* and the great deal that I think and can see as clearly as if I knew it, of Russell's affairs, and I believe that the only step that can save him from *immediate* Bankruptcy, and a Bankruptcy which will prevent him ever rising again, would be an arrangement with his creditors under which his liabilities would be ascertained, the position of each creditor defined and the several works now in hand carried on and no new works undertaken or liability incurred. And if proceeding under inspectors, Mr R. would devote his *mind* and attention to his business – I mean his engineering business – each party providing their own materials and the inspectors attending to the expenditure it is just possible that the several works on hand might be completed and the concern left in a state to carry on business. But unless the creditors are immediately called together and some arrangement made, depend upon it that Bankruptcy, and that not a creditable one, is unavoidable to the ruin of the property in the concern, and I very much fear to the ruin of R.'s character. I think it is much to be regretted that R. seems to have no friend about him to give *strong advice*. I have tried it several times and should not be listened to otherwise I would come and give it, but I think I am better away. R. must feel that he has dreadfully misled me more than anybody and would therefore not be disposed to receive my advice as it would be meant.

That Russell had indeed misled Brunel and betrayed his trust was now becoming the more lamentably apparent with every day that passed. By this time Russell had received all but £40,000 of the sum due to him under the contract. In fact, with the addition of certain sums for extra work, the Eastern Steam Navigation Company had paid him a total of £292,295, a very considerable proportion of which seemed to have vanished into thin air. There was the deficiency of 1,200 tons of iron for which Russell had obtained payment from the Company. Moreover, it would soon be revealed that Russell had not paid for a great deal of the iron which he had received for the ship, for Beale & Company, the suppliers, were his largest creditors. Over and above all this, close examination of the great skeletal hull revealed only too clearly that in his anxiety to assist Russell in his financial difficulties, Brunel had authorized the payment of far more money to him than was warranted by results. Indeed it was stated afterwards that at this time no less than three-quarters of the work on the hull had still to be done. Because he had been so largely responsible for persuading his friends and business associates to subscribe the capital for the venture, we may imagine Brunel's feelings when these consequences of his misplaced trust became known. But worse blows were soon to follow for it is quite clear from his letter to Yates that Brunel still did not realize what was afoot; that he still believed Russell to be merely improvident.

Yates was despatched to Millwall with the object of helping Russell to put his affairs in order, while Freshfield, Russell's solicitor, advised Brunel that his client had been supplied with funds to enable him to carry on for another week. Two days later, on 26 January 1856, Brunel sent Russell a short note in which he said: 'I am anxious as ever to save you from disasters which I think I see though apparently you do not.' Such magnanimity was misplaced. It would soon become apparent that the boot was on the other foot. Because his gesture elicited no response from Russell, Brunel next wrote to Freshfield requesting him to prepare a certificate stating that Russell was no longer in a position to proceed with his contract.

I can see [he wrote] no ground whatever to hope that Russell can suggest any arrangement by which the work can be carried on, nor does he seem disposed to agree to any that I can suggest.

At the same time he requested Freshfield to prepare a lease of that part of Russell's property occupied by the great ship. Most of the building site was upon the Napier Yard which was the Company's leasehold, but the bows were on Russell's land. From Yates at Millwall on the last day of January came the following hurried note: 'Have made arrangements for continuous possession. Fear no hope of Russell continuing works as Mr Martin told me last night they should dishonour his drafts and cheques drawn or due yesterday.'

If Yates supposed he had made such arrangements he was disappointed for the next blow followed swiftly, on 2 February to be precise, in the shape of a letter from Freshfield. 'I prepared a lease of extra ground for the ship and have seen Russell,' he wrote, 'but I am sorry to say that he cannot execute the lease as it appears he mortgaged these particular premises to Martin & Co. last year.' Russell had presumably taken this step at the time of his previous financial difficulty but had not thought fit to inform Brunel, notwithstanding the latter's help on that occasion.

On 4 February Russell suspended payments and on the 11th Yates wrote to Brunel from Millwall: 'Mr Russell had been to the works before I arrived there on Saturday and had peremptorily discharged all the men connected with the works of the ship.' At the same time Russell airily announced that he could not continue work on the great ship contract unless he was paid the fantastic sum of £15,000 a month. This piece of cool effrontery was surpassed, however, in the report which he had prepared for his creditors' meeting which was held at Millwall on 12 February with Samuel Beale in the chair. £122,940 was the total of Russell's liabilities, £100,353 being the estimate of realizable assets. An arrangement very similar to that which Brunel had suggested was proposed whereby existing contracts would be completed and the business liquidated under three inspectors appointed by the creditors. But Brunel had not realized either the extent of the claims of Russell's creditors or the fact that his property was already heavily mortgaged to his bankers. The Eastern Steam Navigation Company had reserved a right of claim on Russell's estate for breach of contract only to discover that, so far as they were concerned, he had no estate. His creditors and his banker had prior claims. Brunel and his colleagues were confronted with the painful fact that Scott Russell was one of those individuals in whose hands money could melt mysteriously away. In this case it was a not inconsiderable portion of the £292,295 paid under the contract. 'If the contracts could be carried on,' Russell's report proposed, 'the mortgagees would allow the use of the premises, and the creditors, by adopting the mode of liquidating under inspection, could thus avail themselves of advantages which would probably lead to an increased dividend and prevent the sacrifice which must otherwise ensue.' If this plan was adopted he stated that he hoped to be in a position to pay 10s. in the £ by June. Now followed his crowning piece of insolence. For the first time Russell displayed his true colours. 'It is not proposed,' he blandly declared, 'to continue the construction of the leviathan vessel of the Eastern Steam Navigation Company, the Contract passing to the Directors. Up to the present time no loss had been sustained in connexion with that steamer; but if the work is continued it would, no doubt, exhibit an unfavourable result.'

As Scott Russell must have realized full well, this cynical repudiation of his contract placed Brunel and the unfortunate Company in an impossible position for which they had no redress. He possessed no assets upon which they could claim, while their great ship lay partly upon property held by his mortgages along with the plant needed for its completion. So, while the skeleton of the leviathan loomed silent and forlorn through the mists of Millwall, the mortgagees moved in. 'Are you aware', wrote Brunel to Yates on 4 March,

that Messrs Martin took possession of the fixed plant on Wednesday or Thursday last and placed a person in charge? It is now being valued and, *I am told*, under the impression that an arrangement is under consideration with the Eastern Steam for a *12 months*' occupation of them.

It seems clear that Martin's, encouraged, no doubt, by Russell, anticipated negotiating a lease with the Company on very favourable terms, whereas the Eastern Steam, not unnaturally, considered that they themselves had a just claim. Yates replied immediately: 'Have given Martin & Co. notice of our claims and rights under legal advice. It is only a guess on their part that we are making an arrangement for occupation.'

Ten days later, fire mysteriously broke out at the yard for the second time. 'There appears to be some evil genius presiding over Millwall,' wrote Yates, perhaps more truly than he realized, 'but I am glad to say the affair of yesterday does not in any way interfere with us as the destruction which is most complete is confined to the saw and planing mills.' It was at this juncture that a scheme was mooted whereby Russell, working under the direct supervision of John Yates, should complete certain work on the ship and Brunel was requested by the directors to procure an estimate of the cost. In reply to Brunel's inquiry Russell promptly quoted the sum of £9,512 10s., but without giving the slightest indication as to how he had arrived at so precise a figure. 'I cannot conceive how you make your calculations,' Brunel replied; 'I must communicate this fruitless result of my endeavours to our Directors. At least will you let me have a statement of what you propose to do for £9,512 10s.'

Meanwhile Yates was evidently inclined to urge the acceptance of Russell's figure but Brunel icily pointed out to him the error of his ways:

I observe that you seem frequently to think it very easy to settle prices and contract sums without any precise definition of what is to be done for the money. I am less able to do so and must either have the information I want or leave the matter to more competent persons. Mr R. asks £9,512 10s. for finishing the work. ... I cannot make out £3,000 worth of work. I must therefore have some definition of what he proposes to do for the £9,512 10s. and this I cannot get the slightest clue to. Whatever you may think, this appears to me a rather serious difficulty. ... It is easy to settle general principles, money matters require details.

Stung by this, Yates at Millwall tried to find out in Russell's absence how the estimate had been arrived at. Hepworth, Russell's assistant, was evasive. He maintained that no record had been kept of the calculations and that the loose papers on which they had been made had been destroyed. Thus confirmed in his suspicions, Brunel advised the directors that Russell's offer should not be accepted.

It is obvious from this that Brunel now had the measure of his man and was not going to be caught a second time, but the fact that he no longer harboured any illusions about Scott Russell was cold comfort. It did not answer the question which banished sleep: how was his great ship to be completed? In a report which he made to the directors later in the year he refers to 'hopes of arrangements with other parties which were unfulfilled'. His correspondence, however, yields no clue to what these hoped for arrangements were or with what parties he endeavoured to make them. Indeed it is difficult to think of any arrangement he could have made under the circumstances. He must have realized only too well that he was caught in a cleft stick and that, whatever he might simulate, Russell knew this as well as he did. Yates seems still to have been more naïve. 'I saw poor Russell at Millwall,' he wrote privately to Brunel on 5 April.

He is terribly taken aback at the discovery for the first time of his real position. What have been his expectations I know not but it is lamentable to see him now. Before the Inspector he professes willingness to do all in his power to forward our views, but without any breach of charity I doubt his sincerity.

Precisely what Yates meant by 'the discovery of his real position' is obscure, but that his sympathy was wasted on Russell is certain.

Meanwhile the situation at Millwall could not have been less happy as representatives of Scott Russell, his creditors, his mortgages and the Eastern Steam Navigation Company wrangled over the corpse of the business or stalked about the yard, eyeing each other as warily as so many predatory tom cats. Poor secretary Yates, Dixon and Hepworth, Russell's chief assistants, and the representatives of the mortgagees and creditors, Martin and Stevens, argued interminably over the stock and plant and the terms under which work on the ship could be continued. Dixon and Hepworth had turned sullen and, when asked to prepare a schedule of materials, refused, muttering that they would take orders from no one but Scott Russell. Martin and Stevens, in their anxiety to salvage all they could from the wreck, were not, it appears, too particular whose property they distrained. 'We see a great deal going on which if we were suspicious men would excite very grave suspicions,' wrote Brunel to Freshfield on 15 April. '... Mr Yates will inform you that parties are marking as Mr Martin's sheer legs and shores about the ship which surely belong to us under contract.' At the same time an apprehensive Yates wrote from Millwall to say that he expected a storm to break at any moment. To this a bitter and exasperated Brunel could only reply: 'I almost wish the storm you refer to would burst and sweep away the ship.'

It must indeed have appeared to Brunel as if, as Yates had suggested, some evil genius infected the very air of Millwall. As though he had set himself to build a latter-day tower of Babel, almost from the moment of conception his great ship had been dogged by misfortune, had been checked and was now brought to a grinding halt by human fallibility and folly. He had suffered before from the envy and the malice of lesser men, nor was he any stranger to disappointment, but never had he confronted a situation so disheartening as this. But in his philosophy there was no such thing as retreat or the acknowledgement of defeat. At long last, after weeks of exhausting argument Russell's mortgagees and creditors grudgingly conceded that the Eastern Steam also had certain rights of claim and as a result an agreement was concluded which allowed the Company occupancy and the use of plant until 12 August 1857. This having been arranged, Brunel decided to pursue the only course open to him – to proceed with construction under his direct supervision. On 22 May he was able to write to Hope, his chairman: 'We propose to commence work on Monday morning,' and to note in his diary on the 26th: 'We have this day recommenced work.'

So, once again, men began to swarm like ants about the hull and the Isle of Dogs re-echoed the clamour of their riveting hammers. As the monster grew sightseers began to flock down to the Napier Yard in such numbers that by the end of June Brunel had to request the directors to restrict admission to certain hours. But not one of those thousands who came to goggle at the wonder of the world had any inkling of what was going on behind the scenes. In charge of operations on the spot was an unhappy partnership consisting of Yates, Dixon and Hepworth, and Brunel's personal assistant, Jacomb. In theory, at all events, Scott Russell had nothing further to do with the great ship from the time of his failure until she was launched. Correspondence and contemporary accounts of the progress of the ship tell us nothing about the human situation at Millwall during this period and our only clue is a paragraph from a rare pamphlet entitled *The Great Eastern Steamship, The Past – The Present* which was published in 1858 by G. Vickers of Angel Court, Strand. This reads as follows:

When the ship passed into the hands of the Company they found three-quarters of the work remained to be done before it was ready for launching. The past being beyond recall, the Company set to work to complete the unfinished task. Every one who had been connected with the previous operations refused to give their assistance except upon the most outrageous terms and for the most exorbitant salaries. There was nothing to do but to submit, for the foremen had the drawings and details of the working plans, they had been trained to the work and their places could only be supplied by men who had received a similar amount of training and experience. A period was named when the whole of the work would be finished. It was exceeded by many months and the Company were shipbuilders for just three times as long as was anticipated.

The foremen referred to were undoubtedly Russell's men, Dixon and Hepworth, who had been in charge of the mechanical engineers constructing the paddle engines and the shipwrights respectively. They were hand in glove with Russell and it is safe to assume that this policy of bare-faced extortion was not essayed on their own initiative, nor solely for their own personal profit. Who did benefit by it and to what extent is a question which remains without a positive answer although tentative conclusions may be drawn from results. From their anxiety to put a term to the Eastern Steam's occupancy of the yard it is evident that Russell's mortgagees and creditors did not. On the other hand, Scott Russell, as we shall see presently, contrived to recoup his fortunes in a most miraculous manner. Exactly how he did so, and how he managed to evade or to satisfy his creditors is a mystery which, not surprisingly, has so far eluded the most painstaking inquiry. But it should be emphasized that while his creditors were entitled to an assignment of his assets *at the time the scheme was agreed upon* in February 1856, they had no claim upon any subsequent fees or profits which he might personally receive. Russell's course of action, indeed, was as astute as it was unscrupulous. The same stroke which enabled him to repudiate his contract with impunity also enabled him to extort moneys from the unfortunate Eastern Steam which his creditors could not touch.

How bitter must these months have been for Brunel. Never, as he had said at the outset, had he embarked on any project on which he had staked so much reputation or to which he had committed himself and those who believed in him so deeply. Yet now he and his associates had been caught in a trap from which there was no escape. Knowing him as we do, the mind shrinks from contemplating the black depths of mortification which he must have plumbed as on the Thames' shore the shape of his 'great babe', as he called her, grew with painful slowness while the precious life blood of capital which should have seen her on the high seas was sucked away by the leeches at Millwall. Happily, in these dark hours he did not lack loyal friends and associates. Considering all things, the unfortunate Board, and in particular the chairman, Hope, stood by him well. There were, too, his own assistant engineers and, needless to say, that indefatigable old warrior, Captain Christopher Claxton, R.N. Then there rallied to the side of the embattled engineer an even more formidable fighter, faithful esquire of many a campaign – Daniel Gooch, whose appointment as his assistant the directors approved in October 1856. When a captain and a chief engineer were appointed to the ship in 1857 Brunel acquired two more loyal admirers and allies. They were Captain William Harrison, late of the Cunard line, and Alexander McLellan. Most gratifying of all was the way in which Robert Stephenson rallied to Brunel.

Throughout their careers the two greatest engineers of the day were in almost constant opposition to one another in public, yet neither ever allowed this to influence in the slightest their long and unbroken friendship. 'It is very delightful,' Brunel had written after an evening spent with Stephenson in May 1846, 'in the midst of our incessant personal professional contests, carried to the extreme limit of fair opposition, to meet him on a perfectly friendly footing and discuss engineering points.' When Stephenson had needed his support and advice at the floating of the first huge tube of his Britannia Bridge across the Menai, Brunel had waived all his engagements and hurried north to be at his friend's side. Now, a sad and lonely widower failing in health, Brunel's great contemporary repaid that debt in full, frequently visiting Millwall or writing encouraging letters when he was too ill to do so. Although education and wealth had given Robert Stephenson a polish which his father had altogether lacked, he never lost those qualities of shrewd appraisal and forthright speech which were part of his north country birthright. He had only to make one visit to Millwall to get the measure of Russell's henchman, Dixon. 'I dislike his face immensely,' he wrote to Brunel afterwards. 'I felt that it was an imperative duty to treat his suggestions irreverently.'

It indicates the extent of Russell's neglect of his contract that notwithstanding all the difficulties and the obstructionism at Millwall, progress on the hull of the great ship was actually faster than it had been before. At the end of June 1857 Brunel was able to report that, with the exception of the sternpost, the screw and the screw shaft, the ship was ready for launching and that this work would be finished in a month. Unfortunately, however, the preparations for launching were far from complete. Owing to Russell's opposition and delaying tactics followed by the protracted dispute over the property it was not until January 1857 that the contract for launching ways and cradles was let to Thomas Treadwell the railway contractor of London and Gloucester. When it became clear that the launching ways could not be completed by the time the Company's agreed period of occupancy expired, Brunel wrote to Yates requesting him to negotiate an extension of time. It is clear from Yates's reply that while it might be to Russell's advantage for the ship to remain at Millwall, his mortgagees and creditors had long since reached the conclusion that it was certainly not profiting them and that the sooner they saw the stern of the ship the better. 'Mr Freshfield and myself', wrote the sorely harassed secretary,

have been for hours today engaged with Messrs Martin and Mr Stevens on the subject of continuing to occupy the premises and plant at Millwall. The utmost we have yet arrived at is that we must peremptorily give up on 15 October, the terms in money not yet agreed upon. This arrangement will necessitate the launch of the ship on the 5th of October at latest, and as the time intervening is only six weeks, not a single hour must be lost. The feeling of the Directors is so strong on this point that nothing short of unforeseen calamity must prevent it.

Five days later, Yates wrote again in desperation. With the unfortunate Company at their mercy it was now Martin's and Stevens's opportunity to turn the screw:

After much difficulty an arrangement has been made with Martin's for the use of the yard from 12 August to 10 October 1857 for £2,500. It is impossible to retain the yard any longer. Launch on 5 October or we shall be in the hands of the Philistines.

But no extortionate demands could move that huge mass of inert metal which towered above the Thames shore and which the public had now christened the *Great Eastern*. Only Brunel could do that, and he was not ready. At the beginning of September he warned Yates that he might not be ready by 5 October and the latter replied:

I do not know what may be the consequences. Stevens appears to me to be eagerly seeking for an opportunity to annoy us and, rely upon it, there will be no quarter given.

Brunel was not ready to launch on 5 October and, just as Yates had forecast, there was no mercy shown. Punctually on 10 October the mortgagees took possession of Russell's yard and those at work on the ship were denied access. This created an impossible situation and it was under such duress that Brunel was forced to declare that he would attempt to launch the ship on the next spring tides on 3 November. It was a decision to which he had been driven against every scruple of his engineering judgement. 'The mortgagees of John Scott Russell', wrote Henry Brunel many years later in his private notebook, '... compelled the Eastern Steam Navigation Company to hurry the launch. Otherwise I.K.B. would not have started on 3 November when none of the tackle had been properly tested ...' A meticulous attention to the smallest detail; a refusal to be rushed into any undertaking until all had been tested and perfected, this had ever been the source of Brunel's strength. But now, like Samson, he had been shorn of it, and it was with sad misgiving that he essayed what was to be his last task, to launch his leviathan and send her to sea.

[14]

The Final Hazard

To the average citizen of Victorian London the Isle of Dogs was *terra incognita*. 'We are bound to confess,' wrote a reporter to the *Illustrated London News* after a visit to the Napier Yard in May 1857, 'that those marshy fields, sparsely studded with stunted limes and poplars, muddy ditches, with here and there a meditative cow cropping the coarse herbage, are not suggestive of the sublime or beautiful.' Of the mean little terraces of cottages and public houses which huddled beside the shipyards between marsh and foreshore another reporter wrote: 'The island is peopled by a peculiar amphibious race, who dwell in peculiar amphibious houses, built upon a curious foundation, neither fluid nor solid. ... The houses, in many cases, drop on one side, at a greater angle than the notorious Leaning Tower of Pisa ...' This was the scene, so different from that of the Isle of Dogs today, which was dominated by the hull of the great ship. As she grew to completion, an iron cliff visible far off across the marshes, speculation as to how such a mass of metal could ever be moved became rife. Ready for launching, the hull weighed over 12,000 tons. Never, since time began, had man attempted to move so great a weight, and there were many who shook their heads and pronounced such a feat impossible, predicting that the great ship would rust away where she lay, a monument to human folly and presumption.

To Brunel the launch represented the greatest technical challenge of his career and to no other problem had he devoted so much time, thought and painstaking experiment. His design of the launching cradles and ways, as constructed by Treadwell, was the outcome of a protracted series of tests made with a weighted cradle sliding on an inclined plane.

With this apparatus the loading per square inch on the sliding surfaces, the nature of these surfaces and the gradient were all varied, while the rate of travel of the cradle and the force required to propel or to restrain it were accurately measured. Brunel's object in these experiments was to determine the layout which would enable the ship to be lowered gently down to low water mark with the minimum application of power either to move or to retard her. As a result of them he decided upon a gradient of one in twelve (which meant a theoretical gravitational force of 1,000 tons) and upon iron sliding surfaces.

The ship was supported in two cradles, each 120 ft wide, in such a way that bow and stern projected respectively 180 ft and 150 ft beyond them while a length of 110 ft was left unsupported between them. Beneath these two cradles inclined ways of the same width extended for 240 ft down to the river, excluding, that is, the portions under the ship. These two launching ways consisted of a triple lattice work of timber baulks upon a piled foundation of concrete 2 ft thick. The lowest layer of timbers, measuring 12 in. × 12 in. was bolted to the piles whilst the uppermost carried standard G.W.R. bridge rails upon which the cradles, shod with inch-thick iron bars laid parallel to the keelplate of the ship, would slide. Altogether there were 9,000 points of contact between the bars and the rails so that the loading on each point would be approximately 1⅓ tons, considerably less than the experimental loading.

Because the hydraulic launching gear which Brunel had originally designed had been ruled out by the Board on the score of expense, the following mechanical equipment was installed. To prevent the ship sliding too fast a huge checking drum was mounted at the head of each slipway. Each drum was 20 ft long by 9 ft in diameter and was secured to a solid base, 20 ft square, of 40 ft piles driven shoulder to shoulder. For the purpose of hauling the ship if she stuck on the ways tackles were rigged at stem and stern. The chains from these tackles passed round sheaves secured to barges moored in the river and back to steam winches on the shore. In addition, to provide a pull amidships, the four 80-ton manually operated crabs or winches which Brunel had previously used at Chepstow and Saltash were mounted on moored barges. Finally, to give the ship an initial start if required there were hydraulic rams at bow and stern. In theory, the total propelling power available, plus the force of gravity, amounted to 2,100 tons, but because Brunel's hand had been forced there was no time properly to test the chain tackles or the moorings. In this department both the time factor and his directors' insistence upon economy had compelled him to take a calculated risk whilst making absolutely certain that the checking tackle was equal to any demands that might be made upon it. He argued that if the ship slid out of control, complete disaster would result, whereas she could come to little harm if she stuck upon the ways.

For directing the launch, Brunel resolved to adopt precisely the same signalling arrangements which had worked so well at the floating of the first truss of the Saltash bridge only two months before. Once again, too, he drafted special instructions which were issued to all those concerned in the operation. 'Provided the mechanical arrangements should prove efficient,' he wrote,

the success of the operation will depend entirely upon the perfect regularity and absence of all haste or confusion in each stage of the proceeding and in every department, and to attain this nothing is more essential than *perfect silence*. I would earnestly request, therefore, that the most positive orders be given to the men not to speak a word, and that every endeavour should be made to prevent a sound being heard, except the simple orders quietly and deliberately given by those few who will direct.

At the same time he wrote to Yates as follows:

The inconvenience we suffer at the present time when we are so pressed by the use and occupation of the yard and shops by

other men than those of the Company is very serious. I must
have sole possession of the whole of the premises on the day of
the launch, no men, even of our own still less strangers, in any
part of the yard except those regularly told off for their respective duties and everybody must be completely under my control.

That he clearly envisaged beforehand that little or no
progress might be made at the first attempt is proved by his
instructions under the heading 'Starting':

A strain being brought upon all the purchases, and the
holding back purchase being slack, if the ship does not move, the
two presses will then be worked; if she does not then move, or if,
when moved, she stops and each time requires the presses, the
attempt will be postponed, and more moving power applied for
the next time.

The attempted launch of the great ship on 3 November
1857 has frequently been misrepresented as a typical piece of
Brunel showmanship which miscarried. Nothing could be
further from the truth. Whereas at Saltash all the arrangements had been so carefully made and rehearsed in the
light of previous experience at Chepstow and the Menai that
the chances of failure were remote, here at Millwall he knew
full well the uncertainty of the outcome. That what he regarded as a tentative and premature experiment should become a public spectacle was the very last thing he desired.
He endeavoured to counter the effects of press publicity by
issuing a press statement himself in which he emphasized
this uncertainty, pointing out that even if he was successful
at the first attempt the proceedings would be quite un-
spectacular. He explained that, properly speaking, the ship
was not going to be 'launched' at all; the most he proposed to
do was to lower the ship slowly down to the end of the ways
where she would remain until the next high tide lifted her
off the cradles. But he underestimated the power of the press.
The public had been so stirred by the growing publicity

which the great ship had received as it neared completion that it was not to be discouraged by one anticlimactic statement or by the thin rain of a raw November morning.

... Men and women of all classes [wrote a reporter] were joined together in one amicable pilgrimage to the East, for on that day, at some hour unknown, the Leviathan was to be launched at Millwall. ... For two years, London – and we may add, the people of England – had been kept in expectation of the advent of this gigantic experiment, and their excitement and determination to be present at any cost, are not to be wondered at when we consider what a splendid chance presented itself of a fearful catastrophe.

Of Millwall itself the same hand wrote:

Across the narrow streets, from public-house to public-house, were stretched broad flowing flags, and every apartment in every house, whether bed-room or sitting-room, if it commanded even a glimpse of the huge vessel stretching along above the tree tops, was turned 'inside out' to accommodate visitors for friendship, relationship or lucre. Bands of music were enlivening the scene at the different public-houses, even at the early hour of ten in the morning, and as the performers were drunk at that period, and miserably out of time and tune, the reader must judge what they were at a more advanced time of the day.

That the occasion might be made the excuse for festivities of this kind Brunel may have expected, but for the scene which greeted him in the Napier Yard itself he was utterly unprepared. Where he had insisted that there must be order and silence there was chaos and pandemonium as inquisitive crowds swarmed about the launching cradles and the big checking drums. If the Thames tide had invaded the yard overnight to sweep away the launching tackle it would have created a situation scarcely less calamitous for the engineer. He had been most cruelly deceived. Apart from the directors and their invited guests, secretary Yates, unknown to Brunel, had sold over three thousand tickets giving admission to the yard. The impoverished board had resolved, whatever their

engineer might say, to cash in upon the public determination to turn the occasion into an entertainment. What
passed between Brunel and the unhappy secretary when his
deception was discovered history does not relate but Henry
Brunel who, as a schoolboy at Harrow, was present at the
time, never, to his dying day, forgot his father's anger. 'In
the midst of all this anxiety,' Brunel wrote afterwards to
Samuel Baker, one of the directors,

I learnt to my horror that all the world was invited to 'The
Launch' and that I was committed to it *coûte que coûte*. It was
not right, it was cruel; and nothing but a sense of the necessity of
calming all feelings that could disturb my mind enabled me to
bear it ...

For the engineer the occasion marked the culmination of
a protracted spell of almost incessant work and anxiety
which had begun on 24 October when, under his direction,
the men, using hydraulic jacks, had begun knocking away
the shores and lowering the ship into her cradles. For the
past few days and nights he had never left the yard, only
pausing to take the shortest of naps on a sofa in the little
office reserved for his use. No wonder our reporter described
him as looking 'like a respectable carpenter's foreman' as he
forced his way through that infuriating throng of sightseers
who had ruined his plans. If any reporter ventured to tackle
Brunel at this moment he would, we may be sure, have received very short shrift, so our informative friend made his
way to a special reserved gallery at the stern of the ship
where he was more cordially received. For, believe it or not,
this gallery was, he wrote, 'the exclusive property of Mr
Scott Russell, whose elegant appearance and general urbanity were a marked contrast to the style and manners of
many other important gentlemen present'. So much for
'poor Russell' whom Yates had found so pathetic, even ab-
ject, a figure less than eighteen months before. Mortgagees
and creditors might inveigh against the great ship which

had brought them no return and merely cumbered their property, the funds of the Eastern Steam Navigation Company might be exhausted and their engineer worn out with work and worry, but here was John Scott Russell flourishing like the proverbial bay tree.

If Brunel had had the advantage of a modern public address system he might have succeeded in imposing order and silence upon the crowd that swarmed under the iron wall of his great ship. As it was he was powerless to do so and the whole proceeding became a grotesque parody of the superbly rehearsed and disciplined operation he had commanded at Saltash. But there could be no postponing the attempt and so with a heavy heart he took up his stand on the high control platform at a little before half-past twelve. The crowd craned their necks to watch the diminutive figure while there sounded from beneath the hull the thudding of sledgehammers as the men knocked away the great wedges which secured the cradles. At a lower level stood Dixon whose duty it was to pass on his orders to the gangs in charge of the two checking drums and the steam winches of the bow and stern overhauls. As soon as he had been advised that the wedges were away, Brunel ordered the checking drum cables to be slackened off and this done a white flag was broken out on the starboard side as a signal to Captain Harrison, who was in charge of the winches on the moored barges, that his men should take the strain. At the same time Dixon was ordered to start up the steam winches. When the strain had been taken up the white flags were waved as a signal to heave hard. For ten seemingly interminable minutes, while the expectant crowd waited for the movement which did not come, there was heard a rumbling noise like a prolonged roll of drums as the straining chain purchases awoke reverberations in the iron hull. Seeing that the power of the tackles was insufficient to start the ship, Brunel now called into action the two hydraulic rams. Their effect was so dramatic that in the general excitement and turmoil no one could agree afterwards as to the exact sequence of events. A sudden shout went up and the ground underfoot trembled as, with great rapidity, the bow cradle slid three feet, whereupon it was apparently stopped by the gang manning the brake lever for the for'ard checking drum. This swift movement of the great ship towards them so terrified Harrison's men on the barges that they forsook their winch handles and one of them, leaping into a small boat, rowed for dear life, convinced that he was about to be overwhelmed. Meanwhile the crew manning the aft checking drum cable winch had apparently become so absorbed in what was going on and so distracted by the crowds which pressed round them that they neglected their charge. Indeed, an elderly Irish labourer by the name of Donovan, who was not a member of the gang at all, was either sitting or standing on one of the winch handles. The next instant the earth quaked again and the mutilated body of Donovan was flying through the air over the heads of the horrified crowds. By the time the gang had collected their wits and strained on the checking drum brake lever the stern of the ship had slid four feet, taking up the slack in the $2\frac{1}{2}$ in. chain cable, and spinning the winch like a cotton reel so that its flailing handles smashed the legs of the unfortunate Donovan (who died shortly afterwards) and injured four others. It is indicative of the chaos which prevailed that immediately after this accident Brunel was prevented from investigating its cause by the press of people. The correspondent of *The Times* maintained that the men had misunderstood an order and let out more slack chain when they should have taken it up. Others declared that the ship was never checked by the drum brakes but stopped of its own accord, a view which, in the light of after events, was most probably correct. One thing was certain, as Brunel freely admitted subsequently when he attended the inquest on Donovan; the crank handles should have been removed from the winches as soon as sufficient slack cable had been hauled out.

After this disaster, Brunel postponed further operations until two o'clock, by which time rain was falling heavily from leaden November skies which held no promise. In the interval Captain Harrison had reported that all his men were in a panicky state and not to be relied upon. Rather than risk another accident, therefore, Brunel ordered all the midship river tackles to be disconnected, which done he signalled the steam winches and the hydraulic rams at bow and stern into action. After a few minutes' effort a gear on the bow winch stripped several teeth and the barge carrying the stern purchase began to drag her anchor. At this juncture Brunel decided to make no further attempt to launch that day so that he could carry out a thorough examination of the tackle, the cradles and the ways before darkness fell. Once again, however, his plans were frustrated for, in the words of his son: 'The whole yard was thrown into confusion by a struggling mob, and there was nothing to be done but to see that the ship was properly secured, and to wait till the following morning.'

Apart from the regrettable accident and the difficulties occasioned by the crowds, nothing had occurred which Brunel had not anticipated, but to the damp and dispirited throngs of sightseers who now struggled and fought their way through the gates of the Napier Yard the whole affair appeared a monumental fiasco and for this they blamed, not the press which had misled them or the Company who had taken their money under false pretences, but the engineer. The papers next day voiced the same opinion, which is not surprising when we remember the number of reporters in that rain-soaked crowd. 'We seem to have been a little unfortunate in our grandiose schemes of late,' remarked *The Times* leader writer pompously. 'The Valentia cable has parted, "Big Ben" has cracked and now *Leviathan* will not budge.'

This popular impression of failure had been heightened by the fact that, instead of waiting until the ship had been driven to the end of the ways and was ready to be floated off, the directors had insisted upon performing a launching ceremony before the operations began, Miss Hope, the chairman's daughter, breaking a champagne bottle on the ship's bows and naming her *Leviathan*. With this futile performance Brunel would have nothing to do. The public had already dubbed the ship *Great Eastern*, a name which he favoured for its association with his two previous ships, but the Board had thought otherwise. When on the morning of the launch Yates had hurried up with a list of suggested names for his opinion the exasperated engineer had replied that so far as he was concerned they could call the ship *Tom Thumb* if they liked. In the event the name *Leviathan* did not stick, for it was as the *Great Eastern* that the ship was eventually registered.

The mortgagees of Scott Russell were naturally no less anxious to see the ship in the water than the directors of the Company. They had grudgingly agreed to Brunel's request for sole possession of the premises on the day of the launch but here was the bow of the great ship still cumbering their property. They could not continue to allow the engineer and his staff unrestricted freedom beyond the boundary of the Napier Yard and consequently the latter were made to feel like so many trespassers so far as the operation of the bow tackles and presses were concerned. The financial implications of this fantastic situation are not revealed by the surviving correspondence, but it is safe to assume that the unfortunate Company was still being charged at the rate of £1,250 per month at the least so long as the ship remained on the slipway. That was the sum agreed when the occupancy which had expired on 10 October was arranged, and Yates's anxiety on that occasion suggests that after that date the penalties exacted may have been still heavier. Even under the very best of circumstances Brunel had set himself a task of a magnitude, difficulty and hazard such as no engineer had ever faced before, but every circumstance seemed to conspire to make his position nightmarish. The insolent defection of the brother engineer whom he had trusted; the condemnation and ridicule of a fickle and uninformed press and public; unscrupulous extortion that was little better than blackmail; understandable but impatient and unreasonable pressure from his directors and, last but not least, the fact that all these circumstances had driven him to attempt the launch with inadequate equipment at the worst season of the year.

From an engineering point of view, Brunel's only worry after the work of 3 November was a possibility that the ship might come to harm. In designing the launching ways he had made no attempt to secure a completely rigid foundation, having arranged for the greater concentration of piling to be round the perimeter of the ways to prevent the ground being extruded from beneath the edges of the concrete apron and so causing unequal settlement. He had argued that a limited resilience in the ways was more likely to ensure an even distribution of the weight. On the other hand the foundation of the building slip was absolutely rigid. Hence his fear that if there was any settlement of the ways under the cradles while the latter rested partly on the ways and partly on the building slip its effect might be to distort the flat bottom plates of the ship in between the bulkheads. It would not be possible now to float the ship off until the next high spring tide on 2 December, but it was important, first to avoid this danger by pushing the ship completely on to the launching ways as quickly as possible and next to get her down to the floating off position without any undue delay in case the ways should prove unable to support so great a static load. To test this second element of uncertainty, Brunel subjected a portion of the ways 10 ft square to a static load of 100 tons which represented double the actual load it would receive. After measuring the deflection he felt confident that even if the ship stuck on the ways for a considerable period there would be no serious settlement.

It was on 19 November that the next attempt was made to move the ship. In the meantime alterations had been made to the gear in the light of experience. The four manual winches from the moored barges had been brought ashore and mounted on piled foundations in the yard, their chain cables being led under the hull between the cradles to purchases on the barges in the same fashion as the steam winches at bow and stern. But Brunel was losing faith in the hauling tackle and had installed two more hydraulic presses, making a total hydraulic power of 800 tons. Two blackboards were now provided beside the cradles on which any movement of the ship could be recorded for Brunel's information as he directed operations from amidships. On this first day's trial no further progress was made. It was found that alterations would have to be made to the piled abutments of the presses to enable them to withstand the increased pressure, while after a few moments' strain the mooring chain of the bow tackle purchase parted. No attempt was made to move the stern because it had already over-run the bow.

The press abutments having been modified and the river tackle again overhauled, the bow rams were brought into action once more on Saturday 28 November. There was a prodigious creaking and cracking of timbers as the presses were pumped up to their full pressure and then suddenly one inch of movement was recorded on the blackboard. Once started, the bows continued to move at the rate of an inch a minute until a foot had been recorded, when Brunel signalled the stern rams into action as well as all the river tackles. As before, the latter failed disastrously. The stern mooring chain and both the bow tackles parted while one of the bow purchase anchors began to drag. Later in the day the moorings of the midships' purchases gave way, but the presses continued to work well and by the time the early darkness brought proceedings to an end the ship had been pushed 14 feet.

Poor Captain Harrison and his men struggled on heroically with the river tackle, laying down fresh moorings or grappling and repairing broken chain cables, hazardous and difficult operations in the best of circumstances but all the more so in the bitter cold and the fogs of that inhospitable winter. No sooner had they repaired one failure than another occurred so that, as was said at the time, there seemed to be a fatality working against them. Such failures would have been understandable if, as has often been assumed, the tackle was subjected to strains beyond its normal capacity, but this was not the case. It was theoretically capable of withstanding the maximum pull which the winches could exert and that it utterly failed to do so was due to a heartbreaking and apparently endless series of defects, mainly flawed chain links. Although the loan of fresh tackle improved matters somewhat, Brunel realized that he could no longer depend on the tackle and that he must revise his plans accordingly. He had originally provided limited hydraulic power merely for the purpose of overcoming inertia if the ship proved stubborn. Once started, he calculated that the tackle should be able to keep the ship moving, as indeed it might have done had it not proved defective. Now it had become obvious that he must rely exclusively upon hydraulic power as he had at first intended, with the difference that in place of the special gear he had originally designed he must improvise as best he could with a motley collection of borrowed equipment. Progress by such methods must be slow even under the best of circumstances. It amounted to jacking the ship down to the water's edge. As soon as the rams of the assembled presses had been extended to their limit they must be retracted and the cumbersome machines shifted and packed, or 'fleeted' as it was called, to new positions, all of which took a great deal of time. The river tackle was not dispensed with altogether but was eventually reduced to the steam winch tackles at bow and stern, the river purchases of which were finally anchored to piled moorings on the further shore. This arrangement would have constituted a great obstruction and danger to passing shipping had the tackles been required to exert a sustained pull, but Brunel had evolved a new technique which called for their momentary use only. If, for example, the for'ard presses failed to start the bow of the ship moving, the stern tackle was hauled taut and then suddenly released. The effect of this sudden release of tension was to send an impulse through the hull sufficient to destroy the inertia.

By noon on Monday 30 November the ship had been moved a total of 33 ft 6 in., and Brunel still entertained a hope that they might get her down the ways in time to float off on the spring tide on 2 December. After lunch on that day the ship made a further short slip of seven inches and then the 10 in. press on the forward cradle burst its cylinder, so putting an end, not only to the day's operations but to any hope of a December launch. It was the 3rd of December and the high tide had passed before the situation was restored and work could begin again. On this day the ship was moved 14 ft and on the next 30 ft, notwithstanding the bursting of two more press cylinders. Despite numerous difficulties of this kind some progress was maintained for the next ten days but it was found that an increasing amount of power was required to start the ship moving. In view of this Robert Stephenson agreed with Brunel that it would be best to suspend operations until a much greater and more reliable concentration of hydraulic power could be assembled.

The blunder made in launching the *Great Eastern* [wrote Scott Russell in later years in his *Modern System of Naval Architecture*] was the determination to try the experiment of launching her in iron ways. The ship slid a few feet until the lubricating stuff was rubbed off and then the rails simply bit one another as the wheels of a locomotive engine bite the rails and this held the ship firmly in its place.

This view, which Russell may have been the first to express, has lived on to this day; so much so that some people think that the use of metal to metal was an example of crass stupidity and folly. Yet there is no evidence, apart from this statement of Russell's, that such 'seizing up', as we might call it, did in fact occur at all. Brunel had been alive to this danger from the outset and indeed his decision to use metal on metal was based on his fear that wood might become 'grain bound' in this fashion. Nowhere in the notes of the results recorded with the experimental apparatus or in the data on the actual launch is there any mention of the metal surfaces seizing to the extent implied by Scott Russell. Furthermore, although the power summoned by Brunel to launch the ship seemed titanic to his contemporaries, it was puny by our standards so that if seizing had occurred to any serious extent the prophecies of the pessimists might have been fulfilled and the ship never launched at all. The inertia of the ship was never very much greater than Brunel had bargained for, while when the presses had overcome that inertia and the ship began to slide it might have been kept moving as he had hoped had suitable tackle been available. As it was he could only muster batteries of presses to move the ship by fits and starts.

It is an ill wind indeed which blows to nobody's advantage. Richard Tangye, in his autobiography, has described the sudden knocking upon the door of their little workshop behind a baker's shop in Birmingham which surprised him and his brothers one dark winter's evening. For them it was literally fortune's knock, for the belated visitor proved to be one of Brunel's assistants with an order for hydraulic presses. The Tangye brothers had but lately moved to Birmingham, where they were practically unknown, but Brunel had long known them and admired their work in Cornwall. The performance of the Tangye presses at Millwall was such that they laid the foundation of a great business and it was the Tangye brothers' boast that 'we launched the *Great Eastern* and she launched us'.

While Brunel awaited the arrival of these new presses he was so inundated with letters from enthusiastic amateurs offering help and advice that in self-defence he had to draft a special standard letter in reply. A Mr Thomas Wright brightly suggested that a posse of 500 troops should circulate round and round the deck at the double to the music of drum and fife (to ensure that they kept step). The vibration thus set up would, he was confident, be quite sufficient to keep the ship moving. The Rev. Flood Page of Woolpit Parsonage, Suffolk, proposed digging a trench up to the bows of the ship and then pushing her into it from the stern, while 'A Berkshire Incumbent' advocated rolling the ship into the river on cannon balls and requested a card of admission to the Napier Yard so that he could put his plan into execution. But perhaps the liveliest idea of all emanated from a certain Mr Parsons, who seriously advised Brunel to turn the ends of the launching cradles into targets and discharge mortars at them. It is consoling to be thus reminded that the lunatic fringe is a hardy perennial and not a phenomenon peculiar to our day and age. Some of these amateur engineers refused to be silenced by the standard letter of acknowledgement; they took great exception to it and wrote again. One of the most persistent was a Mr Thornton, who had been struck by an idea of such brilliance that he hesitated to commit it to paper. To him Brunel was finally goaded into writing: 'I can hardly imagine the cleverest man in the world who is not acquainted with all the particulars suggesting anything useful. Still, I shall pay great attention to anything you may say, particularly if you will tell me what is the difficulty you have assumed to exist.' This silenced Mr Thornton.

While half the cranks in England plagued Brunel with their idiotic notions, the press, which had so lately lauded the great ship in such extravagant terms, now taunted its creator as it had once taunted his father. Even the stale jibes of a dead past were combed from dusty files. 'Why do great companies believe in Mr Brunel?' asked *The Field*.

Is it because he really is a great engineer? If great engineering consists in effecting huge monuments of folly at enormous cost to shareholders, then is Mr Brunel surely the greatest of engineers. There is his and his father's great work, the Thames Tunnel, which our uncles and aunts used to look upon as the eighth wonder of the world. With what an expenditure of treasure, labour and blood, was that most stupendous and useless of 'bores' created! And to what purpose? To add one more shilling sight for holidaymakers, and to afford shelter for some half-dozen gim-crack sellers.

'The Hero of Millwall', quipped *Punch*, '... may be observed, in the evenings, gazing hopefully on the *Leviathan*, ejaculating like another Galileo "*E pur si muove*".' 'Canst thou draw out Leviathan with an hook?' quoted the *Morning Advertiser*. Only *The Engineer* reproved the mockers by pointing out that: 'A brave man struggling with adversity was, according to the ancients, a spectacle the Gods loved to look down upon.'

One thing which Brunel never doubted throughout all those weary weeks of trial, disappointment and obloquy was his ability to launch the ship notwithstanding every handicap under which he laboured. To him it was simply a question of practising that doctrine which he had once preached to Captain Claxton at Dundrum: 'To stick to the one point of attack, however defended, and if the force first brought up is not sufficient, to bring ten times as much; but never to try back upon another in the hope of finding it easier.' So, while amateurs and cranks aired their theories and pessimists continued to shake their heads or mock him, the collection of additional hydraulic presses from various sources went on until by the end of the year there were mustered at Millwall no less than eighteen, nine at the bow and nine at the stern, with a combined thrust of 4,500 tons. They included the great press with a 20 in. cylinder which had been used for lifting the tubes of the Britannia bridge. At last Brunel had under his hand the tools for the job and when, on Tuesday 4 January 1858, he again took his stand on the control platform and signalled his new concentra-

tion of power into action for the first time it was with no
doubt as to the result. Only cruel weather now hampered
and slowed down progress. By day dense fogs suspended
operations because his signals could not be seen and
coordination between the various gangs became impossible. Throughout the long, bitter nights the waterfront of Millwall presented an eerie spectacle. It was as
though an army was encamped there, so numerous were the
fires which warded off frost from the presses and pumps,
while like some monstrous phoenix brooding upon her nest
of flame, the dark shape of the great ship loomed through a
curtain of smoke and fire-reddened river mist. Yet notwithstanding these precautions hours were often wasted in thaw-
ing frozen pipes and pumps before the work of the day could
begin. But through fog, frost or rain, Brunel insisted upon
supervising the whole operation in person. Without his presence not a move must be made, and when the ship began to
near the end of the way he took to spending days and nights
together in the yard as he had done in the previous October.

From that moment when the semblance of his great ship
had first taken shape under his pencil it seemed as if it had
claimed its creator body and soul. It became an all-con-
suming passion to which he dedicated himself utterly without regard for health, for fortune or for family. Slowly but
surely it excluded every other concern from his mind. The
impression he made upon his intimate friends during this
period and their concern for him is vividly conveyed in a
letter written to him by his brother-in-law John Horsley,
now no longer the gay young artist and travelling companion of his youth but a grief-stricken widower who
had just lost a wife and two young children from scarlet
fever.

I would implore you [wrote Horsley] to reflect upon that hour
of death which must come upon you sooner or later, and
whether, at that awful moment, you will be able to look with

satisfaction upon your life, which has been one of almost unparalleled devotion to your profession, to the exclusion, to far too
great an extent, of that which was due to your God and even to
your family, and with an utter disregard of your health . . .

My dear friend, will all this bring you peace at the last? Does it
give you peace now? Will not even the sacrifice you are making
of your health and strength prevent your enjoying (should you
live) what you may have stored up, at the time you may now
propose to yourself to cease from your labours?

But this is surely a consideration for this world. What has that
powerful mind of yours settled within itself for the next?

Oh! my dear Isambard, do not throw this on one side from
weariness at my want of power in writing on this subject. Go to
those who have, talk it over with them, and ask whether your
way of life is such as to give you a reasonable hope of entering
into the Mansions of Heaven.

. . . I do remember a conversation we had at Watcombe, in
which I rejoiced to hear you speak decidedly (and I know and
honour your entire truthfulness and sincerity) of the efficacy of
prayer. May I, therefore, be mistaken in thinking you neglect
private devotion!

If you would only bring your powerful intellect to bear upon
the subject which contains the 'one thing needful' I feel confident of the result.

You have so much virtue and excellence in your nature that,
but for fatal habit, you *must* be a religious man. With my last
breath I would implore you to lay this to your heart.

God Almighty bless these weak words and turn you to Him, is
the heartfelt daily prayer, and will be unto my life's end, of

Your most affectionate,
J. C. Horsley.

Horsley was labouring under great stress of emotion
when he wrote this moving appeal and it was in the belief
that his own death was imminent that he added the super-
scription: 'To be given after my death.' In fact there lay before him, had he known it, a long life of great happiness with
his second wife and by the time he died at the ripe age of
eighty-seven the passing of years had long since softened the

memory of that agony of bereavement out of which he had written. Consequently Brunel never read his letter. But that it would have had the desired effect even if he had is unlikely. To such a pitch had he steeled himself to meet and confound betrayal, hostility, ridicule and every kind of adversity in order to complete his task that no appeal of this kind could reach him. To yield even in the smallest degree to such persuasions would now appear to him an admission of defeat.

That it must have seemed to those nearest and dearest to him that he had become obsessed by his great ship is not to say that this obsession, if such it was, in any way impaired his engineering genius and judgement. That whatever else might fail, that no matter what sickness of body or spirit might secretly afflict him, these splendid powers remained unimpaired was revealed during the last two weeks of January 1858.

From the very outset of the undertaking Brunel had looked upon floating the ship as by far the most difficult and hazardous part of the whole operation. To push the ship down the ways was a straightforward operation in which every factor was under his control. Given the right tackle and hydraulic equipment in the first place there is no doubt that the job would have been carried out without the slightest difficulty. To float off so large a ship in so narrow and shallow a river was quite another matter. As he had emphasized long ago to his assistant Jacomb, there was not an inch to spare. This meant that when presented to the rising tide the hull must be so precisely trimmed that it would float at an exactly predetermined draught fore and aft and also upon a perfectly even keel. This question of trim was one which Brunel obviously could not determine by practical experiment but only by calculation. Hence his anxiety when Scott Russell neglected to give him the necessary data, the centres of gravity and flotation and the weight. But it was not only upon the accuracy of Brunel's calculations that success depended; it was also dependent upon the unpredictable vagaries of wind, weather and tide.

The success of the new presses was such that by 10 January the ship was so far down the ways that she became partially waterborne at high tide. This relieved the ways of part of her weight and progress became even more rapid until on the 14th of the month, Brunel suspended operations for fear that the high tides of the 19th might float her prematurely. When these tides had passed she was pushed on cautiously until the cradles were 25 ft off the end of the ways. Brunel had decided to float off the ship on the spring tides at the month's end and preparations began at once. A firefloat was brought up which pumped water ballast into the double hull to prevent any risk of the ship floating until all was ready. The upright shores on the riverward side of the cradles were removed so that the ship was secured there only by the wedges below them. Because they might otherwise foul paddles or other gear when they floated up with the ship, chains were attached to these wedges and carried on deck so that they could be hauled out of harm's way. The bow and stern haulage tackles were still rigged so that she could be pulled out further if need be and bow and stern mooring cables were also run out to prevent wind or tide drifting the ship. Finally, as the critical time approached, four steam tugs arrived to tow her to her berth on the Deptford side of the river. The *Victoria* and the *Friend to All Nations* stood by at the bow while the *Napoleon* and the *Perseverance* took station at the stern.

30 January was the date upon which Brunel had resolved to make the attempt, but everything depended on the state of the weather and he had arranged for weather reports to be sent to him at Millwall by telegrams from observers at Plymouth and Liverpool. At his instigation also, Henry Brunel had been granted special leave from Harrow to be at his father's side. He travelled by North London Railway to Millwall on the afternoon of Friday the 29th and at once

stepped into an atmosphere tense with expectancy to record fully in his diary a never-to-be-forgotten occasion.

A close watch was being kept on the tides which had so far fallen sadly short of their anticipated heights. The water-level was taken every half-hour by Brunel's gauge and recorded upon a chart in the little office. When darkness fell his father settled himself on the sofa in this office smoking his habitual cigars. Every now and again he would jump to his feet and go out into the yard to ascertain for himself the state of the wind, weather and tide or to confer with Captain Harrison. The wind was rising, the reports from Plymouth and Liverpool were discouraging, but there was some promise of a better tide. If Brunel determined to make the attempt to float on the morning tide he must soon give the order to begin pumping out the 2,700 tons of water ballast which held the ship in her cradles. Despite the unpromising conditions this order was given, but throughout the long hours of that winter's night the weather steadily worsened. While Henry, curled up in a corner, tried to snatch some sleep and the tip of his father's cigar glowed in the gloom beyond the lowered lamp, a rising wind boomed overhead or drove torrents of rain against the windows. By early morning it was blowing a gale from the south-west dead on the starboard beam of the great ship and it became obvious that even if she could be floated under such conditions the tug power available would be insufficient to haul her across the river in the teeth of such a wind. So, at 3.30 a.m. Brunel reluctantly ordered the pumps to stop. It must have seemed to him that even the elements which he had so often successfully defied had now joined a conspiracy to defeat him. Earlier, a petrifying stillness of fog and frost had blinded him and would have paralysed and broken his power had he not fought it with fire. Yet now, when he had need of such a calm, what malice had whistled up the winds?

Through a belated dawn of low-flying cloud and gale-driven rain curtains the firefloat returned to her station with difficulty and began to pump back into the ship the ballast which had been fruitlessly pumped out during the night. It was as well that this was done, for, while the weather still made impossible any attempt to float, the following night's tide, with maddening irony, rose high above normal. At the flood the submerged timbers of the cradles creaked and cracked in protest as 15,000 tons of iron and ballast became no longer an inert mass but an animate thing to tremble and to stir uneasily to the fret of wind or current. It was an anxious hour, but the ballast was just sufficient to hold the ship and when the tide ebbed it was found that she had not shifted.

All that evening and well into the night the rain sluiced down in torrents, but a little after midnight it ceased and the wind went round into the north-east. At the same time a telegram came in from Liverpool reporting a similar improvement in conditions. Encouraged, Brunel determined then that he would launch on the next tide, that is to say, on the morning of Sunday, 31 January. It was his last chance before the next springs. As soon as it was safe to do so he ordered McLellan again to start the pumps. This was at 3.30 a.m., exactly twenty-four hours since they had been stopped.

When his father woke Henry from an uneasy sleep at 6 a.m. it was to see a sky which had cleared at last and was now bright with stars. Now, surely, fortune must favour them. In the first faint dawnlight, Brunel ordered the removal of the bolts securing the cradle wedges and while the men were working at this the sun rose in great splendour. Shortly after this McLellan reported that his pumps were sucking air. The ship was ready; there was nothing more to be done now but to await the coming of the tide which began to flow strongly and unusually early. It had been arranged that all the presses and winches should be manned at 11 a.m., but owing to the exceptional tide Brunel had to send messengers hot-foot to round up the men before this. No sooner were they at their posts than he signalled the presses into action for the last time and they began cautiously to push the cradles farther off the ways as the tide rose. At 1 o'clock Mary Brunel and Sophia Hawes arrived in a carriage from the Barge House and were escorted by Henry to a good vantage point at a corner of the yard. Twenty minutes later the stern of the ship, which had been deliberately advanced farther than the bow, was seen to be afloat. Brunel immediately called upon the for'ard steam winch to haul as quickly as possible and at 1.42 p.m. precisely the great bows lifted gently to the tide. The ship was afloat.

Brunel at once went aboard accompanied by his son, his wife and his sister. No sooner had they done so than the starboard cradle wedges floated up and were hauled clear according to plan. There was then a hitch when the stern mooring cable was let go prematurely in the excitement of the moment and the two tugs at the bow began to haul the ship ahead before she had cleared the cradles. As a result the port paddle wheel fouled the timbers of the for'ard cradle and the ship had to be carefully manoeuvred astern under Brunel's direction. Then, as soon as she had dropped clear of this obstruction, she was swung out into the river to be once more hauled ahead, whereupon, with a swirl and a roar of waters, first the stern and then the bow cradles came shooting out from under her. A few moments later the starboard paddle wheel fouled a moored barge carrying a bow purchase. Brunel at once ordered the barge to be scuttled and in a few minutes this last obstacle sank from sight. The four tugs then successfully shepherded their charge across the river to her Deptford moorings, where she was safely secured. Apart from the fouling of the paddle wheels the whole operation had gone remarkably smoothly, the deceptive ease with which it was performed being the highest tribute to the foresight of the engineer and the accuracy of his calculations. The enforced last-minute postponement from Saturday to Sunday morning undoubtedly favoured Brunel by enabling him to evade those distracting crowds which had thronged the Millwall foreshore ever since the operation began. He had never wanted the undertaking publicized, and cheers which could so quickly turn to jeers held no music for him. On the other hand he must have relished keenly the spontaneous ovation from McLellan and his men which saluted him when, at 7 p.m., he came down the ship's side, his task successfully completed after sixty hours of unsleeping vigilance. The *Great Eastern* was at last in her element, but victory had been purchased at a terrible price. The Company was ruined and his own health was broken. The ship, moreover, was still very far from complete, although £732,000 had been expended.

How great was the ruin of both Brunel's health and his spirits, such scraps of his writing as survive from this period reveal with tragic clarity. 'For the last 3 or 4 months I have had so much to try my temper,' he wrote in February, 'that I am proof against anything, and only fear my becoming too slow and apathetic in this state of mind.' And again: 'I would not trouble you with an invalid's journal ... being weak, I am regularly floored with a concatenation of evils.' In May, he left England with Mary for Vichy in the hope of recovering his lost health. From Vichy he went on to Switzerland, returning to England by way of Holland. While away he worked on his designs for the Eastern Bengal Railway, but his thoughts were always turning to his great ship. He returned in September to find her still lying as he had left her, a useless hulk at Deptford. During his absence the unfortunate Company had endeavoured to raise the £172,000 which it was estimated would be necessary to complete her fitting out by launching a scheme for the sale of annuities, but 1858 was a year of trade recession and financial slump, and the scheme was a miserable failure. There were threats of enforced sale and with the idea of forestalling this several of the directors were in favour of offering the ship for sale by public auction. For this sorry predicament the press, needless to say, made Brunel the scapegoat. The attitude of *The Times*, traditionally hostile to both the Brunels, was typical. In July their financial columnist commented upon the forlorn hulk as follows:

We have already had such specimens as the Thames Tunnel and other enterprises which, however they may have redounded to the profit and glory of individuals have led to the ruin of those who embarked in them on the only principle which should ever be recognized in such cases – namely that of securing an adequate pecuniary return. It may be fine to say that the Thames Tunnel, *Great Eastern* and other analogous constructions excite the wonder of foreigners and should gratify our pride, but in that case [the writer went on with heavy sarcasm] we should consider the establishment of a 'Consolidated Fund for the Dignity of England'.

Fortunately, a section of the Company's board was resolutely opposed to any idea of selling the ship at auction and, at the time of Brunel's return, two members of this party, Campbell and Magnus, had put forward a scheme for the flotation of a new company which would buy the ship and in which the subscribers to the existing Company would be allotted shares in proportion to their original holdings. This proposal was adopted and on 18 November the prospectus of the Great Ship Company was issued, Campbell being nominated as chairman, Yates as secretary and Brunel as engineer. On the 25th of the same month the new Company acquired the hull of the ship for £165,000 and on 17 December the Eastern Steam Navigation Company held its last meeting at which a winding-up resolution was passed unanimously.

The new Company asked Brunel to submit his estimate of the cost of fitting out the ship for sea. After his bitter experience over the original contract with Scott Russell, he agreed not only to submit an estimate but to prepare detailed specifications of all the work to be done. These specifications should be sent out to contractors with the invitations to tender and he stressed to the Board that it was of the utmost importance that the terms of contract should bind the contractor to adhere scrupulously to them. Otherwise he clearly foresaw that the same kind of trouble might easily recur.

In preparing his specifications the need for economy forced Brunel reluctantly to discard many features of his original designs to which he had devoted much time and thought. Among these discards were the two small steamers which were to have been carried in fixed davits just for'ard of the paddle boxes. The harbour facilities which so large a ship requires did not exist at this time and his idea had been that these little ships would act as shore tenders. But the most interesting of all these rejected plans was his design for what he called a 'whirling contrivance' to provide an artificial horizon and enable astronomical observation to be made independent of any motion of the ship. In the course of a long correspondence with Airy, the Astronomer Royal, and Professor Piazzi Smyth of Edinburgh Observatory between 1852 and 1854 Brunel had outlined his plans for what was in essence a gyro compass born years before its time. He proposed a gimbal-mounted platform on which the observer would stand with his instrument. Under the platform there was to be a flywheel weighing 1 cwt housed in a box 2 ft square. The flywheel was to be set spinning by means of pulleys and rubber belt until it had attained sufficient momentum to allow the observation to be made. But alas, Brunel's 'whirling contrivance' was still-born for now it had to be relegated to the ranks of the 'might-have-beens' along with his plan for special charts on reels which would be unwound by the ship's patent log as she steamed.

With the new Company in being and work on the ship about to recommence, the last thing Brunel wished to do was to leave England again. But fate decreed otherwise. Despite the rest and change his health showed no sign of improvement and a consultation with Sir Benjamin Brodie and Doctor Bright disclosed that he was suffering from a chronic form of that disease to which Doctor Bright had recently given his name but which we now call nephritis.

There can be no doubt that the years of constant anxiety and nervous tension culminating in days and nights spent without sleep in the miasmic atmosphere of Millwall and under conditions of exposure to appalling weather had been responsible for the onset of the disease. Just how evil the atmosphere of London's river was at this period is disclosed in a letter written by Sophy Horsley to her sister in Switzerland wherein she describes the Queen's visit to the great ship which took place while Brunel was abroad:

Wm Reid says, on going down to the *Great Eastern*, he took the boat and it was dreadful, the water thick black liquid, and the smell beyond description. The Queen, he says, smelt her nosegay *all the time*; the sight was beautiful, and she and her party remained an hour on board.

The two doctors insisted that Brunel must winter in a warm climate and with the greatest reluctance the sick man agreed. So it came about that a few days after tenders had been invited for completing his ship, he left for Egypt accompanied by Mary, his son Henry and a young physician. He had invited Doctor Parkes, lately superintendent of his Renkioi Hospital, to accompany him, but Parkes was unable to accept the invitation, and the young doctor was a tiresome last-minute substitution. 'He is most obliging,' wrote Mary Brunel to her elder son in England, 'but such a toady to your father! And I wish you could hear Henry's droll imitations of him at night when we go to bed.' They took with them a prodigious quantity of stores with which to stock the dahabiya in which Brunel planned to travel up the Nile. The passage across the Mediterranean by paddle steamer from Marseilles was so rough that all the passengers including the doctor were compelled to retire in distress to their cabins with the sole exception of the invalid. Standing on one of the sponsons with his body securely braced in the angle between the rail and the paddle box, Brunel was far too preoccupied in measuring the degree of pitch and roll, wind velocity and paddle revolutions to realize that all but he had fled.

On Christmas Day he dined in Cairo with Robert Stephenson. Stephenson had been too ill to witness the floating of the great ship and now he too had come to Egypt in the vain pursuit of lost health. When he bade farewell to Brunel as the latter set out on his journey up the Nile, both men may have realized that they were unlikely ever to see each other again. The dahabiya which Brunel had chartered was an iron boat which could not safely ascend the rapids above Assuan. After an excursion on donkey back to Philae, however, he determined to continue up the river. He procured a wooden date boat, enlisted some local labour and in a very short time had fitted it up with three cabins. Judging from the lively description of their subsequent trip up the rapids which he wrote in a letter to his sister Sophia it is unlikely that either Mary or the doctor enjoyed themselves very much. He and Henry, however, quite obviously relished it all hugely.

On leaving Egypt the party went to Naples and from thence to Rome, where they stayed for a time before returning home in May 1859. Meanwhile what was happening to that great ship which was seldom out of his thoughts for long? His chief assistant, Brereton, and the faithful Jacomb were deputizing for him, but the reports which reached him indicated that all was not well. In the circumstances this did not surprise him. On the very day of his departure from England he had learned that the Great Ship Company had received two tenders for fitting out the *Great Eastern*. One of these was from the firm of Wigram & Lucas. The second had been submitted by none other than John Scott Russell. How Russell had regained control of his business and so speedily recouped his fortunes that he was in a position to tender for so large a contract is a question that cannot be answered. The documents which might have supplied that answer have been destroyed. Therefore we have to be content with the bare fact that Russell submitted his tender and, after what had passed, that fact alone may seem a remarkable piece of effrontery. But Russell went further than this. Whereas Messrs Wigram & Lucas agreed to accept Brunel's specifications as the basis of their contract, Russell coolly refused to do so.

It must have seemed to Brunel that no sooner was his back turned than his evil genius had seized a long awaited opportunity to strike. He immediately took what countermeasures he could. From Lyons on 6 December he despatched a long urgent letter to his Board in which he pressed them to close with Wigram & Lucas. Above all, he besought them on no account to arrange any contract except upon the basis of an agreed specification and upon the strictest terms which could admit of no subsequent misunderstanding. His warning was not heeded; the Board of the Great Ship Company accepted Russell's tender and the contract was placed with him apparently upon the vague and ambiguous terms which Russell himself had proposed. Only the director Magnus was vehemently opposed to such a bargain.

As many of those on the Board of the Great Ship Company had served its luckless predecessor in the same capacity, it seems at first thought inconceivable that they should have consented to employ Scott Russell again. But apart from the generalization that men never learn by experience there were practical considerations which counted in Russell's favour. Undoubtedly the fact that their engineer was incapacitated for what might prove to be an indefinite period weighed heavily with them and in Brunel's absence there could be no question but that Russell knew more about the *Great Eastern* than any man in England. Also, he still held all the aces, as he very well knew, in the shape of detail drawings and trained assistants such as Dixon and Hepworth. Even if the contract was given to Wigram & Lucas, they could not complete the work on the paddle engines which had not been shipped as originally planned, but which still lay in Russell's shop at Millwall. Unfortunately, the figure Russell quoted for the fitting out does not appear to have survived, but as it seems to have been one of his maxims to quote low and argue afterwards, it is probable that it was considerably lower than the figure of £142,000 submitted by his rivals. Finally, we must not forget that 'urbanity' upon which the reporter had remarked on the day of the disastrous launch attempt. Russell undoubtedly possessed what is sometimes called 'presence', a very valuable asset indeed, and he could when it suited his purposes, turn on a degree of charm capable of disarming the most hard-headed Victorian business man.

No sooner had Russell landed the contract than history began to repeat itself and we may imagine Brunel's feelings when he stepped on board his ship once more on 18 May 1859 to be confronted by a situation almost identical with that which had prevailed at Millwall in 1855/6. An atmosphere of intense suspicion and mutual distrust enveloped the ship like a fog with Captain Harrison, McLellan and Jacomb in the one camp and Russell and his two men, Dixon and Hepworth, in the other. The inevitable disputes were already raging between Russell and the Company over the former's obligations under his contract. Russell's procedure, whenever he encountered any unexpected snag, was to claim that it was not his but the Company's responsibility. To take but one example, when the auxiliary engines in the paddle room were found to be seriously defective owing to certain errors in design, Russell insisted that he was in no way responsible because the engines were of the Company's, or in other words, Brunel's design. In fact, as was later proved, the detail design of these engines had been left entirely to Russell under the original contract. And so it went on. So serious and irreconcilable did these wretched disputes become that in June, John Hawkshaw, John Fowler and J. R. McLean had to be called in as arbitrators with two famous Thames shipbuilders, Joseph Samuda and John Penn, to give independent expert evidence.

In view of all this, it is no wonder that the atmosphere on board the great ship was, to say the least of it, strained. Conditions less likely to lead to the prompt and efficient completion of the work necessary to make the ship ready for sea could not be conceived. In fact a situation bordering on chaos prevailed and it was immediately obvious to Brunel that if the ship was to make her first voyage in the late summer as had been planned something must be done and done quickly. It was equally obvious to him that, weak as he still was, there was only one man to do it and that was I. K. Brunel. Turning a deaf ear to the warnings of his doctors and the appeal of his friends he once more took command of his great ship, applying himself to the task of restoring order out of chaos with all his old daemonic energy. To this there could be only one end.

On his return Brunel's friends had congratulated him on his improved appearance which they took to be certain evidence of his return to health. But by this time he was very well aware that his disease was mortal and that even if he allowed himself to become a complete invalid the end could not long be postponed. A man of his spirit scorned to take any lease of life on such terms as that. Failure and humiliation in the loss of his powers was the prospect from which he shrank; of death he had no fear. He had now but one ambition left, to complete his leviathan and accompany her when she put out upon her maiden voyage. What better then, he must have reasoned, than to devote what little life remained to him to the fulfilment of this last ambition? He was taking a calculated risk for, as he knew very well, and his doctors told him, to consume so recklessly his last reserves of energy must hasten his end. So he who had confronted and defeated so many enemies now challenged death itself.

Whenever he was able, Brunel was to be found on board the great ship supervising operations himself and checking every particular from engine room auxiliaries to rigging details and cabin fittings. There were some days, however, when physical weakness proved too much even for his will-power to master. But though the body might fail him the spirit did not falter and on such occasions he would lie upon a couch at Duke Street dictating endless letters and memoranda of instruction to his faithful clerk Bennett; letters to William Cory concerning the best coal to use, the price per ton, where and in what quantity it should be delivered and how stowed; letters to Russell and to Blake of James Watt & Co. requesting them to nominate the engineers who would take charge of their respective engines during sea trials; instructions to Westhrops, the foreman of riggers, and to Edward Finch of Bridge Works, Chepstow, about the rubber ring joints of the tubular iron masts which the latter was making: 'I know of nobody', he wrote, 'but Moulton of Bradford, Wilts, who makes the right sort of material.' To Captain Harrison went a reminder that he must move to a deeper mooring before coaling and that it was time that he settled upon his anchorages at the mouth of the Thames, at Weymouth and at Holyhead, from which last port it was hoped that the *Great Eastern* would sail to New York in early autumn.

At the same time, whenever he was away from the ship, Brunel received constant progress reports from Harrison, McLellan, Jacomb and Blake of James Watt & Co., but never a word, be it noted, from Russell or his assistants. Indeed the most significant feature of these reports, especially in the light of after events, is the contrast between the businesslike way in which all progress in the screw engine room was reported and the impenetrable silence which appears to have shrouded the affairs of the paddle engine department. In July McLellan reports satisfactory tests of the screw engine boilers and states that he has agreed that they may be clothed at once. On 1 August he reports a satisfactory first trial of the screw engines for one hour at 12 r.p.m., while Blake supplies Brunel with indicator cards. Meanwhile the only news of the paddle engines is supplied by Jacomb. On 25 July he reported his discovery that Russell and his men had given the paddle engines a first trial at 5 a.m. on the previous Saturday morning without informing anyone of their intention and when neither he, Harrison nor McLellan was on board.[1]

By the end of July Brunel became so ill that it seemed he must surely lose his battle against approaching death. On the 28th he had arranged to meet the arbitrators on board the ship but instead Fowler, Hawkshaw and McLean came to him at Duke Street. By this time the sub-contractor, Crace, had completed his elaborate Gothic decoration of the grand saloon and on 5 August Yates wrote to advise Brunel that his directors had invited the Lords and Commons to a dinner on board on the following Monday and requesting him to meet them on deck at noon. To this Brunel replied regretting that his health would not permit him to attend and adding that he considered such celebrations premature in view of the amount of work still to be done. It was indeed true that the thousands of visitors admitted to the ship by ticket and the dining and wining of celebrities on board seriously delayed completion of the ship. Blake was to write to Brunel, later in the month, complaining bitterly that the 'fêtes and dinners' on board prevented the completion of the superheaters which James Watt & Co. had contracted to install as an extra.

The dinner of 8 August was chiefly remarkable for the speech made by Scott Russell, who was in his element on such an occasion. After acknowledging with a most becoming air of modest magnanimity that credit for the idea of the great ship belonged to Brunel, he went on to quote a most remarkable and entirely fictitious conversation between Brunel and himself which implied that the former had ordered

1. Harrison later gives 2 a.m. as the hour when the test was held (see p. 389).

him to build the ship in precisely the same way that a man might order a new suit from his tailor.

Now, I am not a ship-builder [he quoted Brunel as saying], nor am I an engine-builder, and I now come to you to see if you will devote your mind and attention to the carrying out of this problem to a successful issue. . . . You shall design the ship according to your own lines, make the engines upon your own plan, and construct the ship according to the best of your experience and knowledge.

His speech was not only widely reported in the press, but this extract was printed in a booklet about the ship which was produced for sale to visitors. The latter also included the statement that: 'the merit of the construction of the ship and her successful completion is owing entirely to the untiring energy and skill of Mr Scott Russell'.

If Brunel's attention was drawn to these remarks he disdained to dispute them. He knew his Scott Russell to the bone now and could no longer be surprised or angered by anything which he might say or do. If he was to achieve his last ambition he could not afford to allow Russell's posturings to divert him for one moment but must concentrate every ounce of his failing energies on getting the ship to sea. And if Russell had supposed when he made his speech that he had seen the last of him he was mistaken, for by mid-August Brunel had rallied his strength. He took a furnished house at Sydenham and from this more convenient headquarters was able to visit his ship almost every day. But by this time it must have been obvious to all how desperate the struggle between flesh and spirit had become. The dynamic, self-confident engineer, with the inevitable cigar jutting from his full-lipped mouth, who had posed for the photographer in the Napier Yard only two years before was scarcely recognizable in this shrunken, pitiful figure that moved with painful difficulty about the decks with the support of a stick. At the age of only fifty-three he had suddenly

put on the habit of old age. The mass of dark hair had rece-
ded from the high forehead, the full face had fallen in and
the mouth had pursed to a straight, thin-lipped line of pain.
There were dropsical bags under the dark eyes, but the eyes
themselves had become a redoubt where the forces of life
were fighting their last stubborn rearguard action. Through
them the spirit of the engineer still blazed with defiant, unquenchable courage.

Nothing escaped those vigilant eyes; not for an instant did
he permit his hold upon the reins to relax. Orders and instructions still poured out: 10 August to Blake: 'Let me
know when you work engines next; I should like to attend.'
18 August to Russell: 'I require 6 hour engine trial and indicators.' 25 August to Blake: 'Your funnel stays are weak and
must be attended to at once as we sail on the 3rd.' And so on.
There is no evidence that Russell ever carried out the six-
hour trial of the paddle engines and it is quite clear from
what chief engineer McLellan had to say later that Brunel
regarded Russell's proceedings with the gravest misgivings,
knowing from past experience that he would miss no opportunity to evade his responsibilities.

It was finally announced that the *Great Eastern* would
move down to the Nore to adjust her compasses on Wednesday 7 September and sail from thence to Weymouth.
Brunel had already selected his cabin for the voyage which
would mark the consummation of years of endeavour, but
the race had been too close run and he was never to sail.
Early on the morning of Monday 5 September the engineer
boarded his great ship for the last time. At mid-day that
deadly adversary whom he had kept at bay so obstinately
overtook him and he was suddenly seized by a stroke. Paralysed but still conscious he was carried down her side and
home to Duke Street. His last instructions before this blow
fell upon him were to McLellan. He emphasized, McLellan
said, that neither he nor the Company would approve and
accept the engines until they had had a proper trial at sea.

If all went well they might be accepted at Weymouth. Until then the engines remained the responsibility of the contractors and Brunel particularly impressed upon him that it was his duty during this period of trial to hold a watching brief on behalf of the Company but not to issue any orders or interfere in the working of the engines in any way.

Even after this blow which at the eleventh hour had cheated him of the reward for all his efforts, Brunel rallied. At least he had achieved his goal of sending the ship to sea and by 8 September he was again dictating letters to Bennett. So far as can be discovered, his last letter was dictated and signed on 9 September. It was a request that the men from the Great Western Railway shops at Swindon should be given special passes and time off to enable them to see over his ship while she lay at Weymouth.

Meanwhile that ship which was never for an instant out of his thoughts had moved down river with an escort of tugs to Purfleet, where she lay through the night of the 7th, and from thence to the Nore, where she lay the following night. On board, Brereton and Jacomb deputized for their master. In the screw engine room William Crow, the engineer nominated by James Watt & Co., was in charge under Blake's supervision. In the paddle engine room was Russell's nominee Arnott, although Dixon was also present and Arnott was working to his instructions. Scott Russell stationed himself on the paddle bridge, from which vantage point he issued orders to the paddle engine room, his son acting as runner for him.

In the light of what was to follow it is now necessary to recall the fact that the ship was fitted at the base of each of her five funnels with annular feedwater heaters of the design which Brunel had first adopted on the *Great Britain* after his experience of the fire on the *Great Western* which had so nearly cost him his life. This history of the adoption of these heaters adds a quality of dramatic irony to the tragic conclusion to which events were now swiftly heading and which,

in retrospect, seems to have been ordained from the beginning. So protracted and bitter a conflict as that between
Brunel and Scott Russell seems to infect the very air in
which it rages, creating a tension so oppressive that catastrophe becomes as certain as a lightning flash. Of that final
disaster these feedwater heaters were to be the instrument.

As was explained previously, these heaters performed the
dual function of heating the boiler feedwater and preventing
the saloons from receiving too much heat from the funnels.
On the *Great Eastern* they consisted of a six-inch-wide water
jacket round each funnel which extended throughout the
full height of the saloons. From the top of this casing, which
was level with the upper deck, an open-ended standpipe ran
to the top of each funnel. This enabled a head of water to be
built up sufficient to overcome the boiler clack valves and so
feed the boilers by natural flow. Also, of course, the standpipes prevented any possibility of steam pressure building
up in the heaters. While the ship was fitting out the five casings had been given an hydraulic test to 55 lb per square
inch and, in order to do this, stopcocks had to be fitted temporarily to the standpipe outlets. For reasons which were
never satisfactorily explained, these cocks were not removed
from the two feedwater heaters over the paddle engine boiler
room for which Russell was responsible. The standpipes were
fitted above them, while the cocks were concealed by wooden
casings fitted with small access doors just above deck level
and not, therefore, accessible at all from the boiler room.

An engineer named Duncan McFarlane, who was put in
charge of paddle engine auxiliaries by Russell and was working under Dixon's orders, insisted subsequently that these
two cocks were open at the Nore and that in proof of this he
had seen steam issuing from the mouths of the standpipes. If
McFarlane spoke the truth, some hand closed the cocks
when the ship left the Nore. By what malice or what fatal
aberration was it moved? The cocks served no useful purpose
and should never have been there at all, yet to close them

was as deadly an act as to light the fuse to a charge of dynamite. It is more charitable to assume that McFarlane was mistaken and that through some oversight the cocks had never been opened at all. Yet if this was the case it is strange that no accident happened when the ship was coming down the river.

After leaving the Nore Light on the morning of 9 September the *Great Eastern* was worked up to 13 knots, the screw engines making 32 r.p.m. and the paddle engines 8 r.p.m. As she steamed so splendidly through the Downs she encountered a stiff westerly breeze and seas which made the small ships that had put out to view the wonder of the world disappear in the troughs of the waves as they hastened back to the shelter of Dover or Folkestone. But the privileged passengers and reporters on board were enthralled at the manner in which this leviathan disdained these seas. It was as though the Channel had become a lake, so imperturbably and smoothly did she glide forward upon her course. So far, her owners' boast that the most furious sea could not affect their great ship seemed to be justified by results. But meanwhile down in the paddle engine room, had the passengers known it, things were far from well. McFarlane was having such trouble with his donkey pumps that it was only with the greatest difficulty that he succeeded in maintaining the water level in the boilers. Time and time again they laboured and stopped in a hissing cloud of steam. Finally, in desperation he pushed over the two-way taps which bypassed the feedwater heaters and directed the pump delivery straight to the boilers. At once the pumps thudded into reassuring action. Little did McFarlane realize the consequences of his act. By isolating the heaters in this way they became completely sealed containers in which pressure was now steadily rising. Above him, the passengers who were strolling the carpets of the grand saloon, admiring the elaborately gilded and carved woodwork, little realized the lethal force which was gathering behind the great mirrors, the pillars and the panelling which concealed the funnel casings.

Mercifully, everyone had left the saloon before the explosion occurred. The majority were in the dining saloon, but a party had gathered on deck at the bow of the ship to observe the view of Hastings which was just coming into sight. It was five minutes after six o'clock and the great ship was passing Dungeness Light when the unbelievable thing happened. The little group at the bow stood rooted to the spot, white-faced and unable to credit the evidence of their senses as they beheld the huge forward funnel of the *Great Eastern* suddenly launch itself into the air upon a great cloud of steam. Among them was the correspondent of *The Times*. He wrote:

> The forward part of the deck appeared to spring like a mine, blowing the funnel up into the air. There was a confused roar amid which came the awful crash of timber and iron mingled together in frightful uproar and then all was hidden in a rush of steam. Blinded and almost stunned by the overwhelming concussion, those on the bridge stood motionless in the white vapour till they were reminded of the necessity of seeking shelter by the shower of wreck – glass, gilt work, saloon ornaments and pieces of wood which began to fall like rain in all directions.

The paddle boiler room was instantly filled with scalding steam and there presently stumbled upon the litter-strewn deck nightmare figures, their bodies boiled a leprous white. A passenger ran forward and caught one by the arm only to find the flesh come away in his hand. Another, a trimmer named O'Gorman, leapt screaming straight over the rail to be instantly caught up in the floats of the huge paddle wheel. In a matter of hours three firemen, Adams, McIlroy and Mahon, were dead. To two more, Adams and Edwards, fate was less merciful and they lingered on until the next day. None of the dazed passengers and crew who now stumbled about in the wreckage knew what had caused the explosion, nor did they realize that the second funnel casing above the paddle boilers was miraculously withstanding a fantastic pressure of steam which might blow them to glory at any moment. But fortunately, for a reason which he never satisfactorily explained, Arnott realized at once the cause of the explosion and despatched a greaser named Patrick with a key to open the standpipe cock. A great plume of steam shot up with a roar from the mouth of the standpipe high overhead and the danger was past.

Brunel's designs seemed fated to be subjected to ordeals occasioned through no fault on the part of their creator. That they could survive them at all was an engineering triumph, but from every ot ier point of view these ordeals were unmitigated disasters. Just as the rocks of Dundrum had subjected his *Great Britain* to a racking which would have broken the back of any other ship afloat, so now his *Great Eastern* withstood magnificently an explosion which no other ship could have survived. Her iron bulkheads confined the effects of the explosion entirely to her grand saloon and in the library adjoining not a book had moved upon the shelves and not a mirror shivered. Nor did the great ship for an instant deviate or falter upon her course, but held steadily on for Weymouth through the gathering dusk. Her two for'ard boilers below the fallen funnel were blown down and their fires had been instantly drawn following the disaster. But to allow for any two boilers being taken out of service on voyage, Brunel, it may be remembered, had stipulated that the two boiler rooms should be inter-connected. So the two paddle engine boilers still in steam were now supplemented by steam supplied from the screw engine boiler room. So still she drove on both screw and paddles.

While this drama was taking place in the Channel, the dying engineer lay at Duke Street waiting for news of his ship. It has been said that news of the disaster should have been kept from him, but it is inconceivable that Bennett could have done anything other than break it to him. Brunel had given strict instructions to both Captain Harrison and Brereton that they were to send him reports of the voyage as soon as the ship reached Weymouth, while Brereton was to despatch Jacomb to London by express to deliver a verbal account. By no pretext which would have deceived Brunel could Bennett have withheld the tidings from him when they reached Duke Street. His body paralysed, only the engineer's spirit still rallied him, holding death at bay a while longer for but one reason – to hear news of the success of his great ship. He was so confident of a triumph which must atone for every misfortune. Instead came the news of this crowning disaster. This final stroke was too cruel to be borne. The spirit broken at last, the light in the eyes went out and as night fell on Thursday 15 September he died.

The press which had so often blamed or derided Brunel in life now blossomed forth with fulsome obituaries which need not concern us here; such hollow tributes can add nothing to our knowledge of the man. It is only from the words of friends not intended for the public ear that we can begin to understand the weight of such a loss. Thus in a letter to Claxton, William Patterson, the old Bristol shipbuilder, said of him simply: 'I believe him to have been a very good man, and as an Engineer I do not think we have his equal left.' But the most moving tribute of all came from Daniel Gooch. In his diary he wrote:

On the 15th September I lost my oldest and best friend. . . . By his death the greatest of England's engineers was lost, the man with the greatest originality of thought and power of execution, bold in his plans but right. The commercial world thought him extravagant; but although he was so, great things are not done by those who sit down and count the cost of every thought and act.

So wrote the man who was to do more than any other to honour the memory of his friend and to champion his cause.

Epilogue

THE body of Brunel still lay at Duke Street awaiting its last journey to the grave of his father and mother at Kensal Green when the inquest on the victims of the explosion took place at Weymouth. The issue of *The Times* for 19 September carried at once a long obituary notice and an almost equally long verbatim report of the inquest. Readers who studied both items must have been somewhat puzzled, for whereas the writer of the obituary credited Brunel with 'the conception and design in all its detail of the *Great Eastern*', Russell was quoted as saying in evidence that: 'except as far as the late Mr Brunel was the originator of the idea, I was the builder and designer of the Great Ship'.

The inquest appears to have been a very strange affair indeed. In the first place it attracted no public interest whatever in Weymouth, the coroner's court being practically empty except for the press, while *The Times* correspondent remarked that only the foreman of the jury appeared to pay the slightest attention to the evidence and that several jurymen were reading local newspapers. This was a pity, because the evidence was well worth hearing. Having staked his claim to be the author of the ship, Russell proceeded to deny all responsibility for what had occurred. He was not in charge of the paddle engines, nor were his assistants; they were the responsibility of Mr McLellan, the ship's chief engineer. 'With the trial trip between Deptford and Portland', he went on, 'I had nothing whatever directly or indirectly to do. I went out of personal interest and invited Dixon as my friend.' Asked by the Coroner what, in that case, he was doing upon the bridge, Russell replied that he had volunteered his assistance only when it became obvious to him that the officers in charge were having difficulty in handling the ship. So much so, he added, that he had three times saved the ship from disaster. At this mental picture of the gallant engineer springing to the rescue of his brain-child *The Times* correspondent noted 'murmurs of approbation in court'. This performance must have left the representatives of the Company who were present, the chairman, Campbell, Captain Harrison, McLellan and Brereton, almost speechless with astonishment and indignation and one cannot but admire Russell's histrionic powers. He made only one slip. Asked who was responsible for the screw engines he replied 'Mr Blake'. Fortunately for him, however, he was not taken up on the point, or surely he would have found it hard to explain why, if the Company had assumed responsibility for the paddle engines, the representative of James Watt & Co. should still have been in charge of the screw engines. Russell went on to say that the feedwater heaters were not of his design but were a modification which did not appear on his original detail drawings of the ship. This was true, but his contention that the heaters were a dangerous device the merits of which had never been proved in practice was subsequently emphatically repudiated by Captain Robertson, the Board of Trade inspector, who was present on the ship when the explosion occurred.

Russell's assistant Dixon followed him into the witness box. He stated that the paddle engines, including the 'water casings', had been constructed under his supervision by the old Company and that 'Mr Russell then had nothing to do with the Yard at all'. He contradicted Russell by saying that the heaters were 'an admirable plan when properly carried out'. Asked who was in charge at the time of the disaster he hedged. 'I could not say who was,' he replied, 'or if anyone was in charge at all.'

But the most unsatisfactory witness from Russell's point of view was Arnott, the engineer whom he had nominated to take charge of the paddle engines. Previously, the greaser,

Patrick, had stated clearly that immediately after the explosion Arnott had despatched him with a key under orders to open the standpipe cock at the base of the second funnel, but Arnott denied that he had ever given such an order. 'The reluctance with which witness gave evidence', added *The Times* correspondent, 'caused a sensation in court.' It seems clear that, whoever else may have shared it with him, Arnott certainly knew the answer to the mystery of how the cocks came to be closed.

Said Captain Harrison: 'Mr Brunel informed me that he would not accept the engines unless they worked faster at sea than they had done on the river. My instructions to Mr McLellan were not to interfere with the working of the engines at all.' Of the paddle engines he said: 'Only one trial of four took place in the presence of the Company's officers and engineers, the first trial was made at 2 a.m. without anyone's knowledge. I would not,' he added indignantly, 'have allowed Mr Scott Russell to take up the position he did if I had known that he only volunteered his assistance.'

The evidence of the other witnesses for the Company need not detain us because all corroborated Harrison. The Coroner in his summing up stated that in his view there was evidence of negligence upon someone's part sufficiently serious to warrant a charge of manslaughter. He went on to refer to the unsatisfactory manner in which the witness Arnott had given his evidence. But the bored and uninterested jury were impatient to close the case and returned a verdict of accidental death without retiring.

The report of Russell's evidence provoked three letters of protest in *The Times* from Great Ship Company shareholders who had been on the ship. 'I saw and heard at least a hundred orders given by Russell from bridge to engine room and not one by anyone else,' wrote one. 'Why is Mr Russell so anxious to be released from responsibility before this trial has taken place?' asked another. Meanwhile Captain Claxton received an indignant letter from William Patterson of

Bristol. 'I am sorry to see Mr J. Scott Russell taking all the credit to himself as respects the *Great Eastern,*' wrote the old shipbuilder –

> Mr Brunel spoke of a 1,000 ft ship to me at the time the *Great Britain* was building and at the same time expressed his dislike of the old-fashioned way of framing ships. He would have all the framing in the direction in which the diagonal ribbon lines are in the framing of a wooden ship and this plan of framing he has carried out in the great ship and he has in almost everything then proposed now carried out and I am quite sure that all credit for all arrangements in that ship is due to Mr Brunel. . . .
>
> Russell denied responsibility for the paddle engines at the inquest. Surely he cannot mean to say he had given up those engines in an imperfect state. . . . He calls the turning round of the wheels three or four times at her moorings a trial! I have had experience of trial trips since 1817 when I was first engaged in steam. The engines were always properly tested.

Meanwhile the *Great Eastern* was undergoing repairs at Weymouth which included the removal of the feedwater heaters. The directors had sanctioned this step in deference to strong public prejudice against them which, though quite unreasonable, Russell's statement at the inquest had not unnaturally roused. It was a sad mistake. Not only did thermal efficiency suffer severely but, as Gooch later noted in his diary, when the ship was steaming hard the heat in the saloons became almost unbearable. The repairs completed, the *Great Eastern* sailed for Holyhead but it was considered too late in the year by then for her to make her maiden voyage to New York and so she returned to Southampton Water for the winter. Here the ill fortune which seemed to haunt the great ship struck another blow. Captain Harrison was drowned when his gig was capsized by a sudden squall as he was going ashore from his ship. He was succeeded by Captain Vine Hall.

John Scott Russell had stepped into Brunel's shoes as engineer to the Company, but his tenure of office was brief; so

far as the *Great Eastern* was concerned he had shot his last
bolt, for the views expressed by William Harrison were
widely shared. The argument between Russell and the
Company over the question of responsibility for the paddle
engines dragged on and was made all the more bitter by the
fact that the cost of fitting out had far exceeded the most
pessimistic expectations. Notwithstanding this expenditure,
the state of the ship was still unsatisfactory, the unhappy
Company having been presented, on their ship's return from
Holyhead, with a formidable list of Board of Trade re-
quirements. Too late, the directors of the Great Ship
Company realized ruefully that they had been largely to
blame for what had happened by employing Russell again
and, having done so, for not heeding their engineer's advice
by drawing up an absolutely watertight form of contract. An
influential group of shareholders headed by Sir Benjamin
Hawes and the director, Magnus, thought likewise, and a
circular which Russell sent out to the shareholders in self-
defence seems to have cut little ice. At a special meeting on 7
January 1860, Sir Benjamin called for a committee of inquiry
to investigate the affairs of the Company and in particular
the circumstances attending the placing of the contract with
Scott Russell. Uproar broke out and there followed what a
reporter described as: 'a heated discussion abounding in per-
sonalities'. However, after two adjournments a committee
was appointed with a mandate to inquire: 'Whether the
terms of the Contract with Mr John Scott Russell were wise
and, above all, to inquire in what way £346,000 had been
expended on a ship still not fit for sea.'

It was probably through no fault of its members that the
report of this committee was rather a damp squib. As we
should expect, they lamented the fact that the evidence was
conflicting and that they had lacked the power to obtain evidence under sanction of legal process. They were, however,
'of the opinion that Mr Magnus, in his opposition to the
contracts entered into with Mr John Scott Russell ... was

influenced by a conscientious desire to forward the interests of the Company'. It was a feeble statement but its publication in February was the signal for the directors to resign in a body and for the election of an entirely new board which included Daniel Gooch. Its first action was to dismiss Scott Russell and appoint Gooch engineer. A further sum of £100,000 was raised on debentures to make the ship ready for sea, Gooch going down to Southampton to superintend the work, and on 17 June the *Great Eastern* sailed for New York with Gooch and his family on board.

It would be inappropriate here to follow in great detail the subsequent career of the *Great Eastern*, but because an extraordinary amount of nonsense has been written about the ship and her misfortunes these mishaps must be mentioned briefly in so far as they reflect upon the soundness or otherwise of Brunel's design. That it is possible to point only to two really serious mistakes in a design so huge, so unorthodox and so far ahead of its time is surely remarkable.

The first of these defects was the white-metalled stern bearing of the screw shaft, which proved inadequate and wore itself away with great rapidity despite every effort to improve the lubrication and cooling. Even on her return to Southampton from Holyhead the bearing was in need of repair, and after completing her first two Atlantic crossings the white metal lining had worn to the extent of $2\frac{1}{2}$ in. The *Great Eastern* was put on the grid prepared for her at Milford Haven and a new stern bearing of lignum vitae was fitted. A brass journal, which was shrunk on the screw shaft at the same time, ran in this bearing and the combination was found to be perfectly satisfactory.

The second error in design was the mounting of the lifeboats in fixed davits over the ship's side. Brunel had supposed that the boat deck of his great ship was beyond the reach of any sea, but the North Atlantic was to prove him wrong with consequences which would have been fatal to any other ship. The success of the ship had seemed to be at last assured when she left Liverpool on 10 September 1861 for New York under the command of Captain James Walker, with a large complement of passengers. Three days out, however, she encountered a storm of such severity that a heavy sea carried away one of the boats on her weather side immediately ahead of the paddle box. Swinging at sea level from one of the davit falls it was in imminent danger of fouling the paddle wheel, so Captain Walker ordered the ship astern so that the boat could be cut away without risk. Taken unawares by this sudden manoeuvre the helmsmen lost control of the wheel with the effect that the huge rudder, sucked by the screw, swung over until the quadrant hit its top with such a mighty jar that the vertical steering shaft broke clean in two at its weakest point – just above the ball-bearing which carried its weight. The great ship immediately fell off and lay broadside at the mercy of the storm which proceeded to demonstrate the vanity of the boast that any ship made by man can be the mistress of the North Atlantic. From the stern of the crippled ship there came a terrifying succession of detonations as the useless rudder, now completely unconstrained in its travel, beat itself against the blades of the screw. The screw engines were stopped. The ship might still have been manoeuvred on her paddles but such was the fury of the seas that in a very short time nothing remained of the two great wheels but the bare hubs, and the ship was left completely helpless. She rolled so heavily that her grand saloons were practically wrecked before the crew managed to regain control of the rudder with jury tackle and the ship was able to limp slowly back to Cork. From thence she returned to the grid at Milford, where repairs were made and two new paddle wheels of smaller diameter and stronger construction were fitted. Such was the heavy price that was paid for mounting the boats in so vulnerable a position.

The following summer the *Great Eastern* met with another mishap equally disastrous to her owners but demonstrating this time, not some fresh weakness but rather the excellence of Brunel's design. Early on the morning of 27 August the great ship arrived off Montauk Point at the entrance to Long Island Sound and was slowing down to pick up the pilot when she heeled slightly and there was heard a loud rumbling noise. Any other ship in the world would have gone straight to the bottom, for she had struck a submerged reef which was not marked on the charts. As it was, the ship docked in New York as usual, many of the passengers leaving her without the slightest idea that anything was amiss. But investigation by a diver showed that she had torn a hole in her outer skin 85 ft long and from 4 to 5 ft wide. Her inner skin was perfectly sound, however, and it was most unfortunate that, instead of ordering her home and putting her on the grid at Milford, a great deal of money and time was spent in carrying out repairs under water which proved quite ineffective. 'Repairs cost us £30,000 and pretty well ruined us,' wrote Gooch in his diary. He estimated that between them these two accidents in successive years had cost the Company £130,000. All the original capital was lost and after one more voyage to New York and back the ship was laid up and the bondholders took possession. Gooch, Brassey and Barber, who were the largest bondholders, then brought off a shrewd stroke of business. It had been decided to sell the *Great Eastern* at auction and the trio planned to buy her in and then charter her for cable-laying. To this end they agreed to bid up to £80,000 but to their amazement the great ship was knocked down to Barber for a mere £25,000. Gooch at once chartered the ship to the Telegraph Construction Company for £50,000, they to pay all expenses. So began the most useful period in the career of the *Great Eastern* and one of which her creator would most heartily have approved.

With her cable tanks installed and Daniel Gooch on board, the *Great Eastern* put out from Valentia in July 1865 on her first attempt to lay a trans-Atlantic cable. 'The work has the best wishes and prayers of all who know of it,' wrote Gooch in his cabin –

Its success will open out a useful future for our noble ship, lift her out of the depression under which she has laboured from her birth, and satisfy me that I have done wisely in never losing confidence in her; and the world may still feel thankful to my old friend Brunel that he designed and carried out the construction of so noble a work.

This first attempt, however, was doomed to heartbreaking failure. On 2 August a defect developed in the cable and in the process of hauling it in from a depth of 2,000 fathoms in order to trace the fault, it broke and defied all the crew's efforts to grapple it.

At 1.30 p.m. [wrote Gooch] the cable broke a few yards from the ship and all our labour and anxiety is lost. We are now dragging to see if we can by chance recover it, but of this I have no hope, nor have I heart to wish. I shall be glad if I can sleep and for a few hours forget I live. This is indeed a sad and bitter disappointment. A couple more days and we would have been safe. God's will be done.

This one thing, upon which I had set my heart more than any other work I was ever engaged on, is dead, and all has to begin again, because it must be done and, availing ourselves of the experience of this, we will succeed.

Succeed they did. On the last day of June 1866, to the music of 'Goodbye, Sweetheart, Goodbye' from the band of the guardship and resounding cheers from ships and shore, the great ship swung slowly away from her mooring at Sheerness with her cable tanks replenished and her grappling gear greatly improved. A week later, on the eve of departure from Bantry Bay, Gooch wrote:

I was very much pleased to see the name of the steam tug we have in attendance on us. She came from Cork and is called the *Brunel*. It is rather a curious coincidence that his name should turn up here in the midst of labours in which he has performed an important part. I hope it is an omen of success.

Gooch's hope was justified. It certainly seemed as if, upon

this one occasion, the fates, which with such determination had repeatedly humbled the pride of the great ship, relented. Even the weather, always so capricious off the western coast of Ireland, smiled upon her. Spinning out from the cocoons in her huge holds the thread that was to tie the two halves of the world together, the *Great Eastern* steamed westwards over a calm sea, sailing into a sunset of such splendour that even the spirit of the hard-headed engineer was profoundly moved.

It was [Gooch wrote] as beautiful a scene as mortal eyes could look upon; all along the western horizon there was a streak of yellow light below some dark clouds, and lighted up by the sun behind them the effect was that of a bright, beautiful country and it was hard to believe it was not land.

So the man of the age of steam, standing on the deck of the world's mightiest ship, described that same vision of Hy Brasil that had once moved Brendan the Navigator to sail westwards in his tiny craft from this same shore. Surely Gooch must have felt that the spirit of Brunel went with them, for he knew that by thus vindicating the powers of his great ship he was carrying out his friend's dearest wish.[1]

On 26 July that wish was most triumphantly fulfilled when the *Great Eastern* steamed slowly into the inlet in Heart's Content Bay, Newfoundland, and on the following day, amid scenes of wild enthusiasm the cable was taken ashore by the tender *Medway*. 'The old cable hands seemed as though they could eat the end,' wrote Gooch –

one man actually put it into his mouth and sucked it. They held it up and danced round it cheering at the tops of their voices. It was a strange sight – nay, a sight that filled our eyes with tears. Yes, I felt not less than they did. I did cheer, but I

1. When the American, Cyrus Field, first came to England in connexion with his trans-Atlantic cable scheme, Brunel had pointed to the unfinished hull of his great ship and said to Field: 'There is your ship.'

could better have silently cried. Well, it is a feeling that will last my life – I am glad two of my boys were present to enjoy and glory in their part of so noble a work. They may, long after I am gone, tell their children of what we did.

Gooch then sent through the cable the news of their triumph: 'Gooch, Heart's Content to Glass, Valentia, 27 July, 6 p.m.: Our shore-end has just been laid and a most perfect cable, under God's blessing, has completed telegraphic communication between England and the Continent of America.' Thus for the first time the new world spoke to the old.

To set the seal on the triumph, on 2 September the cable laid the previous year was successfully grappled, found to be live, spliced and carried to Heart's Content. Her mission thus so splendidly accomplished, the *Great Eastern* turned about and steamed away under full power and with all sails set, 'a grand sight' wrote Gooch. So she brought Gooch home to England and the richly deserved honour of a baronetcy.

As a result of this success the *Great Eastern* went on, like some industrious spider, to weave a web of cables round the world; from France to America and then from Bombay to Aden and up the Red Sea. On this last occasion she travelled from England to Bombay via the Cape of Good Hope with so great a load of cable on board that she left England with the enormous displacement of 32,724 tons on a draught of 34 ft 6 in. – figures fantastic at that time.

Her cable-laying exploits over, the Great Eastern Steamship Company, as her owners now styled themselves, refitted her for passenger service but in this capacity she was never a commercial success. She fell upon evil days and finally suffered the ignominy of becoming a showboat, anchored in the Mersey. The record of the Summer Session of the Institution of Naval Architects at Liverpool in July 1886 described her thus:

The first outing was a trip on the Mersey, and the first sight

presented to us was the unfortunate *Great Eastern*. . . . She was
then moored in the River Mersey, and had been transformed as a
speculation by a syndicate into a floating palace, concert hall and
gymnasium, and as we passed the band was playing and the
trapeze was at full work, the acrobats, both male and female,
flying from trapeze to trapeze: a deplorable exhibition which
would have broken the heart of Mr Brunel or Mr Scott Russell
could they have seen it. I believe all the members without exception felt distressed at the sight.

Happily the great ship did not have to endure such
degradation for very long. On 26 August 1888, loyal to the
last, Gooch noted in his diary:

A paragraph in the *Standard* looks like the last of the grand
old ship the *Great Eastern*. I would much rather the old ship was
broken up than turned to any base uses. I have spent many
pleasant and many anxious hours in her and she is now the
finest ship afloat.

In November 1888 she was sold piecemeal at auction, the
firm of Henry Bath & Sons buying the hull for breaking up,
and the old engineer, lone survivor now of those great days of
trial and triumph, paid her his final valediction: 'Poor old
ship,' he mourned. 'You deserved a better fate.'

The breaking up of the *Great Eastern* was remarkable for
two things, for the stubbornness with which her splendid
hull resisted the efforts of her despoilers and for the birth of
a fantastic legend, that the skeletons of a riveter and his boy
were discovered sealed within her double hull. That throughout her chequered career the great ship had been accompanied by two such grisly stowaways was just the kind of
macabre nonsense which a credulous public would swallow
whole without pausing to think. Moreover, to the super-
stitiously inclined it explained her every misfortune. In the
first place, as any shipbuilder will confirm, the possibility of
a riveter and his mate becoming trapped in this way is so
remote as to be almost inconceivable and in the second it is
on record that twice, once at the time of floating off and

again in July 1859 during the fitting out, Brunel gave orders
for the space between the two hulls to be scrupulously
cleaned out. When we recall the tonnage of water ballast that
was pumped in and out of her the reason for these orders will
be obvious. Finally, the firm of Henry Bath & Sons which
broke up the ship is still in being and has no record of any
such discovery, nor do the records of the appropriate coroner's court yield any trace of the inquest which must have
followed such a find. Yet the story persists and no doubt it
will still, like John Brown's body, go marching on in defiance
of all the laws of probability. How it began is a question
which cannot be answered with any certainty, but it is not
difficult to imagine a gang of those stalwart Merseyside ship-
breakers 'telling the tale' to some credulous journalist. Anyone who has ever worked among such men will know how
they delight and excel in just this kind of leg-pulling.

Popularly, if she is remembered at all, the *Great Eastern* is
generally looked upon as a colossal white elephant. For her
cable-laying exploits alone she deserves better than this, but
in justice to the memory of her creator there are other factors to be taken into consideration in making any fair
estimate. First and foremost were the extraordinary
circumstances which have just been related but which have
never been divulged before; they delayed the completion of
the ship until the original purpose for which she had been
designed had disappeared and they sent her to sea ill-found
and staggering beneath an immense and entirely unjusti-
fiable burden of capital expenditure. So great a ship was
not suited to the Atlantic service of that time, for which Brunel, as his notebooks reveal, had in mind smaller and faster
ships,[2] and any hope that she might one day fulfil her in-

2. Under the heading 'American Liner' dated 18/7/54 we find the
following note: 'Wanted a boat to go in 7½ days. Take the distance
from Milford to New York allowing for keeping well clear of Nan-
tucket shoal to be 2,900, say even 2,950 miles, that would require
16½ knots. [*Note continues at foot of next page.*]

tended function by plying to the Far East vanished with the opening of the Suez Canal. Not only did the Suez render unnecessary the prodigious bunker space, but her beam prevented the *Great Eastern* from using the canal and the widening works put in hand in 1888 came too late to help her. Because she was too far ahead of her time, the great ship suffered continually from the lack of adequate harbour facilities and from being undermanned and underpowered. Not only did men lack any experience of handling such a huge vessel but they had none of the electrical devices and power controls which assist the crew of a modern liner. That Brunel appreciated this difficulty is shown by the lengthy instructions he drew up for the guidance of captain and chief engineer and by the amount of thought he devoted to the organization on board. Yet notwithstanding all his care and forethought, many of the misfortunes which befell the ship were partly due to the fact that man's command over his monster was ever precarious. Her most successful commander was Walter Paton, who succeeded Captain John Vine Hall when the latter suffered a nervous breakdown after the maiden voyage to New York.

Although there was built into the hull of the *Great Eastern* the greatest concentration of horse-power that marine engine builders could produce at that date, it was not sufficient for so large a ship and she disappointed in speed. It was hoped that she might be capable of cruising at 20 knots

Say length	535	
beam	63	
mean draft	20' 6	
load	23	
Displacement mean	10,000	tons
Hull etc.	5,000	
Engine	2,000	
Coal	2,000	
	9,000	

leaving spare 1,000 for cargo.'

but she never succeeded in averaging much more than 14 knots on the Atlantic crossing. She was not equalled in length (and not then in beam) until 1899, when Harland & Wolff launched the 28,000 ton *Oceanic* at Belfast for the White Star Atlantic service. The inadequacy of the *Great Eastern's* 10,000 h.p. may be judged from the fact that the *Oceanic* could summon 28,000 h.p. from her two sets of triple expansion engines, the secret being that they were supplied with steam at 192 lb per sq. in. as against 25 lb. But what the *Great Eastern* lacked in power she made up in her ability to manoeuvre. With her inexperienced crews and an almost total lack of the tug services and other harbour facilities such as vessels of her size enjoy today, she would indeed have been a white elephant and might never have survived as she did had not her unique combination of paddles and screw endowed her with manoeuvring powers which no other large ship has ever equalled. By putting her paddle engines astern and her screw engines ahead, for example, she could be made to creep ahead or to remain motionless without losing steerage way because the flow from her screw still acted upon her rudder. If, when her engines were working against each other in this fashion, her helm was put hard over she would rotate about her centre.

Had it not been for this remarkable quality, the *Great Eastern* might have been lost before she ever crossed the Atlantic. While she lay at Holyhead in the autumn of 1859 she was exposed to a terrific gale which was remembered by sailormen for many years after as the '*Royal Charter* Storm', because it caused the loss of the ship of that name. The great ship began to drag her moorings and to make things worse a wooden jetty was demolished and the timbers, floating down upon the ship, began to foul her screw and paddles. Nevertheless, throughout a night of tense anxiety, Harrison and McLellan managed to manoeuvre the ship, using first one set of engines and then the other, and so kept her out of danger. Needless to say, for cable laying such powers of manoeuvre made the ship ideal. Gooch recorded in his diary how even the delirious excitement which accompanied the landing of the Atlantic cable at Heart's Content was suddenly silenced for a moment by the thunder of the great ship's paddle wheels and he remarked the hushed astonishment with which the cablemen watched her slowly pivot upon her axis in that narrow inlet and steam majestically away to her anchorage in the bay.

Only one question relating to the great ship remains to be answered. What of John Scott Russell who, for good or ill, played so large a part in her history? When his association with the ship ended he gave up shipbuilding, became a consulting engineer, and devoted himself to that 'Great Eastern' among books, his *Modern System of Naval Architecture*, whose three volumes must weigh the best part of a hundredweight. He designed the dome for the Vienna Exhibition of 1873, while his last marine work was to design a train ferry for Lake Constance. He died in poverty at Ventnor in 1882. In December 1866, when his re-election to the Council of the Institution of Civil Engineers was imminent, Russell's suitability for that office was questioned as a result of charges alleged against him by Sir William Armstrong and others. A committee of inquiry was set up and all the evidence given and the documents produced were subsequently printed and published in the form of a special report for the information of members. This dispute had nothing to do with Brunel or the *Great Eastern*, but it is mentioned here lest it should be thought that, notwithstanding the quoted evidence, the portrait of Russell which emerges from the preceding pages is not a just one.

Briefly, Sir William Armstrong's charge was this. During the American Civil War Russell had acted as intermediary in the sale of guns, made by Sir William Armstrong at his Elswick works, to representatives of the State of Massachusetts. He had done so in such a way that each side supposed that he was acting for the other. The American representative, Colonel Ritchie, had deposited in London bonds sufficient to cover the full contract price, the understanding being that Russell should draw them and pay Armstrongs by instalments, less his commission, as the guns were finished. The bonds had all been drawn by Russell and the guns all completed but the sum of £5,227 still due to Armstrong had disappeared. The sequel has a painfully familiar ring. The legal representatives of Sir William Armstrong and Colonel Ritchie agreed to act in concert to recover what money they could from Russell and to divide the proceeds between their respective clients. But Russell, they discovered, had no assets upon which they could claim. Finally it was suggested that they should take in settlement 125 copies of Russell's work on naval architecture at £21 per copy. The work had not been published, but Russell wrote an order authorizing the release of the books by his publishers, Day & Son of Lincoln's Inn. This order proved worthless. When it was presented, Day revealed that he had a prior claim on Russell because he had never been paid for the books according to their agreement. So, like so much of the money expended by the unfortunate owners of the great ship, the balance due to Sir William Armstrong had mysteriously melted away.

Anyone who would prefer to form their own estimate of Russell's character is recommended to study the evidence, and particularly Russell's defence, in the report of the Committee of Inquiry. It reveals Russell at the height of that form which he displayed at the Weymouth inquest. But in this case his eloquence did not avail him, for, having considered the findings of the Committee, the Council carried a resolution that John Scott Russell should cease to remain a member of the Council of the Institution. Perhaps the best that can be said of Russell is that he suffered from a form of megalomania, for it would appear that a great deal of his ability to convince others that black was white lay in the fact that he could persuade himself to believe it also.

In the course of Russell's defence before the Civil Engineers' Committee he made the remarkable statement that three times he had risked his all in order to complete the *Great Eastern*. How a man could accomplish such a feat in the space of a mere five years is difficult to understand, and a cynic might reply that he could only do so if he was not making away with his own fortunes but those of others. Of Russell's duplicity Henry Brunel evidently entertained no doubts. On the margin of the page in the report which bears Russell's statement he has written: 'Yes, you rascal, you left filth half an inch thick under the splendid carpets laid on the decks of the *Great Eastern* and perjured yourself in the examination concerning the explosion of the water heaters.'

The disastrous affair of the *Great Eastern* not only brought about the death of her creator but left his family in straitened circumstances. The fact that in the year before his death Brunel had an inventory made of the contents of No. 18 Duke Street suggests that he may have contemplated selling up and removing to some more modest establishment. In the event, however, Mary Brunel continued to live on in semi-retirement at Duke Street for a decade after her husband's death and by all accounts the effect of adversity and impoverishment was to sweeten a character which had been for so many years spoiled by luxury. Among the many engineering associates of Brunel who befriended his widow was Sir William Armstrong who took Henry Brunel into his works at Elswick as an apprentice. So well did Henry acquit himself that when, after ten years, the Duke Street house and its contents were sold by auction he was in a position to buy many of the relics of its former splendour including many fine pieces of French rococo furniture of which his father had been a keen collector. These are now in the possession of the engineer's granddaughter. The fine house itself was later demolished to make way for government offices.[3]

3. Not only No. 18 but the whole of Duke Street disappeared when the Colonial Office was built.

Today, when the Thames tides recede, the timbers of the ways which once launched the *Great Eastern* still appear in the mud below the Napier Yard at Millwall. These timbers, with the few pathetic relics of the great ship which were preserved when she was broken up, and the recently restored *Great Britain* in Bristol, are all that now remain of the three most daring adventures in shipbuilding since man first ventured on the deep seas. Happily, however, Brunel's other works have proved more durable memorials, so durable in fact that no estimate of his achievement would be complete without some account of their subsequent history.

First, then, the Thames Tunnel, that forgotten monument to the tenacity and courage of both the Brunels. Sir Marc Brunel's original plans for the tunnel had included spiral descending carriage ways at Wapping and Rotherhithe, but although the old engineer even tried to solicit the aid of Queen Victoria, this part of the project was never carried out, and its use remained restricted to foot passengers. Thus a work which had been completed at such a heavy price, not only in money but in human life, and which had once been regarded as the wonder of the age became virtually a white elephant until 1865 when the East London Railway Company was formed to acquire it from the old Tunnel Company and adapt it to railway use. This was done by extending the tunnel north and south in order to make connexions with the Great Eastern Railway 49 chains from its Liverpool Street terminus and with the main lines of the London, Brighton & South Coast and the South Eastern Companies at New Cross.

Like its predecessors, the East London Railway Company had its vicissitudes before it was finally absorbed by the London Transport Executive, but it remains to this day the most easterly of all London's rail links across the Thames. Electrified in 1913, a frequent shuttle service operates between Shoreditch and New Cross during the day but after mid-

night the steam locomotive returns to work interchange
goods traffic between the Eastern and Southern regions of
British Railways.

The two great shafts which Sir Marc sunk at Rotherhithe
and Wapping still remain, although the presence of lifts and
pumping machinery within them makes it difficult to appreciate their proportions. There seems no reason to doubt
that the spiral staircases in the shafts are the originals. The
tunnel dips under the river on gradients of 1 in 40, the inclination being such that it is not possible to see the station
lights of Rotherhithe from Wapping or vice versa, but only
their reflections on the highly polished running surfaces of
the rails. Observing this, it becomes easy to understand the
difficulty of keeping the tunnel clear of water as the shield
advanced and how greatly Sir Marc felt the need for the
drainage driftway which he was not allowed to make. It is in
this respect that contemporary engravings of the tunnel are
so misleading.

To those who use the Thames Tunnel today it is merely
an insignificant part of London's labyrinthine underground
railway system, but to walk through it in the silence of midnight after the last electric train had gone to its depot was
for this writer an unforgettable experience. No tunnel in the
world can be so haunted by the echoes and apparitions of
bygone dramatic events; the music of the band of Guards on
that famous night when the side arches, now soot-encrusted
and cluttered with electric cables, were draped in crimson
hangings; the echoing voice of Fitzgerald crying out in his
nightmare; Charles Bonaparte standing up in the boat looking like his uncle at St Helena; the roar of onrushing water
and, framed in a side arch, too fascinated by the sight to
consider his own safety, the shadow of Isambard Brunel.

Although water is continually pumped out of the tunnel it
was interesting to observe that this comes entirely from the
adjoining tunnels built in 1865, flowing down a culvert into
the bottom of the dip under the river. The original Brunel

tunnel is perfectly dry notwithstanding the fact that the crown of the arch is so near the river bed that the throb of propellers can be heard as ships pass overhead. For nearly ninety years the fabric of the tunnel has been subjected to the vibration of constant railway traffic which Sir Marc could never have envisaged, yet so far as is known the tunnel has never required repair and certainly the smooth rendering of hard Roman cement with which the brickwork is faced betrays no sign of such work. Thus, after one hundred and twelve years the Thames Tunnel remains in its perfection, a splendid tribute to its builders.

Like Marochetti's marble on the Embankment, or Norman Shaw's window in the Abbey, the Clifton Suspension Bridge stands today as a memorial to I. K. Brunel rather than of him because it was not possible to complete it in accordance with his original design. Within twelve months of his death members of the Institution of Civil Engineers had banded together to form a company for the purpose of completing the bridge 'as a monument to their late Friend and Colleague, Isambard Kingdom Brunel, and at the same time removing what was considered a slur upon the engineering talent of the country'. Most active amongst those who canvassed support for this venture and chairman of the Company was that most unscrupulous scourge of the broad gauge Captain Mark Huish, now nearing the end of his railway career. One of Huish's co-directors was G. P. Bidder, George Stephenson's friend and Brunel's opponent in the gauge trials, while the engineers were W. H. Barlow and yet another old rival, John Hawkshaw, who had once reported so scathingly on the broad-gauge road. The Company's secretary was none other than Captain Christopher Claxton, R.N., now in his seventieth year but still as active and devoted as ever. It was a happy sequel to Brunel's tragic end that old friends and old opponents alike should thus combine in free association to do him honour.

It was fortunate for the new Company that at this time

John Hawkshaw (later Sir John) was building the new Char-
ing Cross railway bridge, a work which involved the demo-
lition of the Hungerford suspension footbridge which had
been built by Brunel between 1841 and 1845. To replace the
Clifton chains sold to the Cornwall Railway Company for
the Saltash Bridge, the Hungerford Bridge chains were acquired for the bargain price of £5,000. So by an odd twist of
fate a bridge about which Brunel had cared so little that he
only 'condescended' to engineer it, enabled Clifton – 'my first
child, my darling' – to be completed at last. It was the use of
these chains, however, that forced Barlow and Hawkshaw to
make considerable modifications in the design. Lighter than
the originals, they had to be used in triple instead of double
tiers, the platform suspension rods being connected to a link
pin in each chain successively; in other words every third
link pin carried a rod, the rods to the upper chains passing
between the link plates of the chains below. The construction of the bridge platform was much altered and was
five feet narrower than the original design, while the gaunt
stone suspension towers, reminiscent of Cornish mine architecture, were very different from the Egyptian towers which
Brunel had designed with such enthusiasm so many years
before. Nevertheless, in spite of these alterations, when the
bridge was opened with great pomp and circumstance on 8
December 1864, the engineer would surely have approved his
memorial. The broad principles of the grand design were still
Brunel's and today, one hundred and twenty-five years after
it was first projected, in a world satiated with engineering
marvels, the grandeur of it can still uplift the heart. More
surely than any lineaments in stone or pigment, the aspiring
flight of its single splendid span has immortalized the spirit
of the man who conceived it.

   At the beginning of August 1855, the Great Western Railway's great chairman, Charles Russell, resigned, and in the
following spring, like his predecessor William Unwin Sims,
he died by his own hand. Four years after the death of Bru-

nel there passed into law that Amalgamation Act which united the Great Western with the West Midland group of narrow-gauge lines and so fore-ordained the extinction of the broad gauge. This event was immediately followed by the retirement of Charles Saunders and by the resignation of Daniel Gooch as locomotive superintendent. A year later, in September 1864, Saunders died. He was sixty-eight years old. He had worn himself out in the service of the Company to which he had devoted his whole life and his heart was finally broken by the bitter attacks launched against him by the now West Midland directors.

So, within a decade, three out of the four members of that redoubtable broad gauge partnership came to tragic and untimely death. Only Gooch survived, and the Company had by no means seen the last of him. He had been locomotive superintendent of the Great Western for twenty-seven years, but a year after his resignation from that office he returned from his cable-laying exploit, as Sir Daniel Gooch, Bart, to reign for another twenty-four years as chairman. This lifetime of service, almost unique in railway history, was ended only by his death on 15 October 1889. Through these long years the ageing pioneer watched inscrutably, for he became in later life a man of few words,[4] the gradual collapse of Brunel's broad-gauge empire for which he had fought so hard. He saw the broad gauge banished from the Midlands and South Wales but, although he must have known that the end could not be long delayed, he did not live to see it pass from the old main line to the west of England.

It was on the morning of 20 May 1892 that a crowd gathered at Paddington to see the last broad-gauge Cornishman leave for the west, and as the *Great Western* drew away with

4. In twenty years' representation of the Cricklade Division in the House of Commons, Gooch never spoke a word. When he gave up his seat he wrote: 'The House of Commons has been a pleasant club. I have taken no part in any of the debates, and have been a silent member. It would be a great advantage to business if there were a greater number who followed my example.'

her train, that lofty roof reverberated with the sound of their cheering. It was a last tribute to the bravest lost cause in engineering history. When, at 4 a.m. the next morning, this same train arrived back at Swindon from Penzance, the last wheel ceased to roll on the seven-foot gauge. So expeditiously was the work of conversion planned and carried out that almost exactly 24 hours later the first narrow-gauge booked train, the night mail from Paddington, was able to work through to Penzance. Broad-gauge locomotives and rolling stock were either converted or scrapped, the latter fate befalling the famous 8 ft singles of the 'Iron Duke' class with the exception of the *Lord of the Isles*. This engine, together with the historic *North Star*, was preserved at Swindon until the end of 1905, when George Jackson Churchward, to his everlasting shame, issued an edict that both locomotives must be destroyed because they were 'occupying valuable space'. As an example of a churlish contempt for past greatness, this obliteration of the last remaining evidence of the broad gauge and of two of the most famous locomotives in the world was unique in its crass stupidity. This became the more evident when, only a few years later, a great deal of money and time was mis-spent in building at Swindon a meaningless replica of *North Star*.

So, almost overnight, it seemed, the broad gauge vanished. *Punch* celebrated the occasion appropriately with a parody on *The Burial of Sir John Moore*, the accompanying cartoon depicting the internment under cover of night of a broad-gauge locomotive. Above the still open grave looms the shadow of a single ghostly mourner, an unmistakable figure in a tall stove-pipe hat, cigar in mouth and hands thrust in breeches pockets. But although Brunel's way has gone, his works remain. Under the stress of West of England expresses of a speed and weight undreamed of in his day, the famous flat arch bridge at Maidenhead whose collapse his enemies so confidently predicted still stands fast.[5] In the West there still

5. When the main line between Taplow and Didcot was quad-

stands to his memory the great bridge over the Tamar at Saltash, its piers proudly bearing his name. Here the girders of the short land spans have been renewed in steel, but the great main trusses remain to this day as they were when they were built. For this survival we must be thankful that they were constructed in wrought iron which is so much more resistant to corrosion. No bridge built in steel can live so long.

Despite the wonder of the Saltash bridge, however, of all Brunel's works it is, we may be certain, by the old main line from London to Bristol that he would wish to be remembered. It was upon this work above all others that he spent his youthful energies most lavishly and here he was able to express more fully than at any subsequent time those aspirations which he had cherished as a youth. In great cutting and embankment; in the beauty of elliptical arch work where massive permanence is endowed with grace by superb skill; in bell-mouthed tunnels with triumphal portals arching high above the tracks to dwarf the 'Castles' and the 'Kings' of the 1950s no less effectually than the broad-gauge flyers of yesterday; in these things we see the monuments of a brief heroic age of engineering as remote from our world as that of the great medieval cathedrals.

That age did not long survive Brunel. After his death, so soon followed by those of his great contemporaries Robert Stephenson and Joseph Locke, it drew quickly to its close. If we would know the secret of its splendid assurance, its unshakable faith in the equation of material progress with human betterment and also the reason for its sudden end we have only to skim the pages of the Brunel sketch books. To do so is inescapably to be reminded of those earlier and greater notebooks of Leonardo da Vinci, so remarkable is the range and play of intellect and imagination that they dis-

rupled between 1890–92, the fabric of Maidenhead Bridge was not touched and in the construction of the new arches Brunel's original design was faithfully followed.

play. There existed then no problem in architecture, in civil or in mechanical engineering which his mind was not eager to confront and to conquer. It was precisely because Brunel displayed this astonishing versatility to such a degree that he was able to impart that tremendous impetus to the momentum of the industrial revolution which ensured that he could have no successors. He and his generation bequeathed a sum of knowledge which, like his great ship, had become too large and too complicated to be mastered any longer by one mind. Consequently, all scientific and technical development thereafter depended upon specialization to an ever increasing extent. The result has been that while the collective sum of knowledge has continued to increase at a prodigious rate the individual sum has so seriously diminished that, to paraphrase Goldsmith, while machines have multiplied, men have decayed. For just as the machines, by carrying too far the principle of division of labour, degraded the craftsman into a machine minder, so, just as surely and far more subtly, the process of specialization has by perpetual reduction destroyed that catholicity of intellect without which civilization cannot survive.

If we survey the generations of eminent engineers who have followed after Brunel, we cannot find his equal because all lack that catholicity. Nowhere, ever again, do we find so inclusive an intellect. But specialization has produced other effects of much wider import. In Brunel's day engineering, a word which embraced both the civil and the mechanical, that distinction having only just emerged, was still the twin sister of art. Men spoke in one breath of the arts and sciences and to the man of intelligence and culture it seemed essential that he should keep himself abreast of developments in both spheres. But after the mid-century the two sisters became increasingly estranged from each other with consequences disastrous to both. Scientist and engineer lost their sense of proportion as they lost their concern for the humanities. They grew the more arrogant and intolerant as they became more specialized, and as jealous of their petty monopolies of knowledge as some medieval alchemist of his magic formulae. For all their arrogance, however, these latter-day engineers could never dominate the boardroom in the way that Brunel and his great contemporaries had done. For it was a foregone conclusion and a part of the general pattern of disintegration that when scientist and engineer became specialized animals they should also become the tools, first of commercial power and later of the far more terrifying and impersonal power of the State.

So long as the artist or the man of culture had been able to advance shoulder to shoulder with engineer and scientist and with them see the picture whole, he could share their sense of mastery and confidence and believe wholeheartedly in material progress. But so soon as science and the arts became divorced, so soon as they ceased to speak a common language, confidence vanished and doubts and fears came crowding in. No longer mastered by any single mind, scientific and technical development seemed to have acquired a frightening momentum of its own independent of human volition. The more ingenious and complex its manifestations became the more was wonderment tempered by misgiving. In the arts, which are ever the pulse by which we measure the health of a civilization, these misgivings become increasingly apparent throughout the second half of the nineteenth century. It is therefore a mistake to suppose as we sometimes do that the whole century was distinguished by complacent self-confidence and optimism. Beneath the façade of empire building and expanding commerce, what appears to have been a long golden afternoon of peace and prosperity was troubled by signs and portents of the coming storm, being in fact no true peace but an unnatural stillness such as precedes the thunderbolt's ultimate disintegration.

Architecture, which, above all else, should be the bridge between art and science, did not pursue the splendid example which Brunel had set at Paddington. Instead, it turned timidly to the meaningless and dead reproductions of the past, while commercial building ceased to have any connexion with architecture whatever. It is perfectly true that Brunel used past architectural styles with great freedom and in great variety. On the Great Western between London and Bristol alone there are to be found the Moorish, the Egyptian, the Italianate, the Classic, the Tudor and the Gothic, but the spirit in which he used them was at the opposite pole from the escapist, antiquarian romanticism of the later nineteenth century. For Brunel the past was not something to sigh nostalgically about. It was something to admire and to learn from, yes, but he never doubted his capacity to improve upon it and if he failed in his adaptation of old styles to new purposes it was not through timidity but through over confidence in his own powers. It could be said that here again, in this mastery of a whole range of different styles, he surely paved the way for the disintegration which was so speedily to follow.

To remark the lamentable results of the divorce of science from the arts we need to look no farther than the department of biography. Our descendants will discover that there was scarcely a soldier, a politician, a divine, a nobleman, a social reformer or a philosopher of any note in the last century whose life the literary world has not examined and dissected. As for the proportion of biographical writing devoted in recent time to past alumni in the closed world of the arts it is so great that it must surely convince the future scholar of the morbid, self-absorbed condition of the arts. Looking back from his distance in futurity upon the last century and a half he will assuredly see that soldier and statesman, poet and painter, were all so many puppets jerked by the steel threads spun by the engineers and dancing to the tunes they called. Therefore he will be astounded to discover that our literary world displayed no interest whatever in these engineers. They were shunned because they dealt with 'technicalities' and were therefore considered fit subject only for juvenilia or for the arid treatise. That is why only one adult biography has been devoted exclusively to Brunel in all the years which have passed since his death, and that was a work of filial piety published in 1870 which makes no attempt at assessment and was in any case much too close both in relationship and in time to do so. Yet the historian of the future will assuredly see Isambard Brunel as the key character of his century, the archetype of the heroic age of the engineer and the last great figure to appear in this, the twilight of the European Renaissance. In this regard he will wish to know more than the mere circumstances of a career, its immediate triumphs and failures and its impact upon contemporaries. He will wonder how such a man arose at such a time and, most difficult of all, from what vital spring of the spirit he derived his prodigious creative powers.

There are some few men, rare in any age, who are mysteriously endowed with such an excess of creative power that it can be truly said that they are born to greatness. Yet, with such, the precise form in which that power will find its expression will depend upon what we call, not knowing the mysteries of creative purpose, the accidents of birth and history. That Brunel was of this company there can be no question. Creative power such as he possessed will always seek whatever outlet promises to yield it the richest satisfaction. In fifteenth-century Italy, he might have been a great painter, sculptor and architect, in seventeenth or eighteenth-century Europe a master of the baroque, but because he was born in nineteenth-century England he became an engineer. What he would have become had he been born into our world it is idle to speculate. In an age wherein the bureaucrat whom he hated has everywhere triumphed; where mediocrity must prevail in the sacred name of democracy, and decision by committee be everywhere the craven substitute for individual sovereignty, there is no longer any outlet for genius such as this.

What, then, was the mainspring of his inexhaustible energy? An unshakable confidence in his own powers, yes, but to what purpose were they directed? To the making of money? To the gratification of ambition and to earn public applause? Or was there some high religious or philosophical motive? If we could answer this question with certainty we should have solved the profound mystery of the wellsprings of human action. As it is we can only speculate, weighing the possible motives each in turn.

The least percipient of readers will have gathered from the chapters that have gone before that the money-making instinct in Brunel was not strong. When he was at the height of his fame he undoubtedly earned far more than engineers of his father's generation were ever able to do. The scale of his establishment at Duke Street and the luxuries which he lavished upon his wife tell us that. This wealth, however, was to a great extent incidental and fortuitous. Like his brother engineers, his money came to him from fortunate investments rather than from professional rewards. The engineers may have sown the golden seeds but they did not themselves reap the huge private fortunes which were amassed in that age.

When in April 1836 Brunel's fortunes were in the ascendant he wrote in his journal:

> One thing however is not right; all this mighty press brings me but little profit – I am not making money. I have made more by my Great Western shares than by all my profesional work – *voyons* what is my stock in trade and what has it cost and what is it worth?

He made it a principle throughout his life to invest in his own schemes and so to share their financial risks with others. On this occasion his investment had proved fortunate but, as the financial failure of many of his projects indicate, his losses must often have been heavy and more than once we may guess that he sailed uncomfortably close to the financial rocks. In March 1852, for example, he mentions financial

difficulties in a letter to Charles Saunders 'on that most disagreeable of all subjects – money', in which he requests the
settlement of his account for £2,550 for his work on the new
Paddington Station. He explained that this sum included his
own payment of Wyatt's fee for his assistance and pointed
out that if the Great Western had called in an architect
themselves as the London & Birmingham had done at
Euston Square, Paddington might have cost them an additional £20,000 or £30,000. It was shortly after this that he
staked most of his reserves on the most disastrous speculation of all – his great ship. From a money-making point of
view, therefore, Brunel's career cannot be considered to have
been successful and it is abundantly clear that he was of that
company who dislike money and money matters intensely as
distasteful means which they must perforce employ in order
to attain their ends.

The question whether fame was Brunel's spur is a much
more difficult one to answer, for in this respect there are
curious contradictions in his character. His pride, his love of
glory, his ambition to 'distinguish himself in the eyes of the
public' are self-admitted, and yet not once in his career do we
find him deliberately seeking the limelight, while he scorned
public honours. His attitude to the latter he expressed in his
usual forthright style in reply to a suggestion that an
honour conferred upon him in France ought to be recognized in England.

I disapprove strongly [he wrote] of any introduction into England of the system of distinctions conferred by Government upon
individuals, whether engaged in professions, arts, or manu-
factures, whose merits can be so much better and more surely
marked by public opinion. In countries where public opinion is
not so searching and so powerful as in England, the evils of
favouritism may be out-balanced by the advantages of some
means of distinguishing men. I admit the possibility, though I
doubt the fact; but I feel sure that the evils would be far greater
than the advantages in England. The few cases of Knighthood

conferred in England generally follow public opinion, though I should not wish to see this system carried further. Such being my opinion, I could not consistently ask for my own letter of Chevalier de la Légion d'Honneur being recognized here.

Now this declaration is doubly important because it reveals a respect for public opinion which seems unbelievably strange to us today and which helps us to understand Brunel's ambition. The reason why anyone today who aims to 'distinguish himself in the eyes of the public' seems so despicable is simply that since Brunel's day public opinion has been utterly debauched. As long ago as 1925 Belloc could write of the effect of the popular press:

All men of real capacity in this our time seem to be agreed that the peculiar odour of modern success is a thing for a decent man to avoid. It is becoming a sort of habit of the mind throughout the West, that is, throughout civilized Europe today, in such few men as can do some great work, to do it silently and apart, and to disconnect it wholly from immediate renown.

Conversely, you will have monstrous advertisement surrounding the mere name of a man incompetent, and usually thoroughly dishonest as well, because he is athirst for publicity of the modern sort, and loves it . . .[6]

If this was true in 1925 how much more true is it today when to the power of Fleet Street has been added the immensely more potent publicity medium of radio and television, whose huge stage, dedicated to the glorification no less than the edification of the common man, offers for the indulgence of megalomania opportunities unparalleled. From all this babel Brunel was happily spared; only towards the very end of his career was the power of the press beginning to make itself felt to his despite. The public opinion which he respected was critical and well informed, possessing, in fact, his own standards. To seek to win the acclaim of such an audience was no discreditable ambition. This informed opinion represented, we know, only an educated minority,

6. *The Cruise of the Nona*, pp. 161-2.

but we also know from a number of incidents in his life how Brunel also earned and greatly valued the esteem of those miners, navvies, shipwrights and mechanics who laboured under his command. Illiterate, uninformed instead of mis-informed, and with standards hewn in the hard school of experience, these tough, horny-handed men were shrewd judges of a man and his shaping in his chosen trade.

Sweet though such appreciation may have been to Brunel, it would surely be absurd to suppose that his craving for it was the secret of all that mighty and ultimately self-de-structive outpouring of creative energy. If that was true he would certainly have courted applause far more assiduously than he did. Repeatedly during his lifetime he would ignore the most vociferous curtain calls because he was much more concerned to rehearse his next performance. Neither wealth nor the desire for fame, then, give us the key to his life story, nor is there evidence to suggest that his career was inspired by any religious or high moral purpose.

From the letter of John Horsley's which he never received we can infer that Brunel was not a man of deeply religious professions. The impression we get is of a conventional religious belief weakened by the natural scepticism of a ruth-lessly logical and inquiring mind. This impression is con-firmed by what is almost the only reference to religion surviv-ing amongst his writings, a reference to the value of prayer in a letter which he wrote to his son Isambard just after the *Great Eastern* had at last been successfully floated off.

Finally [he wrote], let me impress upon you the advantage of *prayer*. I am not prepared to say that the prayers of individuals can be separately and individually granted, that would seem to be incompatible with the regular movements of the mechanism of the Universe, and it would seem impossible to explain why prayer should now be granted, now refused; but this I can assure you, that I have ever, in my difficulties, prayed fervently, and that – in the end – my prayers have been, or have *appeared* to me to be, granted, and I have received great comfort.

What a world of doubt seems to sound in that one italicized word! And to this doubt there was closely allied a profound pessimism.

Whenever I think seriously [he wrote in his private journal in 1829], it appears to me how ridiculously unimportant *every worldly* occurence is – and how painful it will be to see every event pass off unheeded; time flies, our hopes [are] gratified or blasted ... and [there is] nothing to rest a permanent idea or hope on but our prospects in the next world.

Despite the gaiety, the wit, and the high spirits which so distinguished him as a youth and as a young man, Brunel's was not, fundamentally, a happy disposition. His own writings in his youth confirm this. They suggest that at the core of his being there lay a profound melancholy and that it was to escape from it that he became so addicted to what he called his 'castle building'. In this unhappiness, in a nature so intensely proud and gifted with so vivid an imagination we have, surely, the key to his extraordinary energy. Doubt and pessimism which might have driven weaker natures to apathetic despair or to orgies of self-indulgence drove Brunel into a fury of creative activity. So proud a man could never admit despair nor any defeat. Whatever imagination suggested, pride drove him to undertake and so the *châteaux d'Espagne* of youth became the great achievements of his maturity.

If this assessment be correct then the claim that Brunel was the last great figure of the European Renaissance is no exaggeration. For what was the motive force behind that precocious and prodigal flowering of genius if it was not the proud reaction to a new spirit of doubt and scepticism? Certainly it was not born in tranquillity of spirit. And if Brunel had been asked what he conceived to be the ultimate end to which his efforts were leading he would surely have answered that it was that same 'profane perfection of mankind' to which the arts of the Renaissance testified. It was only after Brunel's passing that 'confusion fell upon our thought'; that the illusion of that high promise stood revealed, and Europe awoke from that dream.

When, at the height of his career, Brunel visited Rome with John Horsley, the latter related afterwards how fascinated his companion had been with the interior of St Peter's. Many years after, when the engineer returned again to Rome with only a few months to live, his son remarked upon the hours which he spent alone in that great church. Whether he received any spiritual consolation during those hours there can be no knowing, but this much is certain: of his captivation by the sight of the profane splendours of that mighty baroque interior and of the great frescoes in the Sistine Chapel there can be no question, so completely was he in sympathy with the spirit of the age that created them. 'He belonged to the genus of deep, violent, colossal, passionately striving natures.' So wrote J. A. Symonds of Michelangelo, but his words would equally well serve as an epitaph to Isambard Kingdom Brunel.

Appendix

DIMENSIONS OF THE THREE GREAT SHIPS

	Great Western	Great Britain	Great Eastern
Length overall	236 ft	322 ft	692 ft
Length between perps.	212 ft	289 ft	680 ft
Breadth of hull	35·3 ft	51 ft	82·7 ft
over paddleboxes	59·8 ft	—	118 ft
Draught	16·7 ft	18 ft	30 ft
Gross register	1340 T	3720 T	18,915 T
Displacement	2300 T	3018 T	32,000 T
Boiler pressure	5 p.s.i.	15 p.s.i.	25 p.s.i.
Heating surface	3840 sq. ft	not known	49,200 sq. ft
Grate area	202 sq. ft	360 sq. ft	2,328 sq. ft.

ENGINES

	Great Western	Great Britain	Great Eastern Paddle	Great Eastern Screw
			Oscillating	Screw
Type	Side-lever direct acting	Vee crank-overhead direct acting	Oscillating	Horizontal direct acting
Makers	Maudslay	G.W.S.S. Co.	Russell	James Watt
No. of cyls.	2	4	4	4
Diameter	73·5 in.	88 in.	74 in.	84 in.
Stroke	7 ft	6 ft	14 ft	4 ft
R.P.M.	15	18	10·75	38·8
N.H.P.	450	1000	1000	1600
I.H.P.	750	1500	3410	4890
Paddle wheel dia.	28·75 ft	–	56 ft	–
Length of floats	10 ft	–	13 ft	–
Screw dia.	–	15·5 ft	–	24 ft
Pitch	–	25 ft	–	44 ft
R.P.M.	–	54	–	38·8

The figures refer to the original engines of the *Great Britain* and the original paddle wheels of the *Great Eastern*. Scott Russell gives the following I.H.P. figures for the latter: Paddles 3676, Screw 3976, but I regard these as highly suspect.

A Note on Sources and Acknowledgements

APART from juvenilia the only previous work exclusively devoted to I. K. Brunel is the *Life* written by his eldest son, Isambard Brunel, which appeared in 1870. Isambard was not himself an engineer and, as he acknowledges in his Preface, his book is to a considerable extent a compilation of material supplied by his brother Henry Brunel and by Captain Claxton, William Froude, William Hawes and other friends and business associates of his late father. The book is a useful source of factual information on Brunel's career but reveals little of his character or his private life and is heavy reading.

E. T. MacDermot's monumental *History of the Great Western Railway* published between 1927 and 1931 is a mine of information upon Brunel's connexion with that Company. In the very few cases where I have found that the facts contained therein differ from older or original sources, I have accepted the older source.

Lastly, in 1938 there was published that charming family tribute, *The Brunels, Father and Son*, by I. K. Brunel's granddaughter, Lady Noble. This is invaluable for the light which it sheds on the family life of the Brunels. These three books, with the addition of Richard Beamish's *Memoir* of Sir Marc Brunel, represent the printed sources upon which I have drawn most freely. Other works consulted which illuminate various aspects of the engineer's life are listed in my bibliography. The recently published book by Mr James Dugan, *The Great Iron Ship*, gives a very racy account of the career of the *Great Eastern*. My chapters dealing with the construction, launch and subsequent mishaps of this ship are based on other sources, mainly original.

It soon became obvious to me that printed sources could do little more than provide me with the skeleton of a fascinating life story and that only by the study of original documents could I hope to bring these bare bones to authentic life. Such success as I have had in this research is due in great measure to the kindness of those who have granted me access to material, and entrusted me with irreplaceable documents. In this respect I am indebted above all to Lady Noble and to her son, Sir Humphrey Noble, Bart, for their great generosity and helpfulness. My thanks are also due to Mr J. Shum Cox, Librarian to the University of Bristol, and to the British Transport Commission Archivist, Mr L. C. Johnson, for access to the Brunel manuscripts in their charge.

This original material may be listed as follows. First, the private diary which Brunel kept intermittently between 1827 and 1829 and which was only recently discovered. It is a brief document but it gives us an immensely valuable glimpse of Brunel's character as a young man and I have therefore quoted freely from it and at length. Next come the two volumes of Brunel's journal in the possession of his great-grandson. These are of great value because they illuminate the period between the stoppage of the Thames Tunnel works and the commencement of the Great Western Railway which, aside from the Clifton Bridge competition, printed sources gloss over. Chapter 3 and a considerable portion of Chapter 4 have been based upon this journal in which Brunel kept an almost continuous record of his activities from 7 March 1830 to 24 January 1834. After this date the pressure of work evidently became too great for him to maintain this record for there are only a few random entries at widely separated intervals, the last dated 3 March 1840.

In his 1870 biography Isambard Brunel gives a few tantalizing quotations which indicate that at some time in the 1840s his father again began to keep a regular journal which he may well have maintained until almost the end of his life. But, alas, no trace survives of what must have been a fascinating record. One longs to know Brunel's private reactions to the tragic affair of the *Great Eastern*, but Lady Noble is most probably correct in her belief that after his death his sons agreed to destroy these self-revelations out of a misguided sense of filial piety, a fate which the early private diary can have escaped only by a lucky chance. This means that the only personal source material available from 1834 onwards consists of Brunel's letters and reports, notebooks of a purely technical character and the wonderful series of sketch books which Lady Noble has presented to the University of Bristol. Fortunately this material, coupled with the writings of such contemporaries as George Henry Gibbs and Daniel Gooch, enable us to form a reasonably clear portrait of the engineer at the height of his fame. Our knowledge of his character in youth also enables us to infer with reasonable accuracy what was going on behind the façade of the public figure.

For Chapter 6, which deals with Brunel's marriage and family life, I have drawn freely upon Lady Noble's book and upon the diaries of Fanny and Sophy Horsley which, edited by Mrs Rosamund Brunel Gotch, were published under the title *Mendelssohn and his Friends in Kensington* in 1934. But I would add that the opinions expressed in that chapter, based on entries in the private diary, are my own.

Also preserved in the Library of the University of Bristol are certain papers and diaries of Henry Brunel. It is upon Henry's diary account that I have based my description of the scenes which attended the floating off of the *Great Eastern*. It differs in a few minor particulars from that given by his brother, who was not present, in his book. It was a combination of idle curiosity and lucky chance which led me to the discovery, amongst a collection of Henry's papers which did not promise to have any bearing upon his father's life, of some notes and transcripts of correspondence concerning the construction and launch of the *Great Eastern* which hinted that there was very much more to this story than had ever been made public. Henry's notes were tantalizingly incomplete but their implication was that his father had been unjustly held responsible for unforeseen difficulties, delay and expenditure which were in fact occasioned by peculiar circumstances which were outside his control. It seems likely that Henry prepared these notes and transcripts from the information of his brother but that the latter considered it imprudent to allude to the matter at all in his biography. As, with the exception of his father, all the protagonists in the drama were still alive at the time he wrote, his decision was probably wise, but the need for such reticence has long since disappeared and it seemed to me that, in justice to the memory of a great man, the true facts of the matter ought to be pieced together and revealed.

This proved a difficult task in which there were many disappointments due to the loss or destruction of documents. For example, documents in the possession of Messrs Freshfield, John Scott Russell's legal advisers, would undoubtedly have shed valuable light on the matter, but the firm's early records were destroyed in the German raid on the City in 1940. However, from the Letter Books of the *Great Eastern*, from the files of *The*

Times and other newspapers and from scraps of information gleaned from other sources I have pieced together the story related in the last three chapters. This I believe to be a substantially true and correct account of the extraordinary events which accompanied the birth of Brunel's great ship. If through my desire that justice should be done to the memory of Brunel the character of John Scott Russell has suffered it was not, I must emphasize, due to any preconceived prejudice on my part. The result of my research was as much of a surprise to me as it may be to the reader. I began with the assumption that Scott Russell was as loyal a partner of Brunel as Claxton, Guppy, Gooch or Charles Saunders. Even when I had discovered Henry Brunel's notes I assumed that the trouble could only have been due to the fact that Russell was financially improvident. It was only after a most careful investigation of all the available facts that there emerged, willy nilly, the strange character who appears in these pages.

It only remains for me to express my most sincere thanks to all those people who gave me such generous assistance in so many different ways by giving me introductions or research permission, by loaning material, carrying out research on my behalf or by imparting so freely their special knowledge. First, as I am sure Brunel himself would have wished, I acknowledge the help of the following Bristolians: Colonel A. R. Boucher, Miss Grinham of the Public Library, College Green, Mr Spittal of the University Engineering School Library and Mr I. H. C. Fraser. Next it gives me particular pleasure to acknowledge the help of one of Sir Marc Brunel's countrymen, M. Henri Delgove of Le Mans, Sarthe, the distinguished French authority on the Brunel family. Thirdly, I acknowledge the courteous assistance given me by those who are today responsible for Brunel's Great Western Railway: Mr M. G. R. Smith, Chief Civil Engineer, Paddington, and the following members of his staff: Mr P. S. A. Berridge, Chief Bridge Assistant, Mr Sidney Stevens, District Engineer, Paddington, Mr H. G. Lakeman, District Engineer, Bristol and Mr E. Tyers. To Mr D. S. M. Barrie of the British Transport Commission I am indebted for a great deal of advice and invaluable assistance and to Mr G. Edrich, Assistant Chief Civil Engineer of the London Transport Executive, for enabling me to inspect the Thames Tunnel. I am similarly indebted to Mr A. E. Tawse of Messrs Joseph Westwood & Company, Napier Yard, Millwall, for permitting me to view that historic yard and for the loan of valuable material. I would also thank the following for their help: Mr J. Foster Petree and Mr C. E. C. Townsend, who together constitute a mine of information on the subject of Thames shipbuilding; Mr Gardiner and Mr H. P. Spratt of the Science Museum, South Kensington; Mr F. C. Ferguson of Portsmouth and his brother Mr J. C. Ferguson of Salisbury, Southern Rhodesia; Mr Rex Wailes and Mr A. S. Crosley of the Newcomen Society; Mr F. G. G. Carr and Mr A. W. H. Pearsall of the National Maritime Museum at Greenwich; Mr W. G. Bennett and Doctor C. de V. Shortt for legal and medical advice respectively, and Messrs Cecil Clutton and Malcolm Gardner for horological information. Others who gave me generous assistance were: Messrs D. C. Barber, W. Leslie-Barrow, A. J. Beamish, N. Burton, Coroner's Officer, Birkenhead, E. C. R. Hadfield, Christopher Jennings, R. B. Kirwan, J. F. Parker, Laurence Pomeroy, J. Rapley, C. W. Reed, Sacheverell Sitwell, and A. W. Whistlecroft. Finally, a special word of thanks to my friend David Cape. The idea of this book was his and but for his infectious enthusiasm and determination it might never have been written at all.

We are indebted to Mrs W. B. Yeats for permission to reproduce 'The Choice' from *Collected Poems of W. B. Yeats*, published by Messrs Macmillan & Co. Ltd.

L. T. C. R.

Bibliography

THE following works were consulted in the preparation of this book and the author is indebted to the writers named:

Abell, Sir Westcott: *The Shipwright's Trade*. London, Cambridge Univ. Press, 1948.

Armstrong, Sir W., and Russell, J. Scott: *Correspondence and Documents Submitted to, with Minutes of Evidence taken before the Council of the Institution of Civil Engineers*. (Privately circulated to members of the Institution.) London, William Clowes & Sons, 1867.

Barrie, D. S. M.: *The Taff Vale Railway*. South Godstone, The Oakwood Press, rep. 1950.

Beamish, Richard, F.R.S.: *Memoir of the Life of Sir Marc Isambard Brunel*. London, Longman, Green, Longman & Roberts, 1862.

Berridge, P. S. A.: 'The Erection and Reconstruction of Large Railway Bridges'. London, British Railways Western Region Debating Soc. *Proceedings*, 1950–51.

Brunel, Isambard: *The Life of Isambard Kingdom Brunel, Civil Engineer*. London, Longmans, Green, 1870.

Cottrell, A .E.: *The History of the Clifton Suspension Bridge*. Bristol, Cottrell, 1928.

Dickinson, H. W., and Titley, A.: *Richard Trevithick Centenary Memorial Volume*. London, Cambridge Univ. Press, 1934.

Geikie, Archibald: *The Life of Sir Roderick Murchison*, vol. 1, 1875.

Gibbs, George Henry: Extracts from the Diary of. London, G.W.R., n.d.

Gooch, Sir Daniel: The Diaries of. London, Paul Trench & Trubner, n.d.

Gotch, Rosamund Brunel: *Mendelssohn and his Friends in Kensington*. Oxford, The University Press, 1934.

Guest, Lady Charlotte: Diaries, Ed. The Earl of Bessborough. London, John Murray, 1950.

MacDermot, E. T.: *History of the Great Western Railways*, 2 vols. London, G.W.R., 1927–31.

MacFarlane, Charles: *Reminiscences of a Literary Life*. London, John Murray, 1917.

Murray, K. A.: 'The Atmospheric Railway Episode'. (K. & D. R.) Dublin, *Irish Railway Record Soc. Journal,* 1954.

Napier, David: *David Napier, Engineer, 1790–1869. An Autobiographical Sketch, with Notes.* Glasgow, James Maclehose, 1912.

Nasmyth, James: *James Nasmyth, Engineer.* An Autobiography, ed. Samuel Smiles. London, John Murray, 1883.

Noble, Lady Celia Brunel: *The Brunels, Father and Son.* London, Cobden Sanderson, 1938.

Pannell, J. P. M.: 'The Taylors of Southampton: Pioneers in Mechanical Engineering'. London, Institution of Mechanical Engineers, *Proceedings,* 1955.

Quartermaine, A. S.: 'Presidential Address'. London Institution of Civil Engineers, *Proceedings,* 1952.

Russell, John Scott: *The Modern System of Naval Architecture.* London, Day & Son, 1865.

Smiles, Samuel: *Industrial Biography.* London, John Murray, ed. 1882.

Smith, Eng. Capt. Edgar C., and Penrose, Charles: 'The Centenary of Atlantic Steam Navigation, 1838–1938'. London, *Newcomen Soc. Trans.,* vol. XVIII, 1937–8.

Spratt, H. P.: *Marine Engineering.* London, H.M.S.O., 1953.

Spratt, H. P.: *Merchant Steamers and Motor Ships.* London, H.M.S.O., 1949.

Spratt, H. P.: *Outline History of Transatlantic Steam Navigation.* London, H.M.S.O., 1950.

Tangye, George: *A Family Cruise.* Printed for Private Circulation, 1904.

Tangye, Richard: *One and All.* An Autobiography. London, S. W. Partridge, 1889.

'Vigil': *Inconsistencies of Men of Genius.* London, John Olliver, 1846.

Whitley, H. S. B.: 'Timber Viaducts in South Devon and Cornwall, G.W.R.'. London, *Railway Engineering,* Oct. 1931.

Woodham-Smith, Mrs Cecil: *Florence Nightingale.* London, Constable, 1951.

Appendix to the Bibliography

Amongst the works on or about Brunel which have appeared since the publication of this book, the following are particularly worthy of note:

Ball, Adrian, and Wright, Diana: *S.S. Great Britain*. Newton Abbot, David & Charles, 1981.

Beaver, Patrick: *The Big Ship: Brunel's Great Eastern – A Pictorial History*. London, Evelyn, 1969.

Beckett, Derrick: *Brunel's Britain*. Newton Abbot, David & Charles, 1980.

Binnie, G. M.: *Early Victorian Water Engineers*. London, Thomas Telford, 1981.

Body, G.: *Clifton Suspension Bridge: An Illustrated History*. Bradford-on-Avon, Moonraker Press, 1976.

Brooke, David: *The Railway Navvy – 'That Despicable Race of Men'*. Newton Abbot, David & Charles, 1983.

Brunel Society: Publications from 1970 include intermittent issues of a *News-letter* and *Gazetteer*, available from the Society at Brunel Technical College, Bristol. See, for example, Farrell, Christopher: 'The Brunels and their Gaz Engine'. *Newsletter* no. 2, May 1977; and Fry, Plantagenet Somerset: 'Brunel's Crimean Hospital'. *Gazetteer* no. 2, September 1984.

Buchanan, R. A.: 'Brunel in Bristol', in McGrath, P., and Cannon, J. (eds.): *Essays in Bristol and Gloucestershire History*. Bristol, Bristol & Gloucestershire Archaeological Society, 1976.

Buchanan, R. A.: 'I. K. Brunel and the Port of Bristol'. London, *Transactions of the Newcomen Society*, vol. 42, 1969–70, 41–56.

Buchanan, R. A.: 'The Overseas Projects of I. K. Brunel'. London, *Transactions of the Newcomen Society*, vol. 54, 1982–3, 145–66.

Buchanan, R. A.: 'The *Great Eastern* Controversy: A Comment'. *Technology and Culture*, vol. 24, no. 1, January 1983, 98–106.

Buchanan, R. A., and Cossons, Neil: *The Industrial Archaeology of the Bristol Region*. Newton Abbot, David & Charles, 1969.

Buchanan, R. A., and Jones, S.: 'The Balmoral Bridge of I. K. Brunel, *Industrial Archaeology Review*, vol. 4, no. 3, Autumn 1980.

Buchanan, R. A., and Williams, M.: *Brunel's Bristol*. Bristol, Redcliffe Press, 1982.

Clark, E. F.: *George Bidder – The Calculating Boy*. Bedford, K. & L. Publications, 1983.

Clements, Paul: *Marc Isambard Brunel*. London, Longmans, 1970.

Clinker, C. R.: *Paddington 1854–1979 – An Official History of British Rail Western Region*. Weston-super-Mare, Avon-Anglia Publications and Services, 1979.

Conder, F. R.: *The Men Who Built Railways* (a reprint of the author's *Personal Recollections of English Engineers*, published anonymously in 1868). Simmons, Jack (ed.), London, Thomas Telford, 1983.

Corlett, Ewan: *The Iron Ship: The History and Significance of Brunel's 'Great Britain'*. Bradford-on-Avon, Moonraker Press, 1975.

Dumpleton, Bernard, and Miller, Muriel: *Brunel's Three Ships*. Melksham, Colin Venton, 1974.

Emmerson, George S.: *John Scott Russell – A Great Victorian Engineer and Naval Architect*. London, John Murray, 1977.

Emmerson, George S.: *The Greatest Iron Ship – s.s. Great Eastern*. Newton Abbot, David & Charles, 1981.

Emmerson, George S.: 'L. T. C. Rolt and the *Great Eastern* Affair of Brunel versus Scott Russell'. *Technology and Culture*, vol. 21, no. 4, October 1980, 553–69.

Emmerson, George S.: 'The *Great Eastern* Controversy: In response to Dr. Buchanan'. *Technology and Culture*, vol. 24, no. 1, January 1983, 107–13.

Gladwyn, Cynthia: 'The Isambard Brunels'. *Proceedings of the Institution of Civil Engineers*, vol. 50, September 1971, 1–14.

Hadfield, Charles: *Atmospheric Railways – A Victorian Venture in Silent Speed*. Newton Abbot, David & Charles, 1967.

Hadfield, Charles, and Skempton, A. W.: *William Jessop – Engineer*. Newton Abbot, David & Charles, 1979.

Hawke, G.: *Railways and Economic Growth*. London, Cambridge Univ. Press, 1970.

Hay, Peter: *Brunel: His Achievements in the Transport Revolution*. Reading, Osprey, 1973.

Mosse, John: 'Bristol Temple Meads'. *BIAS Journal 4*, Bristol, 1971.

Noble, Humphrey: *Life in Noble Houses*. Newcastle, 1967.

O'Callaghan, John: *The Saga of the Steamship Great Britain*. London, Rupert Hart-Davis, 1971.

Pudney, John: *Brunel and His World*. London, Thames & Hudson, 1974.

Pugsley, Sir Alfred (ed.): *The Works of Isambard Kingdom Brunel – An Engineering Appreciation*. London and Bristol, Institution of Civil Engineers and University of Bristol, 1976, and Cambridge Univ. Press, 1980.

Rowland, K. T.: *The Great Britain*. Newton Abbot, David & Charles, 1971.

Simmons, Jack (ed.): *The Birth of the Great Western Railway – Extracts from the Diary and Correspondence of George Henry Gibbs*. Bath, Adams & Dart, 1971.

Simmons, Jack: See Conder, F. R., above.

Tames, Richard: *Isambard Kingdom Brunel*. Aylesbury, Shire Publications, 'Lifelines 1', 1972.

Toppin, David: 'The British Hospital at Renkioi 1855'. *The Arup Journal*, vol. 16, no. 2, July 1981.

Totterdill, J. W.: '. . . a peculiar form of construction'. *Journal of the Bristol and Somerset Society of Architects*, vol. 5, 1961, 111–2.

Tudor, Geoffrey: 'To the spanner born'. *The Times Higher Education Supplement*, 19 August 1988, p.13 (on Henry Brunel).

Vignoles, K. H.: *Charles Blacker Vignoles – Romantic Engineer*. Cambridge, 1982.

Walker, Charles: *Thomas Brassey – Railway Builder*. London, Muller, 1969.

Index